Richard Courant

Dirichlet's Principle, Conformal Mapping, and Minimal Surfaces

Reprint

Springer-Verlag New York Heidelberg Berlin

AMS subject Classifications (1970): 49 F10, 30 A38, 31 B25, 3 A24

Reprint 1977

ISBN-13:978-1-4612-9919-6 e-ISBN-13:978-1-4612-9917-2
DOI: 10.1007/978-1-4612-9917-2

Softcover reprint of the hardcover 1st edition 1950

NY/3014-543210

DIRICHLET'S PRINCIPLE, CONFORMAL MAPPING, AND MINIMAL SURFACES

R. COURANT
INSTITUTE FOR MATHEMATICS AND MECHANICS
NEW YORK UNIVERSITY, NEW YORK

with an Appendix by
M. SCHIFFER
PRINCETON UNIVERSITY AND UNIVERSITY OF JERUSALEM

1950

INTERSCIENCE PUBLISHERS, INC., NEW YORK
INTERSCIENCE PUBLISHERS LTD., LONDON

To

Otto Neugebauer

Preface

It has always been a temptation for mathematicians to present the crystallized product of their thoughts as a deductive general theory and to relegate the individual mathematical phenomenon into the role of an example. The reader who submits to the dogmatic form will be easily indoctrinated. Enlightenment, however, must come from an understanding of motives; live mathematical development springs from specific natural problems which can be easily understood, but whose solutions are difficult and demand new methods of more general significance.

The present book deals with subjects of this category. It is written in a style which, as the author hopes, expresses adequately the balance and tension between the individuality of mathematical objects and the generality of mathematical methods.

The author has been interested in Dirichlet's Principle and its various applications since his days as a student under David Hilbert. Plans for writing a book on these topics were revived when Jesse Douglas' work suggested to him a close connection between Dirichlet's Principle and basic problems concerning minimal surfaces. But war work and other duties intervened; even now, after much delay, the book appears in a much less polished and complete form than the author would have liked.

It was felt desirable to include a report on some recent progress in the theory of conformal mapping: fortunately Professor M. Schiffer, who had a most active part in those developments, agreed to write a summary of the material; the result is the comprehensive appendix which will certainly be considered as a highly valuable contribution to the volume.

In a field which has attracted so many mathematicians it is difficult to achieve a fair accounting of the literature and to appraise merits of others. I have tried to acknowledge all the sources of information and inspiration of which I am conscious, and I hope that not too many omissions have occurred.

A first draft of the book was completed eight years ago, supported by a grant from the Philosophical Society and with the help of

Dr. Wolfgang Wasow. Assistance for the present publication was partly provided under contract with the Office of Naval Research. On the scientific side the book owes much to Professor Max Shiffman, who has been concerned with the theory of minimal surfaces ever since a good fortune brought him as a student to my seminar on the subject. Carl Ludwig Siegel read the manuscript carefully and gave much valuable advice. Avron Douglis, Martin Kruskal, Peter Lax, Imanuel Marx, Joseph Massera, and others have unselfishly devoted time to scrutinizing the manuscript, reading proof, and preparing the bibliography. The drawings were made mainly by George Evans, Jr., Beulah Marx, and Irving Ritter. Edythe Rodermund and Harriet Schoverling gave outstanding secretarial help. The strenuous responsibility for the editorial work and for the supervision of all the steps from preparing the manuscript to the final printing was in the competent hands of Natascha Artin. Without the collective help of all these friends the book could hardly have appeared at this time.

Naturally, a word of thanks must be added for the understanding and patient publisher whose interest has been most encouraging.

The book is dedicated to Otto Neugebauer as a token of friendship and admiration.

R. Courant

New Rochelle, New York
April 1950

Contents

III. Plateau's Problem—*Continued*

V. Conformal Mapping of Multiply Connected Domains—*Continued*

Introduction

A new era in the history of mathematics opened when Gauss proved the fundamental theorem of algebra. Abandoning the futile attempts of his predecessors to solve algebraic equations of higher degree by root extraction, he took a step of general significance by proving merely the existence of the roots in question. For the first time it was clearly understood that the primary task in a mathematical problem is to prove the existence of a solution. To find procedures by which the solution can be explicitly obtained is a further question, distinct from that of existence. Since the beginning of the last century this distinction has played a clarifying role contributing greatly to progress in all fields of mathematics.

Among the existence proofs which dominated the mathematical thinking of this period, the most celebrated and consequential were those based on extremum problems of the calculus of variations and suggested by actual or imagined physical experiments. Bernhard Riemann's geometric function theory, published in his doctoral thesis (1851) and in his memoir on Abelian functions (1857), is the outstanding example of the power of such an approach.

To describe the physical reasoning underlying Riemann's conception, let us consider a surface S in space with or without boundaries, of any topological structure. This surface S we assume to be covered by a thin uniform sheet conducting electricity. Imagine a stationary electric current over S, generated by the connection of arbitrary points on the surface with the poles of electric batteries. The potential of such a current will be the solution of a boundary value problem of a differential equation—just that type of boundary value problem obtained from a variational problem. In general this variational problem consists of seeking, among all possible flows, that which produces the least quantity of heat. If the existence of the mathematical function corresponding to such a minimal condition is assumed, the existence theorems of Riemann's function theory follow almost immediately.

Mathematically, such a minimum problem can be formulated as

that of minimizing an integral of the form

$$D[\phi] = \iint_G (\phi_x^2 + \phi_y^2)\, dx\, dy,$$

in which the domain G of integration and the range of functions ϕ admitted to competition are specified according to the particular function-theoretical theorem to be proved.

Already some years before the rise of Riemann's genius, C. F. Gauss and W. Thompson had observed that the boundary value problem of the harmonic differential equation $\Delta u = u_{xx} + u_{yy} = 0$ for a domain G in the x, y-plane can be reduced to the problem of minimizing the integral $D[\phi]$ for the domain G, under the condition that the functions ϕ admitted to competition have the prescribed boundary values. Because of the positive character of $D[\phi]$ the existence of a solution for the latter problem was considered obvious and hence the existence for the former assured. As a student in Dirichlet's lectures, Riemann had been fascinated by this convincing argument: soon afterwards he used it, under the name "Dirichlet's Principle," in a more varied and spectacular manner as the very foundation of his new geometric function theory.

This theory, which was to have so profound an influence on the subsequent development of mathematics, greatly impressed many mathematicians. Immediately it was felt that progress of lasting importance had been made. Thus, when Weierstrass found a flaw in Dirichlet's Principle and, in 1869, published his objection, it came as a shock to the mathematical world. His criticism can be simply summarized: The positive definite character of $D[\phi]$ implies the existence of a greatest lower bound d. Riemann as well as his predecessors had taken it for granted that this bound d is a proper minimum, actually attained by an admissible function $\phi = u$; yet this is just the point which requires careful scrutiny. While continuous functions of a finite number of variables always possess a least or greatest value in a closed region, there are variational problems where no minimum or maximum is attained although a greatest lower or least upper bound may exist.

By no means could this objection be easily answered. Riemann died without having had time to consider Weierstrass' criticism, and the efforts of others to save Dirichlet's Principle remained unsuccessful. Yet mathematicians refused to give up the results gained by Riemann's method. Substitute methods had to be invented to prove his existence theorems, and so Dirichlet's Principle became indirectly the stimulus for a far-reaching development of new methods in analysis. Yet the fascination of Dirichlet's Principle itself persisted: time and again attempts at a rigorous proof were made. Finally, fifty years after Riemann, D. Hilbert succeeded. In a famous publication (1900), he established some of Riemann's existence theorems by proving directly that the corresponding minimum problem actually has a solution. Since Hilbert's pioneering achievement, the theory has been both simplified and extended. Today Dirichlet's Principle has become a tool as flexible and almost as simple as that originally envisaged by Riemann. It has, moreover, been the starting point for the development of the so-called direct methods of the variational calculus, a development equally important to both pure and applied mathematics.

During the last decades it became apparent that another of the classical extremum problems of analysis and geometry is intimately connected with Dirichlet's Principle. Since the early period of the calculus of variations, the problem of determining the surfaces of minimal area spanned in a given curve or subject to other boundary conditions has been attacked by many of the great mathematicians. Again physical experiments, such as those carried out by the Belgian physicist Plateau,[1] lead immediately to the intuitive conviction that such problems can be solved. If a closed contour of wire is dipped into a soap solution, the liquid forms a film which, by virtue of the laws of surface tension, assumes as position of equilibrium the shape of a minimal surface spanned in the contour. But empirical evidence can never establish mathematical existence—nor can the mathematician's demand for existence proofs be dismissed by the physicist as useless rigor. (Only a mathematical existence proof can ensure that the mathematical description of a physical phenomenon is meaningful.)

At any rate, to solve "Plateau's problem," that is, to prove the existence of a minimal surface spanned in a given contour, appeared for a long time to be a most difficult task. Investigations by Riemann, Weierstrass, H. A. Schwarz, G. Darboux, and others linked the problem with the theory of harmonic functions and conformal mapping and provided explicit solutions in significant cases. It was

[1] See also Plateau [1], as well as Courant [13] p. 385, where various other types of significant experiments are discussed.

not, however, until 1939 that the first general existence theorems concerning Plateau's problem were proved, independently, by Tibor Radó and Jesse Douglas. The natural sequel was the use of Dirichlet's Principle as the basis of an attack on a variety of classical as well as new problems concerning minimal surfaces, with a view to developing at the same time general methods of the calculus of variations. Such a program has been pursued by the author and his collaborators. It is applicable not only to problems of minimal surfaces with fixed or free boundaries furnishing a stable equilibrium, as represented by soap film experiments, but also to problems concerning unstable minimal surfaces. The latter were first successfully attacked by M. Shiffman and, independently, by M. Morse and C. Tompkins.

The present book is based on previous publications and on new material. It represents an attempt to develop Dirichlet's Principle together with applications to conformal mapping and to the theory of minimal surfaces, and to give access to a wide, still largely unexplored field connected with many topics in analysis and geometry.

CHAPTER I

Dirichlet's Principle and the Boundary Value Problem of Potential Theory

1. Dirichlet's Principle

1. *Definitions.* We consider domains G in the x, y-plane whose boundaries γ consist of Jordan curves. Functions $\phi(x, y)$, $\psi(x, y)$, $u(x, y)$, $\cdots$ in G will be called piecewise smooth if they are continuous in G and if they have continuous first derivatives in G except on a finite number of arcs with continuous tangents and at a finite number of isolated points. (For convenience we shall in general assume that such arcs of discontinuity are either circular or straight.)

"Dirichlet's integral"

$$D[\phi] = \iint_G (\phi_x^2 + \phi_y^2)\, dx\, dy \tag{1.1}$$

for piecewise smooth functions ϕ is defined as an improper integral.[1] This means

$$D[\phi] = \lim_{n\to\infty} \iint_{G_n} (\phi_x^2 + \phi_y^2)\, dx\, dy,$$

where G_n denotes a sequence of closed subregions of G converging to G in the sense that each point of G lies in almost all the regions G_n. (A region G_n may consist of several domains.) Without restricting the generality of our definition, we may furthermore assume that the regions G_n form a monotone sequence, i.e. that G_{n+1} contains G_n.

In polar coordinates r, θ,

$$D[\phi] = \iint_G \left(\phi_r^2 + \frac{1}{r^2}\phi_\theta^2\right) r\, dr\, d\theta \tag{1.1a}$$

[1] We might instead admit functions continuous in G and possessing first derivatives square integrable in the Lebesgue sense, but for our purposes little would be gained by the use of this wider class of functions.

or, with the abbreviation $ds = r\,d\theta$,

$$D[\phi] = \iint_G (\phi_r^2 + \phi_s^2)\,dr\,ds. \tag{1.1b}$$

Dirichlet's integral for a subdomain S of G will be denoted by $D_S[\phi]$.

2. *Original Statement of Dirichlet's Principle.* We describe Dirichlet's Principle in the not quite precise form in which it was originally used. On the boundary γ of the domain G, continuous boundary values $\bar{g}$ are prescribed. Consider the problem of minimizing $D[\phi]$. Admitted to competition are all functions ϕ, continuous in $G + \gamma$ and piecewise smooth in G, whose boundary values coincide with $\bar{g}$. Dirichlet's Principle states: The problem is solved by a uniquely determined admissible function $\phi = u$ for which $D[\phi]$ attains its minimum value d. This function u possesses continuous first and second derivatives in G and is harmonic, i.e. satisfies the differential equation

$$\Delta u = u_{xx} + u_{yy} = 0. \tag{1.2}$$

Thus the boundary value problem of (1.2) is reduced to the solution of Dirichlet's variational problem: to find the admissible function that minimizes $D[\phi]$.

3. *General Objection: A Variational Problem Need Not be Solvable.* At first glance, the preceding statement seems to be an immediate consequence of the classical calculus of variations, since $\Delta u = 0$ is Euler's equation for the integral $D[\phi]$. Yet this observation alone cannot be accepted as proof of Dirichlet's Principle. Certainly there exists a greatest lower bound d for $D[\phi]$, since the set of possible values of $D[\phi]$ has zero as a lower bound; but it is not evident that d is actually attained by an admissible function u. As a matter of fact, any naive assumption that a "reasonable" variational problem always has a solution is easily refuted by examples to the contrary:

a) Consider the problem of minimizing

$$I[\phi] = \int_0^1 [1 + \phi'^2(x)]^{\frac{1}{4}}\,dx$$

for all continuous functions $\phi(x)$ having piecewise continuous first derivatives in the interval $0 \leq x \leq 1$ and satisfying the boundary conditions

$$\phi(0) = 1, \qquad \phi(1) = 0.$$

The value 1 is evidently a lower bound for this integral. It is in fact, the greatest lower bound: for on substituting in $I[\phi]$ the particular admissible function

$$\phi(x) = \begin{cases} \dfrac{\delta - x}{\delta}, & 0 \leq x \leq \delta < 1, \\ 0, & x > \delta, \end{cases}$$

we find

$$I[\phi] = \int_0^\delta (1 + \delta^{-2})^{\frac{1}{2}}\, dx + \int_\delta^1 dx = \delta^{\frac{1}{2}}(1 + \delta^2)^{\frac{1}{2}} + 1 - \delta < 1 + \delta^{\frac{1}{2}},$$

where δ may be arbitrarily small. But there is obviously no admissible function for which $I[\phi]$ assumes the value 1.

b) In the following problem the non-existence of a minimum is geometrically obvious: Among all surfaces $z = f(x, y)$ bounded by a circle K in the x, y-plane and passing through a fixed point P above this plane, find the one with the smallest area. Consider a surface that is identical with the x, y-plane except for a small disk under the point P, where it rises in the form of a steep cone with vertex at P. By making the base of this cone sufficiently small, we can bring the area of the surface arbitrarily close to the area of K; but there is, of course, no admissible surface of the same area as K.

c) There are even variational problems for which a least upper, as well as a greatest lower, bound exists, although neither is attained by an admissible function. Examples are furnished by extremum problems concerning

$$I[\phi] = \int_0^1 \frac{dx}{1 + (\phi')^2},$$

for all continuous functions ϕ having piecewise continuous first derivatives in $0 \leq x \leq 1$ and satisfying the boundary conditions $\phi(0) = 0$, $\phi(1) = 1$. Obviously $0 < I[\phi] < 1$; but while this is a strict inequality, the limits 0 and 1 can be approximated as closely as desired. For if $|\phi'|$ is sufficiently large throughout the interval except at a finite number of points of discontinuity, $I[\phi]$ will be arbitrarily near its lower bound zero. On the other hand, if $|\phi'|$ is zero except in a finite number of intervals of sufficiently small total length, the integral will be arbitrarily near 1. Now any admissible curve can be approximated as closely as desired by curves of either type, consisting in the one case of a polygon whose sides are all nearly perpendicular to the x-axis, in the other of horizontal lines with steep connecting segments the sum of whose projections on the x-axis is kept sufficiently small. Therefore, in an arbitrarily small neighborhood of any admissible function ϕ there are other functions for which the integral is as close to 0 or 1 as desired.

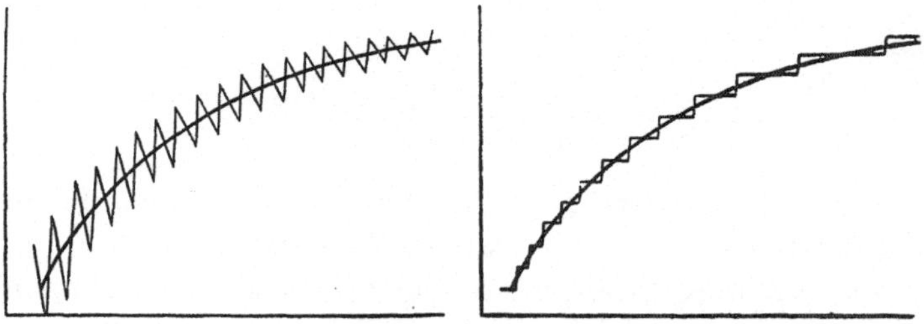

Figure 1.1

4. *Minimizing Sequences.* Let $I[\phi]$ be a variational integral. Consider a sequence of admissible functions $\phi_1, \phi_2, \cdots$, such that the values $I[\phi_1]$, $I[\phi_2]$, $\cdots$, tend to the greatest lower bound d of $I[\phi]$: such a sequence is called a "minimizing sequence." Whenever the set of possible values of $I[\phi]$ is bounded from below, the existence of a minimizing sequence is assured. However, the preceding examples show that a minimizing sequence need not converge, and that, if it converges, the limit function need not be admissible. That Dirichlet's integral is not exempt from this difficulty is shown by the following example:

Consider the minimum problem for $D[\phi]$ in a circle of unit radius, and require the admissible functions to vanish on the boundary. This minimum problem is solved by $\phi \equiv 0$ and by no other function; $d = 0$ is the minimum value, not merely the greatest lower bound. Now define in polar coordinates r, θ a sequence of admissible functions ϕ_n:

$$\phi_n = \begin{cases} c_n \log \rho_n, & r \leq \rho_n^2, \\ c_n \log (r/\rho_n), & \rho_n^2 \leq r \leq \rho_n, \\ 0, & \rho_n \leq r \leq 1, \end{cases}$$

where the c_n are constants. Formula (1.1a) yields

$$D[\phi_n] = 2\pi c_n^2 \int_{\rho_n^2}^{\rho_n} \frac{1}{r^2} r\, dr = -2\pi c_n^2 \log \rho_n .$$

If we choose $\rho_n = e^{-n}$ and $c_n = -n^{-2/3}$, $D[\phi_n]$ tends to zero, i.e. ϕ_n is a minimizing sequence. The value $n^{1/3}$ of ϕ_n at $r = 0$ tends to infinity, and hence the functions ϕ_n do not converge. Incidentally, by the same construction one can obtain a minimizing sequence for Dirichlet's integral which diverges in an everywhere dense point set.

Thus a minimizing sequence cannot in general be expected to yield the solution of the problem by a mere passage to the limit. The essential point in the "direct variational methods" is to introduce appropriate additional constructions that produce convergence.

5. *Explicit Expression for Dirichlet's Integral over a Circle. Specific Objection to Dirichlet's Principle.* The general objections of the preceding section seem accentuated by a fact first discovered by Hadamard: not only is the possibility of solving Dirichlet's variational problem not obvious, but Dirichlet's minimum problem actually has no solution in some cases in which the boundary value problem for the differential equation $\Delta u = 0$ *can* be solved. Thus the idea of reducing the latter problem to the former seemed even more discredited. Here is Hadamard's example:

Let K be a disk of unit radius about the origin, and again introduce polar coordinates r, θ. On the circumference κ of K, continuous boundary values $\bar{g} = \bar{g}(\theta)$ are assigned. Consider the—not necessarily convergent—Fourier series for $\bar{g}$:

$$\bar{g} \sim \frac{a_0}{2} + \sum_{\nu=1}^{\infty} (a_\nu \cos \nu\theta + b_\nu \sin \nu\theta).$$

For $r < 1$, the solution of $\Delta u = 0$ satisfying the boundary condition $\bar{u} = \bar{g}$ on κ is given by the convergent series

$$u(r, \theta) = \frac{a_0}{2} + \sum_{\nu=1}^{\infty} r^\nu(a_\nu \cos \nu\theta + b_\nu \sin \nu\theta). \tag{1.3}$$

This form of the solution, as well as the equivalent representation by *Poisson's integral*,

$$u(r, \theta) = \frac{1}{2\pi} \int_0^{2\pi} \bar{g}(\alpha) \frac{(1 - r^2)\, d\alpha}{1 - 2r \cos(\alpha - \theta) + r^2}, \tag{1.3a}$$

will be used frequently. For $r < 1$ the derivatives of (1.3) are obtained by termwise differentiation. We have $D[u] = \lim_{\rho \to 1} D_\rho[u]$, where $D_\rho[u]$ is Dirichlet's integral for the circle of radius $\rho < 1$ about the origin. From (1.3) we easily evaluate

$$D_\rho[u] = \pi \sum_{\nu=1}^{\infty} \nu(a_\nu^2 + b_\nu^2)\rho^{2\nu}.$$

This implies that for every N

$$\pi \sum_{\nu=1}^{N} \nu(a_\nu^2 + b_\nu^2)\rho^{2\nu} \leq D_\rho[u] \leq \pi \sum_{\nu=1}^{\infty} \nu(a_\nu^2 + b_\nu^2),$$

where the right side may be a divergent series. By letting ρ tend to 1, we infer immediately:

Dirichlet's integral (1.1a) for the harmonic function u over the unit circle is given by the series

$$D[u] = \pi \sum_{\nu=1}^{\infty} \nu(a_\nu^2 + b_\nu^2) \tag{1.4}$$

and exists if, and only if, this series converges.

As pointed out by Hadamard, there exist continuous functions $\bar{g}(\theta)$ for which the series (1.4) diverges; for instance, the function defined by the uniformly convergent Fourier expansion

$$\bar{g} = \sum_{\mu=1}^{\infty} \frac{\sin \mu!\theta}{\mu^2} \tag{1.5}$$

would yield $D[u] = \pi \sum_{\mu=1}^{\infty} \frac{\mu!}{\mu^4}$. With boundary values such as (1.5), therefore, the boundary value problem of $\Delta u = 0$ can certainly not be reduced to a variational problem for Dirichlet's integral: Dirichlet's Principle for such a case is invalid. There is no full equivalence between the variational problem and the boundary value problem.

6. *Correct Formulation of Dirichlet's Principle.* An obvious way of avoiding the last difficulty is to restrict the prescribed boundary values from the outset in such a manner as not to exclude the possibility of solving the variational problem. While such conditions are not necessary for the boundary value problem, they are essential for a meaningful variational problem. Accordingly, it will be assumed that the prescribed boundary values $\bar{g}$ on γ are those of a function g, defined in $G + \gamma$, for which $D[g]$ *is* finite; in other words, we explicitly assume that there exists at least *one* admissible function g with a finite Dirichlet integral. Thus we are led to the following formulation:

Dirichlet's Principle: Given a domain G whose boundary γ consists of Jordan curves. Let g be a function continuous in $G + \gamma$, piecewise smooth in G, and with finite Dirichlet integral $D[g]$. Consider the class of all functions ϕ continuous in $G + \gamma$, piecewise smooth in G, and having the same boundary values as g. Then the problem of finding a function ϕ for which $D[\phi]$ attains a minimum d has a unique solution $\phi = u$. This function u is the solution of the boundary value problem for $\Delta u = 0$ with the prescribed boundary values $\bar{g}$ on γ.

2. Semicontinuity of Dirichlet's Integral. Dirichlet's Principle for Circular Disk

It is easy to prove Dirichlet's Principle in the preceding formulation if γ is a circle. In the proof we use a fundamental fact which will now be formulated for domains G without restriction to circular disks:

Lemma 1.1 (Semicontinuity of Dirichlet's integral for harmonic functions): If a sequence of harmonic functions u_n converges to a harmonic function u, and if the convergence is uniform in every closed subdomain of G, then

$$D_G[u] \leq \liminf. D_G[u_n]. \tag{1.6}$$

Proof: For any closed subdomain G' of G the assumed convergence of the u_n implies, by Harnack's theorem,[2] the uniform convergence of the derivatives of u_n to those of u. Hence

$$D_{G'}[u] = \lim_{n\to\infty} D_{G'}[u_n] \leq \liminf. D_G[u_n].$$

We obtain the inequality immediately by letting G' tend to G.

To prove Dirichlet's Principle for the disk K bounded by the unit circle κ, let g be any admissible function for which $D[g] < \infty$ and consider the "harmonic polynomials"

$$u_n = \frac{a_0}{2} + \sum_{\nu=1}^{n} r^\nu(a_\nu \cos \nu\theta + b_\nu \sin \nu\theta),$$

[2] Harnack's theorem states: If a sequence of harmonic functions converges uniformly in a domain, their derivatives converge uniformly in every closed subdomain and the limit function is again harmonic. The proof is a simple application of Poisson's integral.

where a_ν, b_ν are the Fourier coefficients of the given boundary function $\bar{g} = g(1, \theta)$. While the Fourier series for $\bar{g}$ need not converge, the polynomials u_n are regular and harmonic everywhere. For $\zeta_n = g - u_n$ we have $D_K[\zeta_n] < \infty$ (see §3). With the notation

$$D[\phi, \psi] = \iint (\phi_x \psi_x + \phi_y \psi_y)\, dx\, dy \tag{1.7}$$

we may write

$$D[g] = D[u_n] + D[\zeta_n] + 2D[u_n, \zeta_n].$$

By Green's formula—which is applicable to the polynomials u_n, though not necessarily to u—we find

$$D[u_n, \zeta_n] = \int_0^{2\pi} \zeta_n \frac{\partial u_n}{\partial r}\, d\theta \Big|_{r=1} - \iint_K \zeta_n \Delta u_n\, dx\, dy.$$

The second term on the right side vanishes because $\Delta u_n = 0$. The first term on the right is also zero; this follows immediately if we substitute for u_n its explicit expression, and observe that the first $2n + 1$ Fourier coefficients of ζ_n vanish, i.e. that

$$\int_0^{2\pi} \zeta_n\, d\theta = 0,$$

$$\int_0^{2\pi} \zeta_n \cos \nu\theta\, d\theta = 0, \qquad \int_0^{2\pi} \zeta_n \sin \nu\theta\, d\theta = 0, \qquad \nu = 1, 2, \cdots, n.$$

Hence $D[u_n, \zeta_n] = 0$, and therefore

$$D[u_n] = D[g] - D[\zeta_n] \leq D[g].$$

In K the polynomials u_n tend to the harmonic function u with boundary values $\bar{g}$, i.e. $\lim_{n\to\infty} u_n = u$, and the convergence is uniform in every closed subdomain of K. The semicontinuity of Dirichlet's integral for harmonic functions leads therefore to

$$D[u] \leq \text{lim inf. } D[u_n].$$

This and the preceding inequality imply

$$D[u] \leq D[g],$$

hence the minimum property of $D[u]$ is proved.

To prove uniqueness, we note first that

$$D[u, h] = 0 \tag{1.8}$$

for every function h with $D[h] < \infty$ which is piecewise smooth in K and vanishes on κ. For if $D[\phi]$ attains a minimum for $\phi = u$, then in the expansion[3]

$$D[u + \epsilon h] = D[u] + 2\epsilon D[u, h] + \epsilon^2 D[h]$$

the coefficient of ϵ must vanish. Equation (1.8) for $h = g - u$ implies

$$D[g] = D[u + h] = D[u] + D[h].$$

Thus $D[u] < D[g]$ unless $D[h] = 0$, i.e. unless h has the constant value zero.

3. *Dirichlet's Integral and Quadratic Functionals*[4]

Dirichlet's integral $D[\phi]$ is a quadratic, non-negative functional of a function $\phi(x, y)$. The function ϕ may range over a "linear function space," i.e. a set of functions which, with two functions ϕ, ψ, also contains their linear combinations $\lambda\phi + \mu\psi$ with constants λ and μ. Another quadratic functional important for us is

$$H[\phi] = \iint_G \phi^2 \, dx \, dy.$$

We denote quadratic functionals such as $D[\phi]$ or $H[\phi]$ by the neutral symbol $Q[\phi]$. Some of the essential properties of non-negative quadratic functionals are:

a)

$$Q[\phi] \geq 0. \tag{1.9}$$

b) If ϕ and ψ are two functions for which $Q[\phi]$ and $Q[\psi]$ exist, there exists also a functional $Q[\phi, \psi]$, called the *symmetric bilinear form*, for which

$$Q[\lambda\phi + \mu\psi] = \lambda^2 Q[\phi] + \mu^2 Q[\psi] + 2\lambda\mu Q[\phi, \psi], \tag{1.10}$$

[3] The existence of $D[u + h]$, $D[u, h]$, and $D[g - u]$ is assured by the existence of $D[u]$, $D[h]$, and $D[g]$, as follows from the general fact, proved in §3, that the existence of $D[\phi]$ and $D[\psi]$ implies the existence of $D[\phi + \psi]$ and $D[\phi, \psi]$.

[4] See Courant [2], Chapter VII.

with arbitrary constants λ, μ. Obviously

$$Q[\phi] = Q[\phi, \phi]. \tag{1.11}$$

We may define the bilinear form by

$$2Q[\phi, \psi] = Q[\phi + \psi] - Q[\phi] - Q[\psi]. \tag{1.12}$$

For $Q = D$ the bilinear form is given by (1.7), for $Q = H$ by $\iint \phi\psi \, dx \, dy$.

c) From (1.9) and (1.10) follows
Schwarz' Inequality:

$$\{Q[\phi, \psi]\}^2 \le Q[\phi]Q[\psi], \tag{1.13}$$

which expresses the non-negative character of the ordinary quadratic form in λ and μ represented by (1.10).
d) Equivalent to (1.13) are the
Triangle Inequalities:

$$\sqrt{Q[\phi]} + \sqrt{Q[\psi]} \ge \sqrt{Q[\phi + \psi]}, \tag{1.13a}$$

$$\sqrt{Q[\phi]} - \sqrt{Q[\psi]} \le \sqrt{Q[\phi - \psi]}. \tag{1.13b}$$

The name "triangle inequality" is suggested by the interpretation of $\sqrt{Q[\phi - \psi]}$ as the "*distance*" between the functions ϕ and ψ in the "*linear function space*" under consideration. We may call it a "Q-metric" in the function space. Our concepts likewise apply to integrals Q_S extended over subdomains S of G.

Since integrals for G are defined as improper integrals, a proof is needed for the existence of $Q[\phi + \psi]$, $Q[\phi - \psi]$, and $Q[\phi, \psi]$ as a consequence of the existence of $Q[\phi]$ and $Q[\psi]$. The triangle inequality for any closed subregion S where the integrals are proper yields

$$\sqrt{Q_S[\phi + \psi]} \le \sqrt{Q[\phi]} + \sqrt{Q[\psi]},$$

$$\sqrt{Q_S[\phi - \psi]} \le \sqrt{Q[\phi]} + \sqrt{Q[\psi]}.$$

By letting S tend to G and applying (1.12) we complete the proof.

The triangle inequality is instrumental for establishing lemmas on convergence of quadratic functionals for a sequence of functions $\phi_1, \phi_2, \cdots$ in the linear function space. We decompose the domain

G into a closed subregion G' and a "boundary strip" G^*, so that $G = G' + G^*$, and state:

Lemma 1.2: Suppose $Q[\phi_n] < \infty$ and

$$Q[\phi_n - \phi_m] \to 0$$

as $n, m \to \infty$. Then there exists for every $\epsilon > 0$ a small "boundary strip" G^* such that

$$Q_{G^*}[\phi_n] < \epsilon$$

for *all* n. In other words, the lemma implies that the contribution to $Q[\phi_n]$ from a suitably small neighborhood of the boundary can be made uniformly small.

Proof: By virtue of the triangle inequality (1.13b) applied for $\phi = \phi_n$, $\psi = \phi_m$, with m fixed, the assumption $Q[\phi_n - \phi_m] \to 0$ implies the convergence of $Q[\phi_n]$ for n tending to infinity. Hence $Q[\phi_n]$ is bounded. We choose N so large that

$$Q[\phi_n - \phi_m] < \frac{\epsilon}{4}$$

for $n, m \geq N$. Furthermore, we choose a closed interior part G' so large that, for the boundary strip $G^* = G - G'$, we have

$$Q_{G^*}[\phi_\nu] < \frac{\epsilon}{4}$$

for $\nu \leq N$. Then the statement of the lemma follows from (1.13a) if we set

$$\phi = \phi_n - \phi_N, \qquad \psi = \phi_N .$$

A consequence is the

Corollary to Lemma 1.2: Suppose $Q[\phi_n]$ exists for all n, and

$$Q[\phi_n - \phi_m] \to 0.$$

If ϕ is a function for which

$$Q_{G'}[\phi_n - \phi] \to 0$$

for all closed subdomains G' of G, then $Q[\phi]$ exists and

$$Q[\phi] = \lim_{n\to\infty} Q[\phi_n]$$

as well as

$$Q[\phi_n - \phi] \to 0.$$

Proof: By (1.13b) we have

$$Q_{G'}[\phi] = \lim_{n\to\infty} Q_{G'}[\phi_n].$$

Since $Q_{G'}[\phi_n] \leqslant Q[\phi_n]$ and since $Q[\phi_n]$ is uniformly bounded, the existence of $Q[\phi]$ follows. Now for a given ϵ we choose a strip G^*, again with $G = G' + G^*$, such that

$$Q_{G^*}[\phi] < \frac{\epsilon}{8}, \qquad Q_{G^*}[\phi_n] < \frac{\epsilon}{8}$$

for all n. Choose a number N so large that for $n > N$

$$Q_{G'}[\phi - \phi_n] < \frac{\epsilon}{2}.$$

Then

$$Q[\phi - \phi_n] = Q_{G^*}[\phi - \phi_n] + Q_{G'}[\phi - \phi_n] < \epsilon$$

for all $n > N$, i.e.

$$Q[\phi - \phi_n] \to 0$$

and therefore

$$Q[\phi_n] \to Q[\phi],$$

as $n \to \infty$.

In making use of lemma 1.2, we shall occasionally replace G by a subdomain S of G.

4. *Further Preparation*

1. *Convergence of a Sequence of Harmonic Functions.* As example b) of §1, 3 shows, we cannot derive the convergence of a sequence ϕ_n merely from relations such as $H[\phi_n] \to 0$, $D[\phi_n] \to 0$, or $D[\phi_n - \phi_m] \to 0$ as $n, m \to \infty$. However, such relations do ensure

convergence if we restrict ϕ_n to the narrower class of harmonic functions:

Lemma 1.3: If $H_S[w_n] \to 0$ for a sequence of functions w_n harmonic in a domain S, the functions w_n, and consequently the derivatives of w_n, tend to zero uniformly in any closed subdomain S' of S.

Proof: Given a closed subdomain S', we can find a positive number h such that the disk K with radius h about any point P in S' lies entirely in S. By the mean value theorem for harmonic functions, we have

$$w_n(P) = \frac{1}{\pi h^2} \int_K w_n(x, y)\, dx\, dy,$$

hence by Schwarz' inequality

$$[w_n(P)]^2 \leq \frac{1}{\pi h^2} H_S[w_n].$$

This inequality implies the statement of the lemma for the functions w_n. The statement concerning the derivatives follows automatically. Moreover, we can now prove the two further lemmas:

Lemma 1.3a: Let u_n be a sequence of functions harmonic in S for which

$$H_S[u_n - u_m] \to 0$$

as n and m tend to infinity. Then in each closed subdomain S' of S the functions u_n tend uniformly to a harmonic function u and consequently also the derivatives of u_n tend uniformly to the derivatives of u.

Lemma 1.3*b*: Denote again by S' a closed subdomain of S and by u_n a sequence of functions harmonic in S. Assume that $D_S[u_n - u_m] \to 0$ as m, n tend to infinity, and that in addition the difference $| u_n - u_m |$ becomes arbitrarily small for sufficiently large n and m—either at a fixed point P of S' or even only in a suitable point P_{nm} of S' which may vary with n and m. Then the functions u_n tend uniformly to a harmonic function u in S'. Consequently also the derivatives of u_n tend uniformly to the derivatives of u in S'. Furthermore

$$D_S[u_n] \to D_S[u]$$

and

$$D_S[u_n - u] \to 0.$$

The proofs of lemmas 1.3a and 1.3b follow immediately—for an arbitrary closed subregion S'—from lemma 1.3 and from Harnack's theorem (see footnote 2, p. 11) and then for S by lemma 1.2 and its corollary.

2. *Oscillation of Functions Appraised by Dirichlet's Integral.* The analogue of Dirichlet's integral for functions $\phi(x)$ of one variable x is simply $\int_a^b \phi_x^2(x)\,dx$. By Schwarz' inequality, for $a \le x_1 \le x_2 \le b$,

$$[\phi(x_2) - \phi(x_1)]^2 = \left[\int_{x_1}^{x_2} \phi'(x)\,dx\right]^2 \le |x_2 - x_1| \int_a^b [\phi'(x)]^2\,dx.$$

Thus in one dimension the oscillation of a function can be appraised by means of Dirichlet's integral. Such a simple fact is no longer true in two dimensions, as seen from example b) in §1, 3. However, similar estimates for the oscillation in two dimensions are possible at least along selected individuals of given families of curves.

First we consider concentric circles in G about a center O. Let r, θ be polar coördinates about O, $s = r\theta$. For a circular disk K: $r \le a$ in G, define the function

$$M(a) = D_K[\phi] \ge \int_0^a dr \int_0^{2\pi r} \phi_s^2\,ds,$$

where ϕ may at first be assumed to have continuous first derivatives in K.

Let a and $b = a - h < a$ be constants. By the mean value theorem for integrals, there exists an intermediate value $r = r_0$ in the range $b \le r \le a$ such that

$$\int_0^{2\pi r_0} \phi_s^2\,ds \le \frac{M(a)}{h}.$$

As above, we appraise the oscillation of ϕ between two points P_1, P_2 on a circular arc $r = r_0$ of length $|s_2 - s_1|$:

$$[\phi(P_2) - \phi(P_1)]^2 \le \frac{M(a)}{h} |s_2 - s_1| \le 2\pi M(a) \frac{a}{h}.$$

Now we choose the width h of our ring $b \leq r_0 \leq a$ such that a/h is a fixed positive constant (for example 2). Since $M(a) \to 0$ as $a \to 0$, we can, for given a, always find circles $r = r_0$ with $a/2 \leq r_0 \leq a$ on which the variation of ϕ is appraised by Dirichlet's integral; this variation is arbitrarily small for sufficiently small a.

These facts may be formulated in a slightly more general way for piecewise smooth functions ϕ without restricting the center O to a position in G. Then the circles r = constant may have arcs in G and arcs outside G. Again we denote by $K = K_a$ the part of G lying in $r \leq a$ and by $M(a) = D_K[\phi]$ the corresponding Dirichlet integral of ϕ. We state

Lemma 1.4: For $b < a$ and $a - b = h$ there exists a circle $r = r_0$ with $b \leq r_0 \leq a$ such that on each connected arc of this circle lying in G the inequality

$$(1.14) \qquad |\phi(P_2) - \phi(P_1)|^2 \leq \frac{M(a)}{h} |s_2 - s_1| < \frac{2\pi a M(a)}{h}$$

holds, where $|s_2 - s_1| < 2\pi a$ is the length of the arc between P_1 and P_2. In particular, for sufficiently small a there exist circles $r = r_0$ with $a/2 \leq r_0 \leq a$ such that the variation of ϕ on each arc of the circle lying in G is arbitrarily small.

Proof: We proceed as above, except that the definition of $D[\phi]$ as an improper integral necessitates some caution in splitting the integral $D_K[\phi]$ into simple integrals. For this purpose we consider K as the limit of a sequence of closed sets K_n in G bounded by polygons, so that ϕ_x and ϕ_y are discontinuous at most along a finite number of arcs and isolated points in K_n. If $L_r^{(n)}$ is the set of circular arcs in which the circle of radius r intersects K_n, we can split D_{K_n} as above and have

$$\int_0^a dr \int_{L_r^{(n)}} \phi_s^2 \, ds \leq D_{K_n}[\phi] \leq M(a);$$

for fixed n there is a value $r_0 = \rho_n$ for which

$$\int_{L_{\rho_n}^{(n)}} \phi_s^2 \, ds \leq \frac{M(a)}{h},$$

and

$$|\phi(P_2) - \phi(P_1)|^2 \leq 2\pi \frac{aM(a)}{h}$$

on each arc of $L_{\rho_n}^{(n)}$. As n tends to infinity we may assume that ρ_n tends to a value ρ, and we obviously have the same inequality for every closed arc L on $r = \rho$ that lies in G. This proves our lemma, if we set $h = a/2$.

In the same way the following modifications of the lemma can be proved:

Lemma 1.4a: Denote by S the part of G between $y = a$ and $y = a + h$. There exists a value y_0, $a \leq y_0 \leq a + h$, such that on each connected segment in G on the line $y = y_0$ we have

$$|\phi(x_2, y_0) - \phi(x_1, y_0)|^2 \leq |x_2 - x_1| \frac{D_S[\phi]}{h}.$$

Lemma 1.4b: Consider a family of concentric squares $x = \pm r$, $y = \pm r$ about the origin O. The statement of lemma 1.4 remains true if in the inequality (1.14) the factor 2π is replaced by 8.

Lemma 1.4c: Suppose the domain G in which ϕ is defined extends to infinity, $D[\phi]$ remaining finite. Let K_a denote the part of G outside the circle $r = a$, so that $M(a) \to 0$ as $a \to \infty$. Then again for each a there exists a value r_0, with $a \leq r_0 \leq 2a$, such that inequality (1.14) holds on each connected arc of $r = r_0$ lying in G. The right side of the inequality tends to zero as a tends to infinity, if, say, $h = a$. In a similar way lemma 1.4*b* holds for large squares instead of small ones if G is an infinite domain.

3. *Invariance of Dirichlet's Integral under Conformal Mapping. Applications.* In the next two articles we add a few observations for use in the later chapters.

Theorem 1.1: Dirichlet's integral is invariant under conformal mapping.

Proof: A conformal mapping $x = x(x', y')$, $y = y(x', y')$ may transform the point (x, y) into the point (x', y') and the domain of integration G into a domain G'. The Cauchy-Riemann equations imply

$$\phi_{x'}^2 + \phi_{y'}^2 = (\phi_x^2 + \phi_y^2) \frac{\partial(x, y)}{\partial(x', y')},$$

hence

$$\iint_G (\phi_x^2 + \phi_y^2)\, dx\, dy = \iint_{G'} (\phi_{x'}^2 + \phi_{y'}^2)\, dx'\, dy'.$$

This equation expresses the invariance of Dirichlet's integral and, of course, also implies the invariance of the mixed integral $D_G(\phi, \psi)$.

As a consequence, Dirichlet's Principle is proved for all domains which we can map conformally onto a circle. (From now on we will on occasion take the liberty of using the word "circle" to mean either the circumference or the disk, provided that no confusion can arise.) We recall a few elementary conformal transformations:

a) *The Semicircle:* The transformation

$$\frac{1 - \zeta}{1 + \zeta} = \left(\frac{1 - z}{1 + z}\right)^2$$

maps the interior of the upper half of the unit circle in the z-plane onto the upper half of the ζ-plane. The mapping

$$\eta = \frac{\zeta - i}{\zeta + i}$$

transforms this upper half-plane into the interior of the unit circle in the η-plane.

b) *The Circular Lune:* More generally, a circular lune in the z-plane with the vertices z_1, z_2 and the angle α can be transformed into a lune in the ζ-plane with the vertices ζ_1, ζ_2 and the angle $\alpha\lambda$ by the transformation

$$\frac{\zeta - \zeta_1}{\zeta - \zeta_2} = \left(\frac{z - z_1}{z - z_2}\right)^\lambda.$$

If λ is chosen equal to π/α, the lune in the ζ-plane is a circle.

c) *The Circular Sector:* The center of the circle of a circular sector in the z-plane may be assumed to be the origin. If the angle of the sector is α, the transformation $\xi = z^{\pi/\alpha}$ maps the sector onto a semicircle in the ξ-plane, which can then be mapped conformally on the unit circle.

d) *The Ellipse:* The interior of an ellipse may be seen to be equivalent to a circular disk by elementary conformal transformations.

4. *Dirichlet's Principle for a Circle with Partly Free Boundary.* In Chapters II and VI we shall need a generalization of Dirichlet's Principle to the case in which the boundary values are left free on

a part of the circumference of the circle. On such an arc the solution of the minimum problem will be seen to satisfy a "natural boundary condition": the normal derivative vanishes, or, what is equivalent, the harmonic function v conjugate to u attains constant boundary values. Since the circular disk is conformally equivalent to a semi-circular domain in such a way that the free arc is transformed into the bounding diameter, it is sufficient to prove

Theorem 1.2: Let H be the domain $x^2 + y^2 < 1$, $y > 0$. Denote by μ its bounding semi-circular arc, by λ its (open) bounding diameter $y = 0$, $-1 < x < 1$. Let $g(x, y)$ be a function continuous in H and on μ, with piecewise continuous first derivatives in H, and assume $D_H[g] < \infty$. Then there exists a function u, harmonic in $H + \lambda + \mu$, with the same boundary values as g on μ, with vanishing normal derivative on λ, and such that $D_H[u] \leq D_H[g]$, the equality sign holding only for $u = g$.

Proof: If g is assumed to be continuous on the whole boundary $\lambda + \mu$, the statement follows immediately from Dirichlet's Principle for the circle; for we may then obtain a piecewise smooth function in the full circle by extending g symmetrically into the lower half-circle, i.e. by defining $g(x, -y) = g(x, y)$. The harmonic function in the full unit circle with the boundary values of g and of its symmetric extension is then likewise symmetric and, therefore, has a vanishing normal derivative on λ. Under the less restrictive conditions of the lemma, which does not stipulate continuity of g on λ, we have to supplement this reasoning. Instead of H, we consider first the subdomain H_ϵ defined by

$$x^2 + y^2 < 1, \qquad y > \epsilon.$$

The domain H_ϵ can be mapped on H by an explicit elementary conformal transformation (see §4, 3) which also establishes a continuous biunique correspondence between the boundaries. For $\epsilon \to 0$ this mapping tends uniformly to the identity. It transforms the values of g in H_ϵ into a function g_ϵ which is continuous in $H + \mu + \lambda$ and tends in $H + \mu$ uniformly to g as $\epsilon \to 0$. According to the previous remark, the lemma is proved for the function g_ϵ and for the harmonic function u_ϵ which has the same boundary values as g_ϵ on μ and whose normal derivative vanishes on λ; i.e. we have

$$D_H[u_\epsilon] \leq D_H[g_\epsilon].$$

By the invariance of the Dirichlet integral under conformal mapping we have

$$D_H[g_\epsilon] = D_{H_\epsilon}[g] \leq D_H[g],$$

so that

$$D_H[u_\epsilon] \leq D_H[g].$$

For $\epsilon \to 0$, u_ϵ tends to a harmonic function u, uniformly in $H + \mu + \lambda$, as follows by symmetric extension of g_ϵ and u_ϵ to the full circle.[5] In view of the semicontinuity of Dirichlet's integral, we therefore obtain for $\epsilon \to 0$ the inequality

$$D_H[u] \leq D_H[g].$$

As previously shown in the proof of Dirichlet's Principle for the circle (see §2), the equality sign is excluded except for $u = g$.

5. *Proof of Dirichlet's Principle for General Domains*

1. *Direct Methods in the Calculus of Variations.* Weierstrass' theorem, which states that a continuous function $f(x)$ in a closed interval always possesses a minimum and a maximum, is based on two facts: 1) the *compactness* of the set of real numbers in a closed interval, i.e. the existence, in each infinite subset, of sequences convergent to an element of the set; 2) the continuity of the function $f(x)$. For the existence of a *minimum* alone the second property could be replaced by the less restrictive condition of *lower semicontinuity*: $f(x) \leq \liminf f(x_n)$ for x_n tending to x.

In the calculus of variations the independent variable is not a number x ranging over a closed interval, but an admissible function $\phi(x)$ or $\phi(x, y)$, ranging over a set or a "space of functions." The relevant questions are therefore: a) Is such a space or set of functions compact? b) Is the functional $I[\phi]$ or $D[\phi]$, whose minimum we seek, lower semicontinuous in the space? Of course, the meaning of "compactness" in a function space depends on the meaning of "convergence," which in turn depends on the meaning of "neighborhood" or of "distance" in the space.

[5] For if a sequence of functions is harmonic in a domain G and continuous on the boundary γ, and if the boundary values converge uniformly, the functions converge uniformly in $G + \gamma$ to a function regular and harmonic in G and continuous in $G + \gamma$.

It is entirely possible to attack Dirichlet's variational problem, which we shall henceforth call "Variational Problem I," by seeking affirmative answers to these two questions. However, to proceed along this line we would have to enlarge the original function space to a "Hilbert space" by the addition of "ideal elements," and at the end would be faced with the problem of proving that the solutions found in the larger function space are really contained in the narrower space of functions with continuous second derivatives. For our specific purposes it seems preferable to proceed more directly, exploiting from the outset the individual features of Dirichlet's problem and using the ordinary convergence concept. The proof of semicontinuity presents no serious difficulty, in view of lemma 1.1. We shall have to cope with the fact that sets of minimizing sequences are not compact in the sense of ordinary (uniform) convergence, so that appropriate limiting processes must be devised to obtain the desired solution from a minimizing sequence.

The proof of Dirichlet's Principle proceeds in three steps:

1) Construction of a harmonic function u in G.
2) Proof that $D[u] \leq d$, where d is the greatest lower bound of $D[\phi]$ for all admissible functions.
3) Proof that u attains the prescribed boundary values.

2. *Construction of the Harmonic Function u by a "Smoothing Process."* Consider a minimizing sequence ϕ_n of admissible functions in G. As has been seen in § 1, 3, the functions ϕ_n need not converge. However, the "distance," in the "D-metric," between the functions of the sequence tends to zero, as will be shown presently:

Lemma 1.5: Let $\zeta_n(x, y)$ be any sequence of variations piecewise smooth in G, vanishing at the boundary, and possessing a common upper bound M for $D[\zeta_n]$; then

$$D[\phi_n , \zeta_n] \to 0 \tag{1.15a}$$

for any minimizing sequence ϕ_n. More precisely, if $D[\phi_n] = d_n$ we have

$$D[\phi_n , \zeta_n]^2 \leq (d_n - d)M. \tag{1.15}$$

This is the relation which in our direct approach replaces the vanishing of the first variation in the ordinary variational calculus.

Proof: Since $\phi = \phi_n + \epsilon\zeta_n$ is an admissible function in our variational problem, we have

$$D[\phi] - d = d_n - d + 2\epsilon D[\phi_n, \zeta_n] + \epsilon^2 D[\zeta_n] \geq 0,$$

hence all the more

$$d_n - d + 2\epsilon D[\phi_n, \zeta_n] + \epsilon^2 M \geq 0.$$

Here the left side is a quadratic polynomial in ϵ, and (1.15) simply expresses its non-negative character.

As a consequence we state

Lemma 1.5a: For a minimizing sequence

$$D[\phi_n - \phi_m] \to 0 \tag{1.16a}$$

as n and m tend to infinity. More precisely,

$$D[\phi_n - \phi_m] \leq (\sqrt{d_n - d} + \sqrt{d_m - d})^2. \tag{1.16}$$

Proof: Setting $\zeta_n = \phi_n - \phi_m$ and $M = D[\phi_n - \phi_m]$ we have by (1.15)

$$D[\phi_n, \phi_n - \phi_m] \leq \sqrt{d_n - d}\,\sqrt{D[\phi_n - \phi_m]}.$$

Interchanging n and m and adding, we find that

$$D[\phi_n - \phi_m] \leq \sqrt{D[\phi_n - \phi_m]}\,(\sqrt{d_n - d} + \sqrt{d_m - d}).$$

Cancellation of the factor $\sqrt{D[\phi_n - \phi_m]}$ immediately gives (1.16).

Corollary to Lemma 1.5a: If $\phi_n^{(1)}$ and $\phi_n^{(2)}$ are two minimizing sequences, then

$$D[\phi_n^{(1)} - \phi_n^{(2)}] \to 0$$

as $n \to \infty$; for the mixed sequence $\phi_n^{(1)}, \phi_n^{(2)}$ is again a minimizing sequence.

On the basis of lemma 1.5a, Dirichlet's Principle for the circle can be used to construct the solution u of the general problem by a "smoothing process." To that effect let K be a circle in G with boundary κ, and consider in K the harmonic function u_n having the same boundary values as ϕ_n on κ. The sequence of admissible functions ψ_n defined by

$$\psi_n = \begin{cases} u_n & \text{in } K, \\ \phi_n & \text{in } G - K, \end{cases}$$

is then again a minimizing sequence: for $D_K[u_n] \leq D_K[\phi_n]$ in consequence of Dirichlet's Principle for the circle. Therefore we have

$$d \leq D[\psi_n] \leq D[\phi_n].$$

The replacing of the minimizing sequence ϕ_n by the sequence ψ_n will be referred to as a "smoothing" of ϕ_n in K.

By lemma 1.5a, $D[\psi_n - \psi_m] \to 0$ and, *a fortiori*,

$$D_K[u_n - u_m] \to 0,$$

as well as

$$D_K[\phi_n - u_n] \to 0.$$

From these relations the convergence of u_n in K can be inferred by means of lemmas 1.3b and 1.4. The hypothesis of lemma 1.3b is

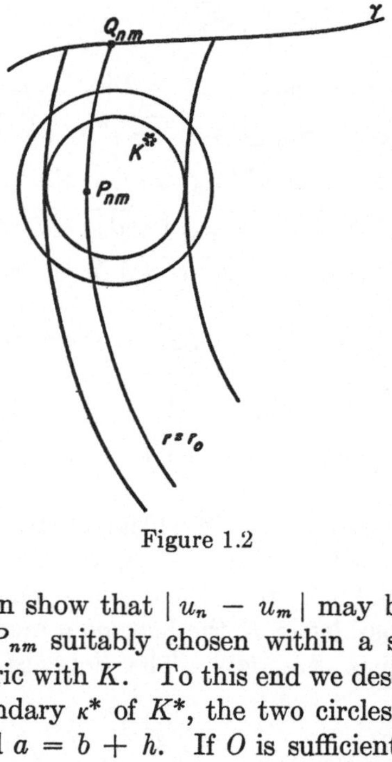

Figure 1.2

satisfied, for we can show that $|u_n - u_m|$ may be made arbitrarily small at a point P_{nm} suitably chosen within a smaller disk K^* of radius $h/2$ concentric with K. To this end we describe about a point O outside the boundary κ^* of K^*, the two circles tangent to κ^*; let their radii be b and $a = b + h$. If O is sufficiently distant from κ^*, all circles about O that intersect κ^* also intersect the boundary γ of G (since G is bounded). By lemma 1.4 these circles include a circle

K_0 of radius $r = r_0$ on which, along any connected arc in G, the oscillation of $p = \psi_n - \psi_m$ is not more than $\sqrt{2\pi(a/h)D[p]}$.

Now let P_{nm} be a point of K_0 lying in the disk K^*, Q_{nm} the nearest point common to K_0 and γ. Since all functions ψ_n have the same boundary values, $p(Q_{nm})$ is zero, and therefore $p(P_{nm}) \leq \sqrt{2\pi(a/h)D[p]}$. By lemma 1.5a, $D[p] \to 0$; consequently $p(P_{nm}) \to 0$ as $n, m \to \infty$. Finally, lemma 1.3b ensures the uniform convergence of the u_n to a harmonic function u, in every closed subdomain S of K. Note also that

$$D_S[u - \phi_n] \to 0. \tag{1.17}$$

Since the derivatives of u_n converge uniformly in S to those of u we have $D_S[u - u_n] \to 0$. Furthermore, $D_S[u_n - \phi_n] \to 0$; hence (1.17) is established by the triangle inequality.

This smoothing process can be applied to any circle in G; it leads in every such circle to the definition of a certain harmonic function u. We assert that this construction defines a uniquely determined function in the *whole* domain G.

For the proof we show that the functions u_1 and u_2 resulting from the smoothing in two *overlapping* circles K_1 and K_2 are identical in the common part K_1K_2. Suppose the sequence ϕ_n gives rise to the minimizing sequences $\psi_n^{(1)}$ and $\psi_n^{(2)}$ by smoothing in the circles K_1 and K_2, respectively. The mixed sequence $\psi_1^{(1)}, \psi_1^{(2)}, \psi_2^{(1)}, \psi_2^{(2)}, \cdots$ is also a minimizing sequence, and therefore $D[\psi_n^{(1)} - \psi_m^{(2)}] \to 0$, in consequence of the corollary to lemma 1.5a. If K' is a circle in K_1K_2, we have, *a fortiori*, $D_{K'}[\psi_n^{(1)} - \psi_m^{(2)}] \to 0$. The functions $\psi_n^{(1)}$ and $\psi_m^{(2)}$ are harmonic in K', and converge to u_1 and u_2, respectively. From the preceding argument it follows that the mixed sequence also converges to a harmonic function u in K'. Hence u_1 and u_2 are identical with u in K', therefore also identical everywhere in K_1K_2.

Remarks: a) The limit function u does not depend on the particluar minimizing sequence ϕ_n from which it was derived by the smoothing process. For let ϕ_n' be another minimizing sequence; then the smoothing of the mixed minimizing sequence $\phi_1, \phi_1', \phi_2, \phi_2', \cdots$ in any circle K leads to a sequence of functions $u_1, u_1', u_2, u_2', \cdots$, harmonic in K, that converge to a harmonic function u, the common limit of u_n and u_n'.

b) Instead of using a fixed circle K for the smoothing of the minimizing sequence, we may use a different concentric circle K_n for the smoothing of each function ϕ_n, provided the radius r_n of K_n remains greater than a fixed positive number, say $r_n > h$. The uniform convergence of u_n to a harmonic function u in a concentric circle of radius less than h is then proved exactly as before.

3. *Proof that* $D[u] = d$. Consider a closed subdomain G' of G and represent it as the sum of a finite number of non-overlapping closed domains S, each of which can be enclosed in a circle K in G. Then, according to (1.17), $D_S[u - u_n] \to 0$, and therefore

$$D_{G'}[u - u_n] \to 0.$$

Hence, by lemma 1.2 and its corollary,

$$D[u] = d.$$

4. *Proof that u Attains Prescribed Boundary Values.* The function u, so far defined in G only, can be extended continuously to the boundary γ and assumes the prescribed boundary values $\bar{g}$ on each Jordan curve α of γ.[6] More precisely: with arbitrarily small prescribed ϵ we have $| u(P) - g(A) | < \epsilon$ for all points P of G whose distance $\overline{PA}$ to a point A on α is less than a suitably small quantity $\delta = \delta(\epsilon)$. This statement expresses the convergence of $u(P)$ to the prescribed boundary values.

Since the boundary values $\bar{g}$ are uniformly continuous on α, it is sufficient to prove the statement in a weaker form: For given ϵ there exists a value $\delta(\epsilon)$ such that $| u(P) - g(R) | < \epsilon$ if the distance $\overline{PR}$ of P to a point R on α nearest to P is less than $\delta(\epsilon)$.

For the proof we set $\overline{PR} = 2h$, intending to fix a proper upper bound δ for $2h$. First we require δ to be small enough to ensure that P has a distance greater than 2δ from every boundary curve except α.

Now we consider the subregion S_h of G consisting of all points P with a distance from α not exceeding $4h$. Because of lemma 1.2 there is a number $\sigma(h)$ such that $\sqrt{4\pi D_{S_h}[\phi_n]} \leq \sigma(h)$ for all n, with $\sigma(h) \to 0$ for $h \to 0$. As a consequence of lemma 1.4 applied to S_h instead of to G, there exists a circle about P with radius r_n, $h \leq r_n < 2h$, on whose circumference the oscillation of ϕ_n is less than $\sqrt{4\pi D_{S_h}[\phi_n]} \leq \sigma(h)$. Similarly, there exists a circle about R with radius ρ_n, $h \leq \rho_n < 2h$, on whose circumference the oscillation of

[6] As to isolated boundary points, see end of article.

ϕ_n is again less than $\sigma(h)$. This circle intersects the circle about P at a point Q_n and, if h is chosen sufficiently small, also intersects α at a point R_n such that an arc of the circle connecting Q_n and R_n lies in G.

Therefore since $\phi_n = g$ on α,

$$|\phi_n(Q_n) - g(R_n)| \leq \sigma(h). \tag{1.18}$$

Let u_n be the harmonic function obtained by smoothing ϕ_n in the circle about P; then the value $u_n(P)$ coincides with some value of ϕ_n on the circle and hence cannot differ from $\phi_n(Q_n)$ by more than $\sigma(h)$:

$$|u_n(P) - \phi_n(Q_n)| \leq \sigma(h). \tag{1.19}$$

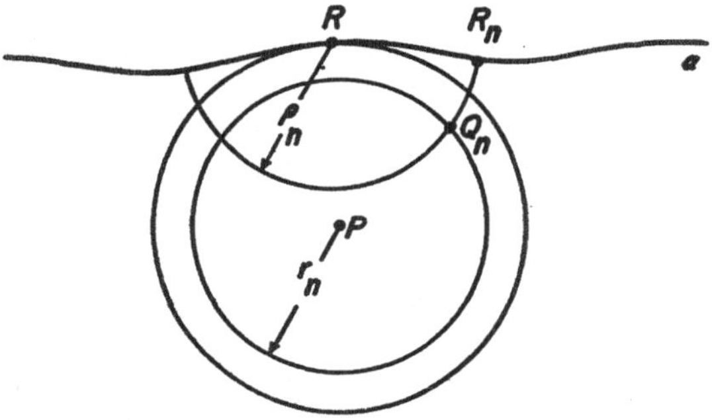

Figure 1.3

By the uniform continuity of g on α there is a positive quantity $\eta(h)$, tending to zero with h, for which

$$|g(R) - g(R_n)| < \eta(h), \tag{1.20}$$

since the distance between R and R_n is less than $2h$. Combining (1.18), (1.19), and (1.20), we find

$$|u_n(P) - g(R)| \leq 2\sigma(h) + \eta(h)$$

Finally we choose δ so small that for $2h < \delta$ the condition $2\sigma(h) + \eta(h) < \epsilon$ is satisfied.

Thus u is recognized as the solution of the boundary value problem. By article 3, $D[u] = d$; consequently u, being admissible in the variational problem, solves this problem as well. The uniqueness proof is exactly the same as in the case of the circle. Hence Dirichlet's Principle, as stated in §1, 6, is established.

It should be observed that our proof becomes invalid if α consists of an isolated boundary point, for then the construction of the arc R_nQ_n breaks down. Assumption of boundary values can be expected only on connected non-degenerate point sets α of γ.

5. *Generalizations.* We can remove two inessential restrictions made so far: a) the condition of boundedness for G; b) the assumption that the boundary consists of Jordan curves.

a) The restriction to bounded domains is eliminated by the following argument: Boundedness was used solely for establishing a connection between the values of the function $\phi_n - \phi_m$ in K and the boundary values by means of circular arcs (see article 2). This connection may, however, be equally well established for an unbounded domain B as long as a boundary continuum T of positive diameter $2h$ is known to exist. In this case we can find a circle A intersecting T on which $| \phi_n - \phi_m |$ is arbitrarily small, for sufficiently large n and m. If A intersects K as well, our previous reasoning remains valid; if A does not meet K, we may assume that A contains K, since otherwise this situation could be obtained by inversion of the plane in A. Finally, the values of $| \phi_n - \phi_m |$ in K can be referred to the (small) values on A, again by the same construction as in article 2.

b) For a boundary arc counted twice, a "slit" in the plane along which we distinguish opposite edges, our reasoning and our results remain literally valid, if we permit g and u to have different boundary values at opposite points of the boundary slit.

To give general definitions for the case of multiple boundary points, one might proceed as follows:

A sequence of points P_n in G is said to be convergent if P_n and P_m can be joined in G by a polygon of arbitrarily small diameter, for n and m sufficiently large. A convergent sequence whose limit point is not in G is said to define a boundary point R. Two convergent sequences P_n, P'_n defining boundary points are said to define the same boundary point if the mixed sequence $P_n P'_n$ is also convergent.

With these definitions the concepts of boundary values and continuity on the boundary become obvious and no change in the reasoning or in the results will be needed if, for admissible functions ϕ, boundary values are understood as limits of $\phi(P_n)$ as P_n tends to R.

Finally it may be stated that the results and the proofs remain valid also if G is a *Riemann surface composed of a finite number of plane domains*, each domain containing only one branchpoint of finite order.

6. *Alternative Proof of Dirichlet's Principle*

Many alternative proofs for Dirichlet's Principle are available. One of these variants, which is given in the following sections, avoids use of the solution of the problem for the circle and, by its slightly different approach, illuminates some of the preceding lines of reasoning.

1. *Fundamental Integral Inequality.* As a preparation, instead of lemma 1.4, we formulate an important inequality due to H. Poincaré:

Lemma 1.4d: For all piecewise smooth functions ζ with vanishing boundary values there exists a constant k depending on G alone such that

$$H[\zeta] \leq k\, D[\zeta]; \tag{1.21}$$

k may be chosen as l^2, where l is the diameter of G.

Proof: To avoid complications due to the boundary we consider instead of G a subdomain G_δ, with polygonal boundary γ_δ, such that G_δ tends monotonically to G for $\delta \to 0$. For a given positive ϵ we have $|\zeta| < \epsilon$ on γ_δ if δ is chosen sufficiently small. Through every point P in G_δ draw the line y = constant in the direction of increasing x up to the first point P' of intersection with γ_δ. Set $\zeta(P') = \epsilon(P)$; then $\epsilon(P)$ is a piecewise continuous function of P with

$$|\epsilon(P)| < \epsilon. \tag{1.22}$$

From

$$|\zeta(P) - \zeta(P')| = \left| \int_{P'}^{P} \zeta_x \, dx \right|$$

we find by Schwarz' inequality that

$$[\zeta(P) - \epsilon(P)]^2 \leq l \int_{P'}^{P} \zeta_x^2 \, dx \leq l \int_I \zeta_x^2 \, dx;$$

here I is the whole intersection of y = constant with G_δ, and l is the diameter of G as before. Integrating over I we find

$$\int_I [\zeta(P) - \epsilon(P)]^2 \, dx \leq l^2 \int_I \zeta_x^2 \, dx.$$

Next we integrate with respect to y over the range of values corresponding to points in G_δ, obtaining

$$\iint_{G_\delta} [\zeta(P) - \epsilon(P)]^2\, dx\, dy \leq l^2 D[\zeta],$$

and consequently by the triangle inequality and (1.22)

$$\left(\iint_{G_\delta} \zeta(P)^2\, dx\, dy\right)^{1/2} \leq \left(\iint_{G_\delta} \epsilon^2\, dx\, dy\right)^{1/2} + (l^2 D[\zeta])^{1/2}.$$

Since ϵ can be chosen arbitrarily small, and since any fixed closed subregion of G is contained in G_δ for sufficiently small δ, the lemma follows almost immediately.

For the analysis of boundary values we need a slight refinement of (1.21), obtained by the same method. As before we decompose G into a closed interior region G' and a boundary strip G^*, so chosen as to contain all points of G whose distance from the boundary does not exceed $4h$. Then

$$D_{G^*}[\zeta] < \sigma(h),$$

where $\sigma(h)$ tends to zero with h.

We consider a continuous boundary component c of G whose diameter exceeds $4h$ and suppose that all other boundary components have a distance from c exceeding $4h$. Let T be a straight segment of length $4h$ extending from a point R on c into the interior of G.[7] We consider all circles about R whose radii vary between 0 and $4h$, and single out those arcs of the circles that lie in G and intersect T. These arcs form a simply connected subdomain L of G; L is a subdomain of G^*, so that

$$D_L[\zeta] < \sigma(h).$$

Now we state:

Lemma 1.4e: Independently of the choice of R we have

$$H_L[\zeta] < 64\pi^2 h^2 \sigma(h). \tag{1.23}$$

The proof is parallel to that of lemma 1.4d with the modification that we now consider circles instead of the straight lines y = constant.

2. *Solution of Variational Problem I.* To construct the solution of Dirichlet's variational problem we again start with an arbitrary

[7] The existence of such segments is obvious, since each straight ray from an interior point has a first point of intersection with γ.

minimizing sequence $\phi_1, \phi_2, \cdots$ and make use of the following facts proved before in §5:
(1.16a) $D[\phi_n - \phi_m] \to 0$ for n and m tending to infinity.
(1.15a) $D[\phi_n, \zeta] \to 0$ for any admissible variation ζ.
(1.24) $D_{G^*}[\phi_n - g] < \sigma(h)$, where G^* denotes a boundary strip of width h and where $\sigma(h)$ is a quantity, not depending on n, that tends to zero with h. We proceed by the following steps:
1) We choose a fixed positive value α and consider the region G_α—possibly consisting of several pieces—comprising all points P in G whose distance from the boundary exceeds α. To any admissible

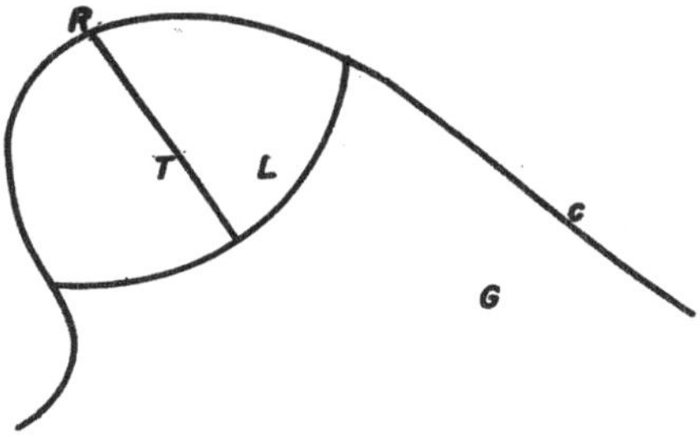

Figure 1.4

function ϕ in G_α a smooth function ω is associated according to the relation

$$\omega(P) = \frac{1}{\pi\alpha^2} \iint_K \phi \, dx \, dy, \tag{1.25}$$

where the integral is extended over the circular disk K of radius α about P. This mean value ω is obviously a continuous function of P in G_α. Furthermore, the first derivatives of $\omega(P)$ exist, are continuous, and are given by

$$\pi\alpha^2\omega_x = \iint_K \phi_x \, dx \, dy, \qquad \pi\alpha^2\omega_y = \iint_K \phi_y \, dx \, dy. \tag{1.26}$$

For since $\omega(x + \xi, y) = \omega(P')$ is expressed by (1.25), but with the integration extended over a disk K' of radius α about P', we have

$$\frac{1}{\xi} \pi\alpha^2[\omega(x + \xi, y) - \omega(x, y)]$$
$$= \frac{1}{\xi}\left\{\iint_{\Lambda_1} \phi(x, y) \, dx \, dy - \iint_{\Lambda_2} \phi(x, y) \, dx \, dy\right\},$$

where Λ_1 is the portion of K' exterior to K and Λ_2 the portion of K exterior to K', each having area less than $2\xi\alpha$. As $\xi \to 0$ the right side tends to the limit $\int_\kappa \phi\, dy$ (where κ is the circle bounding the disk K) as becomes immediately evident if the double integral of the continuous function ϕ is split into an integral with respect to y and an integral with respect to x. Expressing the contour integral by the integral of ϕ_x over K, we obtain the result stated in (1.26); a similar procedure yields the result for ω_y. That ω_x and ω_y are continuous follows immediately; for

$$\pi\alpha^2[\omega_x(x + \xi, y) - \omega_x(x, y)] = \iint_{\Lambda_1} \phi_x\, dx\, dy - \iint_{\Lambda_2} \phi_x\, dx\, dy,$$

and the square of the right side is by Schwarz' inequality not larger than $4\xi\alpha D[\phi]$ since each lune has area smaller than $2\xi\alpha$.

We identify ϕ with the functions ϕ_n of our minimizing sequence and apply (1.16a), as well as the relation

$$H[\phi_n - \phi_m] \to 0, \tag{1.16b}$$

which follows from the fundamental inequality (1.21). As a consequence, the expressions

$$\pi^2\alpha^4(\omega_n - \omega_m)^2 = \left[\iint_K (\phi_n - \phi_m)\, dx\, dy\right]^2 \le \pi\alpha^2 H[\phi_n - \phi_m],$$

$$\pi^2\alpha^4\left(\frac{\partial\omega_n}{\partial x} - \frac{\partial\omega_m}{\partial x}\right)^2 = \left[\iint_K \left(\frac{\partial\phi_n}{\partial x} - \frac{\partial\phi_m}{\partial x}\right) dx\, dy\right]^2 \le \pi\alpha^2 D[\phi_n - \phi_m],$$

$$\pi^2\alpha^4\left(\frac{\partial\omega_n}{\partial y} - \frac{\partial\omega_m}{\partial y}\right)^2 = \left[\iint_K \left(\frac{\partial\phi_n}{\partial y} - \frac{\partial\phi_m}{\partial y}\right) dx\, dy\right]^2 \le \pi\alpha^2 D[\phi_n - \phi_m],$$

converge to zero uniformly at all points in G_α; hence there exists in G_α a limit function $u(P) = u(P; \alpha)$ such that

$$u(P) = \lim_{n\to\infty} \omega_n(P) \tag{1.27}$$

and

$$u_x = \lim_{n\to\infty} \omega_{n_x}, \qquad u_y = \lim_{n\to\infty} \omega_{n_y}, \tag{1.28}$$

convergence being uniform in G_α.

2) Next we show that the limit function $u(P)$ is independent of the

radius α. For the proof we make use of (1.15a), choosing suitable variations ζ. With r denoting distance from P, let

$$\zeta = \begin{cases} 0, & \text{for } r \geqslant \alpha, \\ \frac{1}{2}\left[1 - \frac{r^2}{\alpha^2}\right], & \text{for } r < \alpha. \end{cases}$$

Then (1.15a) becomes

$$D_K[\phi_n, \zeta] = -\iint_K \phi_n \Delta\zeta \, dx \, dy + \int_\kappa \phi_n \frac{\partial \zeta}{\partial r} \, ds \to 0,$$

where s denotes arc length on κ. Since $\Delta\zeta = -2/\alpha^2$ and $\partial\zeta/\partial r = -1/\alpha$ on κ we obtain

$$\text{(1.29)} \qquad \frac{1}{\pi\alpha^2} \iint_K \phi_n \, dx \, dy - \frac{1}{2\pi\alpha} \int_\kappa \phi_n \, ds \to 0;$$

this relation shows that $u(P)$ is also the limit of the mean value of ϕ_n over the circle κ.

Furthermore we choose a variation

$$\zeta' = \begin{cases} 0, & r \geqslant \alpha, \\ \log \frac{r}{\alpha}, & t \leqslant r < \alpha, \\ \log \frac{t}{\alpha}, & r < t. \end{cases}$$

Relation (1.15a) leads immediately to

$$\text{(1.30)} \qquad \frac{1}{2\pi\alpha} \int_{r=\alpha} \phi_n \, ds - \frac{1}{2\pi t} \int_{r=t} \phi_n \, ds \to 0,$$

and (1.30) shows that the mean value of ϕ_n over a circle is in the limit independent of the radius.

Hence the definition of u, u_x, u_y is independent of α and these functions are continuous everywhere in G.

3) We now show that the function u attains the prescribed boundary values, or that the function $\zeta = u - g$ has the boundary values zero. For the proof, consider a point P in G at a distance $2h$ from the nearest point on the boundary. Suppose that R is one of the points nearest to P so that $RP = 2h$, and that R lies on a boundary component c of diameter exceeding $4h$. The disk K with radius h about

P lies in the region G^* and in particular in the domain L attached to R, G^* being a boundary strip and L a subregion of G^* for which $D_L[\phi_n] < \sigma(h)$, as in article 1. Hence by (1.24) and by lemma 1.4e of article 1 we have

$$\left[\frac{1}{\pi h^2}\iint_K (\phi_n - g)\,dx\,dy\right]^2 \leq \frac{1}{\pi h^2} H_L[\phi_n - g]$$
$$< \frac{1}{\pi h^2}\, 64\pi^2 h^2 \sigma(h) = 64\pi\sigma(h).$$

The right side is arbitrarily small for sufficiently small h. The left side, for fixed h and sufficiently large n, differs arbitrarily little from the square of the difference between $u(P)$ and the mean value of the continuous function g over the disk K. Since this mean value approaches $g(R)$ as $h \to 0$, it follows that $| u(P) - g(R) |$ is arbitrarily small for sufficiently small h. The proof is complete.

4) To show that u is admissible in Dirichlet's variational problem we have to ascertain that $D[u]$ is finite. Beyond that, we shall prove that $D[u] = d$. It suffices to show that $D_{G'}[u] \leq d$ for every closed subregion G' of G. To this end we make use of

Lemma 1.6: Every closed bounded region G' in the plane can be covered with a finite number of non-overlapping circular disks in such a way that the part of G' not covered by disks has an arbitrarily small area. Moreover, the maximum radius of the covering circles may be chosen arbitrarily small.

Proof: Let ϵ be an arbitrarily small positive quantity. We first cover G' by a finite number of non-overlapping squares with edges shorter than 2ϵ, so that the remaining part of G' has an (upper) area less than $\epsilon^2/2$. In each of these squares we consider the inscribed circle. The remaining parts of the squares are again covered with smaller squares so that the total residual area is less than $\epsilon^2/4$. We add to the set of covering circles the circles inscribed in these new squares, and proceed as before. By continuing this process sufficiently long, we obtain a paving of G' by non-overlapping circles that leaves free as small an area as we may prescribe; for each circle covers more than $\frac{3}{4}$ of the area of the circumscribed square, and hence the area r_m remaining after m steps is appraised by

$$r_m < \frac{A}{4^m} + \frac{\epsilon^2}{2}\left[1 + \frac{1}{2} + \cdots + \frac{1}{2^{m-1}}\right] < \frac{A}{4^m} + \epsilon^2,$$

where A is the area of G. Obviously, by starting with sufficiently small squares and by taking m large enough we can satisfy the stipulations of the lemma.

On the basis of this lemma the integral

$$I = \iint_{G'} f(x, y)\, dx\, dy$$

of a continuous function $f(x, y)$ can be represented as follows: For a given ϵ we choose a covering of G' by circles of radii $\rho_1, \rho_2, \cdots, \rho_m$ such that the remaining area is less than ϵ and the oscillation of f in each of the circles is less than ϵ. Denoting by f_i the value of f at the center P_i of the circle K_i we have

$$I_\epsilon = \sum_{i=1}^{m} \pi\rho_i^2 f_i \to I$$

for ϵ tending to zero, as follows from the elementary definition of a two-dimensional integral.

By Schwarz' inequality, the value of ω_{n_x} at the point P_i is appraised by

$$[\omega_{n_x}(P_i)]^2 = \frac{1}{\pi^2\rho_i^4}\left|\iint_{K_i} \phi_{n_x}\, dx\, dy\right|^2 \le \frac{1}{\pi\rho_i^2}\iint_{K_i} \phi_{n_x}^2\, dx\, dy.$$

Thus we find that

$$\sum \pi\rho_i^2[\omega_{n_x}^2(P_i) + \omega_{n_y}^2(P_i)] \le \sum \iint_{K_i} (\phi_{n_x}^2 + \phi_{n_y}^2)\, dx\, dy \le D[\phi_n].$$

Letting n tend to infinity, we have

$$I_\epsilon \le d,$$

where I_ϵ now refers to the integrand $u_x^2 + u_y^2$. Hence

$$D_{G'}[u] \le d,$$

and the relation $D[u] \le d$ and therefore $D[u] = d$ is established: u solves the variational problem. Consequently $D[u, \zeta] = 0$ for any admissible variation ζ. Replacing ϕ_n by u in (1.29) and (1.30) we find

$$u(P) = \frac{1}{\pi\alpha^2}\iint_K u\, dx\, dy$$

for a disk K of arbitrary radius α about P. Likewise we have

$$u_x(P) = \frac{1}{\pi\alpha^2} \iint_K u_x \, dx \, dy,$$

$$u_y(P) = \frac{1}{\pi\alpha^2} \iint_K u_y \, dx \, dy.$$

Hence it follows by the reasoning above that u_x and u_y possess continuous derivatives, given by

$$u_{xx} = \frac{1}{\pi\alpha^2} \int_\kappa u_x \, dy,$$

etc., with notation as above. Finally we apply the classical formalism of the variational calculus and conclude from the relation $D[\zeta, u] = 0$ and from the continuity of u_{xx} and u_{yy} that u is harmonic.

7. *Conformal Mapping of Simply and Doubly Connected Domains*

The solution of the boundary value problem for harmonic functions leads readily to the conformal mapping of simply and doubly connected domains on the interior of a circle or a circular ring, respectively. We first consider doubly connected domains:

Let G be bounded by two Jordan curves γ_1 and γ_2, and let U^* be the harmonic function in G with the boundary values 0 on γ_1 and -1 on γ_2. If V^* is the harmonic function in G conjugate to U^*,

$$U^* + iV^* = F^*(x + iy) = F^*(z)$$

is analytic as a function of the complex variable $z = x + iy$ in G. The curves γ_1 and γ_2 belong to the family of curves $U^* = \text{constant}$, and the lines $V^* = \text{constant}$ are the orthogonal trajectories of the family. The function V^* is not single-valued, but increases by a "period" k if the point z in G is made to describe a closed circuit Γ that encircles γ_2 in the positive sense. The period

$$k = -\int_\Gamma [U_y^* \, dx - U_x^* \, dy] = \int_\Gamma \frac{\partial U^*}{\partial n} \, ds$$

(where $\partial/\partial n$ denotes the normal derivative, s arc length on Γ) is independent of the specific choice of Γ but does depend on the shape of G. The level curves $U^* = \text{constant} = c$, $0 > c > -1$ are closed, simple (analytic) curves in G. For there can be only one subdomain $U^* > c$ of G adjacent to γ_1: no domain $U^* > c$ can reach γ_2 where

$U^* = -1$, and there can be no closed subdomain $U^* \geq c$ inside G, since this would imply $U^* \equiv c$. Consequently the system of curves $U^* = c$ is homotopic to γ_1 and γ_2, and this system, together with its orthogonal trajectories, is described by the lines in Figure 5. Obviously $k \neq 0$.

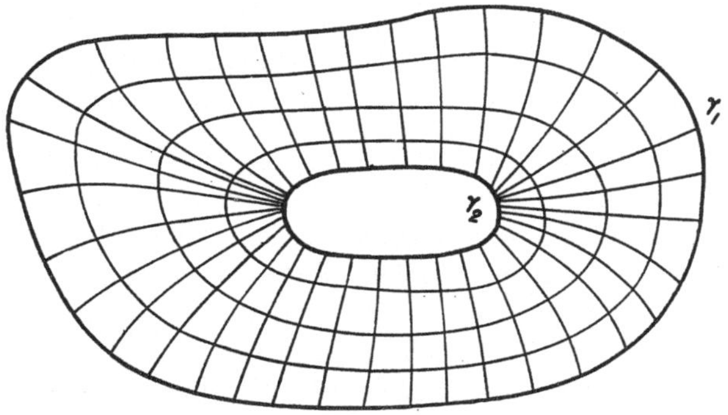

Figure 1.5

For the function

$$U + iV = F(z) = \frac{2\pi}{k} F^*(z)$$

the period is $2\pi i$ and hence

$$\zeta = u + iv = e^{F(z)} = f(z)$$

maps G onto the circular ring

$$e^{-2\pi/k} < |\zeta| < 1$$

of the ζ-plane.

In the limiting case when γ_2 degenerates into the origin O, the function U^* would become identically zero (since the boundary value -1 at the isolated boundary point O is without influence on the solution), and $f(z)$ could not be directly obtained as above.[8] To produce

[8] In the physical interpretation of analytic functions our transformations correspond to vortex motions of a two-dimensional liquid in G such that the lines U = constant or U^* = constant are the streamlines.

It is not difficult to obtain the mapping of a simply connected domain G on the circle $|\zeta| < 1$ as a limiting case by letting γ_2 shrink to a point O, which will then be mapped on the origin $\zeta = 0$ (see, e.g., theorem 2.2).

the period 2π for the conjugate function V, the function U must have the form

$$U = \log r + w(x, y),$$

where r is the distance of the point (x, y) from the origin and w is a regular harmonic function. For such functions U Dirichlet's integral is infinite, and a treatment by Dirichlet's Principle will have to aim not at U but at the regular function w, whose boundary values, by the condition $U = 0$ on γ_1, coincide with those of $-\log r$, hence are known.

The function

$$f(z) = e^{F(z)} = e^{U+iV}$$

maps G onto the unit circle in such a manner that O is transformed into the origin. Thus the problem of conformal mapping of simply connected domains on a circle can likewise be subordinated—in a slightly less direct way—to Dirichlet's Principle.

By linear transformations of the complex plane, the interior of a circle or of a circular ring may be mapped onto the exterior of one or of two circles, respectively; in the latter case the radii of both circles may be chosen as unity.

For domains of connectivity higher than two, reduction of the problem of conformal mapping to a boundary value problem is no longer possible in quite so simple a manner. We shall obtain results of much wider scope by a different approach which, incidentally, will be largely independent of the general results of this chapter.

8. Dirichlet's Principle for Free Boundary Values. Natural Boundary Conditions

Except for §4, 4 we have so far assumed that the boundary values of admissible functions ϕ were prescribed on the boundary γ. Now we consider the same variational problem as before except that we do not impose boundary conditions on the admissible functions for a portion λ of the boundary $\gamma = \lambda + \mu$, while the values of ϕ remain prescribed on the complementary part μ. If such "free boundaries" λ occur in our variational problem, nothing is changed in the preceding reasoning. The solution u exists, is harmonic, and attains the prescribed boundary values on those connected parts μ of γ where boundary values are imposed, isolated points of μ being again excluded.

As we found in §4, in the special case where G is half of a circular disk, μ its circular and λ its straight boundary, the solution u on the "free" part λ of G is characterized by the fact that the conjugate harmonic function attains constant boundary values on λ. As follows from §7, any simply connected domain G can be mapped conformally on a half-disk in such a manner that a prescribed segment of γ passes into the bounding diameter. By this mapping u and its conjugate v are again transformed into a pair of conjugate harmonic functions; hence the result is immediately generalized to such domains.

In the following chapter we shall, however, be particularly interested in related facts for multiply connected domains G. In preparation we prove

Theorem 1.3: Let the boundary γ of a k-fold connected domain consist of at least one boundary line α formed by Jordan arcs and of connected boundaries $\beta_1, \beta_2, \cdots, \beta_m$. Consider $D[\phi]$ for all functions ϕ piecewise smooth in G and assuming prescribed boundary values on α. Then the solution for the variational problem of minimizing $D[\phi]$ is a harmonic function u that assumes the prescribed values on α and satisfies the conditions

$$\int_{\beta_i'} \frac{\partial u}{\partial n}\, ds = 0, \qquad i = 1, 2, \cdots, m.$$

Here β_i' is an arbitrary polygon bounding, with β_i, a two-fold connected subdomain of G; $\partial/\partial n$ again denotes the normal derivative, s arc length on β_i'. Since $\partial u/\partial n = \partial v/\partial s$, this relation expresses the fact that the conjugate function v is single-valued in the two-fold connected domains under consideration. Moreover, the conjugate function v possesses constant boundary values on each boundary β_i. This fact constitutes what we call the "natural boundary condition" for free boundaries, implying that v is single-valued in each of the two-fold connected boundary strips. That v is single-valued in the whole domain G is then assured if α consists of only one boundary curve.[9]

[9] Without making use of the natural boundary conditions, we may immediately establish that v is single-valued by the following reasoning based on the minimum property of $D[u]$. The minimum property is equivalent to the condition that $D[u, \zeta] = 0$ for any function ζ continuous in $G + \alpha$, piecewise smooth in G, vanishing on α, and yielding a finite value $D[\zeta]$. Let σ be an arbitrary polygon in G which separates α from some of the boundaries

Proof: We may accept the existence of the solution u of the variational problem, established by the same considerations as for completely fixed boundary values; hence only the natural boundary conditions for the boundaries β_i need be ascertained. Obviously the function u also furnishes the minimum for the integral $D_S[\phi]$ extended over a doubly connected strip S between β_i and a Jordan curve β_i' so drawn that it bounds with β_i a doubly connected subdomain of G. Admissible for this minimum problem are all functions ϕ continuous in $S + \beta_i'$, piecewise smooth in S, and having the same values as u on β_i'. We map S conformally on a ring $R: a < r < 1$ about the origin, in such a way that $r = a$ corresponds to β_i' and $r = 1$ to β_i. Retaining the notation u, v for the images of u and v we have, because of the invariance of Dirichlet's integral under conformal mapping, $D_R[u] \leq D_R[\phi]$, which means that u also solves the corresponding minimum problem for R. Proceeding as in §4, 4, we replace $\phi(r, \theta)$ in R by $\phi_\epsilon(r, \theta) = \phi(r', \theta)$ with $r' = r - \epsilon \dfrac{r - a}{1 - a}$, and extend ϕ_ϵ by reflection into the ring $\bar{R}: 1 \leq r \leq 1/a$, defining there $\phi_\epsilon(r, \theta) = \phi_\epsilon(1/r, \theta)$. For the harmonic function $w(r, \theta)$ in the ring $R + \bar{R}$ with boundary values $u(a, \theta) = u(1/a, \theta)$ we have $D_R[w] = D_{\bar{R}}[w] \leq D_R[\phi_\epsilon]$. Since $D_R[\phi_\epsilon] \to D[\phi]$, we find

$$D_R[w] \leq D[\phi].$$

The function w is determined solely by the prescribed boundary values, hence does not depend on the individual function ϕ admissible in the variational problem for R; consequently w solves this problem.

$\beta_1, \beta_2, \cdots, \beta_m$, leaving α on the inside. We pick a small positive constant h and consider in the polygon a strip S consisting of points P whose distance $d(P)$ from σ is less than h. Now let $\zeta = 1$ outside and on σ, $\zeta = [h - d(P)]/h$ in S, and $\zeta = 0$ otherwise, in particular near α. Then $D[\zeta] < \infty$ and $D[u, \zeta] = D_S[u, \zeta] = 0$, so that, by Green's formula,

$$\int_\sigma \frac{\partial u}{\partial n} ds = \int_\sigma \frac{\partial v}{\partial s} ds = 0,$$

where s denotes arc length on σ and n the interior normal. This relation exhibits the single-valuedness of v under analytic continuation along σ.

It should be noted that the proof becomes invalid for analytic continuation of v along curves σ which do not separate G as stipulated, a remark which will become relevant in the generalization of our theory to Riemann surfaces not of genus zero.

Since, by the same reasoning employed before in §2, the solution of this minimum problem is uniquely determined, we have $w = u$. Therefore it is proved that u is a harmonic function regular not only in R but in the whole ring $R + \bar{R}$; because of $\partial w/\partial r = \partial u/\partial r = 0$ for $r = 1$ it is further shown that the conjugate harmonic function v is constant on $r = 1$. Under conformal mapping of the ring R onto the strip S, the functions u and v remain conjugate and the constant character of the boundary values of v is preserved.[10] Thus our theorem is proved.

[10] If β_i is not analytic the analytic extension of u and v outside G naturally ceases to be possible.

CHAPTER II

Conformal Mapping on Parallel-Slit Domains

1. Introduction

1. *Classes of Normal Domains. Parallel-Slit Domains.* As shown in Chapter I, every simply connected domain G in the plane (except the full plane or the plane with one point removed) can be mapped conformally on the unit circle; every doubly connected domain G (except the plane with two points removed) can be mapped onto a circular ring whose outer circle may be chosen as the unit circle, the inner circle depending on G and possibly degenerating to a point.

We shall now turn to the general problem of mapping conformally all domains of given topological structure (e.g., all k-fold connected plane domains) onto individuals of a simple special family which depends on a finite number of parameters alone. Such a class $\mathfrak{N}$ of domains will be called a *class of normal domains.* It is of course desirable to reduce the number of parameters in the class $\mathfrak{N}$ as far as possible so that no two domains in $\mathfrak{N}$ are conformally equivalent. The family $\mathfrak{N}$ then contains one and only one conformal representative of every individual of the set of domains G. As implied in our previous results, for simply and doubly connected plane domains the minimal numbers of such parameters or "moduli" are 0 and 1, respectively. We shall see that for k-fold connected plane domains, with $k > 2$, the minimal number is $3k - 6$.

For convenience the term "class of normal domains" is somewhat loosely used; superfluous parameters may be retained, to be eliminated afterwards. For example, as we shall see in Chapter V, k-fold connected domains can be mapped onto the exterior of k circles, a domain depending on $3k$ parameters. By linear transformation of the complex plane the number of parameters may be reduced: one of the circles may be chosen as the unit circle and the center of another prescribed.

In the present chapter we shall be concerned with another class $\mathfrak{N}$: "*parallel-slit domains*" or, more briefly, "*slit domains.*" Such slit domains consist of the whole plane of the complex variable $w = u + iv$ except for straight segments v = constant, the "boundary slits." (In the case of infinite connectivity, these may be infinite in number.) Also admitted are slits which degenerate into points.

Likewise we shall consider "*half-plane slit domains*," consisting of the half-plane $v > 0$ (instead of the full plane) except for a number of parallel slits. The line $v = 0$ is then one of the boundaries.

By the following physical argument one is led to the conjecture that such k-fold connected slit domains constitute classes of normal domains. Suppose that G is a k-fold connected domain in the plane of the complex variable $z = x + iy$. Consider a potential flow in G coming from a dipole O at the interior point $z = 0$. This flow may be characterized by an analytic function $w = u + iv = f(z)$ with the singularity $1/z$ at $z = 0$. In this representation, the curves u = constant are equipotential lines, v = constant streamlines of the flow. The boundary curves of G, which we may visualize as smooth, are parts of streamlines v = constant. It is plausible that the streamlines are smooth analytic closed curves through O, along which u varies monotonically from $-\infty$ to $+\infty$, with the exception of those k streamlines $v = c_1, v = c_2, \cdots, v = c_k$, which reach the boundaries, split there into two branches, and pass around the boundary in different directions until they meet again to lead back to the source at O. Each streamline $v = c$, except for $c = c_1, c_2, \cdots, c_k$, is then mapped biuniquely onto a full line $v = c$ in the w-plane. Of these exceptional streamlines, the parts coinciding with the boundary are mapped onto slits, i.e., straight segments $v = c_1, v = c_2, \cdots, v = c_k$ in the w-plane, in such a way that the two edges of the slit correspond to the two branches of the streamline along the corresponding boundary curve of G. If the dipole O is placed on a boundary curve of G with its axis in the direction of the curve, we obtain a half-plane slit domain. This domain may be chosen as the upper half-plane by mapping the boundary curve onto the entire line $v = 0$.

In the preceding consideration the choice of 6 real parameters for a given domain G is left open, since we may place the dipole at an arbitrary point z_0 of G, choose the direction and intensity of the dipole arbitrarily, and modify the mapping function by the addition of an arbitrary constant. In other words, we may specify a mapping function in the form

$$u + iv = w = f(z) = \frac{a}{z - z_0} + b + (z - z_0)R(z)$$

where $R(z)$ is regular and analytic in G, $a \neq 0$, b are arbitrary complex parameters, and z_0 is an arbitrary point in G. Since a k-fold connected slit domain depends on $3k$ parameters, namely the initial points and lengths of the k slits, we are led to the conjecture that for $k > 2$ a reduction of the number of parameters to $3k - 6$ is possible.[1] For half-plane slit domains this number of "moduli" is more readily

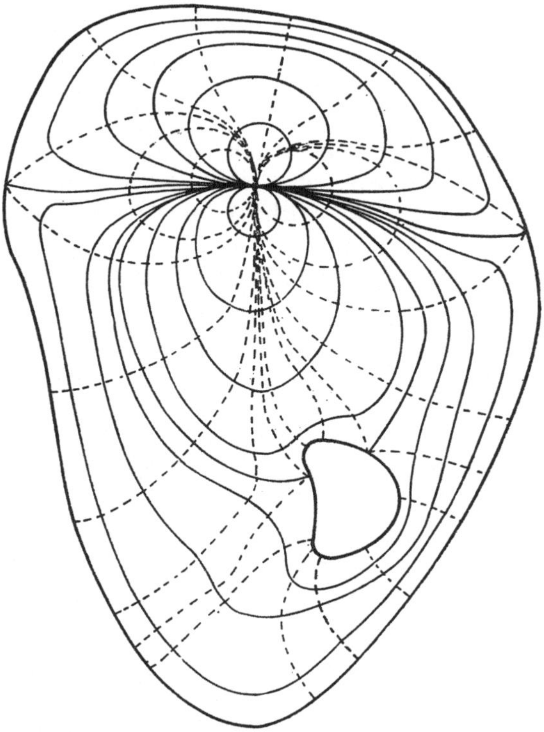

Figure 2.1. Dipole flow in doubly connected region.

visualized. Since $v = 0$ is a boundary line, only a real additive constant b is admissible, and z_0 and a are similarly restricted; thus only three parameters remain at our disposal for reducing to $3k - 6$ the $3(k - 1) = 3k - 3$ parameters in a k-fold half-plane slit domain. We may, for example, fix these parameters by making the point

[1] However, as shown in Chapter I, the number of moduli is not $3k - 6$ for $k = 1$ or $k = 2$, since for these cases a slit domain admits, respectively, three-parametric or one-parametric transformations into itself.

$w = \infty$ correspond to a given boundary point of G and by fixing the initial point of another slit as the point $w = i$.

While the first objective of the present chapter is the proof that these slit domains form a class of normal domains for k-fold connected plane domains G, we shall extend the mapping theorem greatly to Riemann domains G with finite Euler characteristic of arbitrary topological structure. For domains not of genus zero, however, the concept of slit domain must be modified. The corresponding topological structure of the slit domains is introduced by means of pairs of "inner slits" whose boundaries are coordinated by simple rules of "identification" (see §7).

2. *Variational Problem: Motivation and Formulation.* Our object is to characterize the dipole potential u by a minimum problem for the Dirichlet integral $D[\phi]$. Without loss of generality it may be assumed that the dipole is at the origin and that the singularity there is of the form $x/(x^2 + y^2)$. We aim at a complex mapping function $w = u + iv = f(z) = \dfrac{1}{z} + R(z)$, $R(z)$ being regular in G. The Dirichlet integral of u over G is infinite; therefore we must seek to characterize the potential by a modified minimum problem. A clue is provided by the following argument. As seen in Chapter I, §8, the boundaries γ of G are automatically streamlines $v =$ constant, for a potential u obtained by minimizing $D[\phi]$ with the values of ϕ left free on γ. Consequently the following property of u is plausible: If κ is an analytic curve in G enclosing the origin, say a circle, K the interior of κ, and K^* the subdomain $G - K$ of G, u furnishes the minimum of $D_{K^*}[\phi]$ for all functions ϕ piecewise smooth in K^*, continuous in $K^* + \kappa$, and having on κ the same values as u. In itself this statement does not yet characterize u as the solution of a variational problem, since the minimum property of u is formulated with reference to the values of the function u itself on an arbitrary small curve κ. It leads, however, to the following, as yet only tentatively conjectured property of u:

The potential u is characterized by the prescribed singularity $x/(x^2 + y^2)$ at the origin and by the variational condition

$$D[u, \zeta] = 0, \tag{2.1}$$

valid for an arbitrary function ζ which is piecewise smooth in G, and identically zero inside an arbitrarily small circle κ about the origin.

That such a function u exists, and that it provides the mapping of

G on a slit domain, will be proved in the next sections. Here we merely prove

Lemma 2.1: A function u, with the prescribed singularity at O but otherwise regular in G, is uniquely determined—except for an additive constant—by the condition (2.1), valid for any function ζ which is piecewise smooth in G and vanishes identically in a neighborhood of the origin, and for which $D[\zeta] < \infty$.

Proof: Suppose the functions u_1 and u_2 both satisfy the condition (2.1). Then the function $W = u_1 - u_2$ is regular and harmonic in G, even at the origin, and satisfies the condition $D[W, \zeta] = 0$ for arbitrary piecewise smooth ζ with $\zeta = 0$ inside a circle κ.

We remove this restriction by proving that $D[W, \eta] = 0$ for any function η, piecewise smooth in G, for which $D[\eta] < \infty$. Let a be the radius of κ and r, θ polar coordinates about the origin O. We define the function

$$h = \begin{cases} \eta, & r \leq a, \\ \dfrac{2a - r}{a}\,\eta(a, \theta), & a < r < 2a, \\ 0, & r \geq 2a. \end{cases}$$

Then $D[h] < \infty$ if we assume (without loss of generality) that

$$\int_0^{2\pi} [\eta_\theta(a, \theta)]^2\, d\theta < \infty.$$

Set $\zeta = \eta - h$, so that ζ satisfies the requirements for condition (2.1), whence $D[W, \zeta] = 0$. Furthermore $D[W, \eta] = D[W, \zeta] + D[W, h] = D[W, h]$; but by Green's formula

$$D[W, h] = -\iint\limits_{r<2a} h\Delta W\, dx\, dy + 2a \int\limits_{r=2a} h \frac{\partial W}{\partial r}\, d\theta = 0,$$

since W is harmonic and $h = 0$ for $r = 2a$. Hence $D[W, \eta] = 0$.

In particular we may now set $\eta = W$ and obtain

$$D[W] = 0.$$

This equation implies $W_x^2 + W_y^2 = 0$, so that $W =$ constant; thus our statement of uniqueness is proved.

To formulate a genuine variational problem solved by the function u, we avoid the singularity by aiming not directly at u but at a function $U = u - S$, where S is a suitably chosen piecewise harmonic function having the prescribed singularity. The advantage of obtaining, as natural boundary conditions, constant boundary values for v is retained by stipulating $S = 0$ except inside a small circle κ of radius a about the origin. Accordingly S is chosen as a function discontinuous at $r = a$. Specifically we choose $S = \frac{x}{x^2 + y^2} + R(x, y)$ inside κ, R being a regular harmonic function. Consider functions

$$\phi = \Phi + S$$

piecewise smooth in G except at O, such that Φ is piecewise smooth except for the discontinuity on κ. Among all admissible functions Φ we seek a solution U of the problem

$$D[\Phi] = \text{min.} = d. \tag{2.2}$$

To ensure that $U + S = u$ satisfies condition (2.1) and is thus independent of the radius a of κ, we must choose the singularity function S properly. Note that condition (2.1) is satisfied if the normal derivative $\partial S/\partial n$ of the singularity function vanishes on κ.[2] For the solution U of our minimum problem necessarily satisfies the condition

$$D[U, \zeta] = 0 \tag{2.3}$$

for arbitrary ζ, piecewise smooth in G, for which $D[\zeta] < \infty$.

Let κ' be any smooth closed curve inside κ enclosing O, T the ring between κ' and κ, and K' the domain bounded by κ'; assume, furthermore, that ζ vanishes throughout K'. Then, with $K'^* = G - K'$,

$$0 = D[U, \zeta] = D_{K'^*}[u, \zeta] - D_T[S, \zeta].$$

Now

$$D_T[S, \zeta] = -\int_{\kappa'} \zeta \frac{\partial S}{\partial n}\, ds + \int_{\kappa} \zeta \frac{\partial S}{\partial n}\, ds - \iint_T \zeta \Delta S\, dx\, dy = 0,$$

since $\zeta = 0$ on κ', $\partial S/\partial n = 0$ on κ, and $\Delta S = 0$ in T. Condition (2.1) is established for arbitrarily small curves κ'. Consequently lemma 2.1 guarantees that u is independent of the radius a and the

[2] The converse likewise holds, as may easily be verified.

singularity function S. Incidentally, κ as well as κ' need not be a circle. What matters is solely the relation $\partial S/\partial n = 0$ on κ.

We now assume that the disk $r \leq a$ lies entirely in G and define the singularity function[3] S by

$$S = \begin{cases} 0, & r > a, \\ \dfrac{\cos\theta}{r} + \dfrac{r\cos\theta}{a^2} = \dfrac{x}{x^2+y^2} + \dfrac{x}{a^2}, & r \leq a. \end{cases} \tag{2.4}$$

Independently of the motivating arguments presented so far, we formulate

Variational Problem II: Among all functions $\Phi = \phi - S$, with S defined by (2.4), for which ϕ is piecewise smooth in G except at O and Φ is smooth in a neighborhood of O, find the function $\Phi = U$ which minimizes $D[\Phi]$.

The preceding argument then establishes

Lemma 2.2: Except for an additive constant, the solution U of variational problem II defines the function $U + S = u$ uniquely and independently of the specific choice of the circle κ and the singularity function S. The function u satisfies condition (2.1) of lemma 2.1. We may fix the additive constant by demanding $\Phi = 0$ at the origin.

It remains to be shown that the solution U of this variational problem exists.

2. *Solution of Variational Problem II*

1. *Construction of the Function u.* First we ascertain that there exist admissible functions Φ with $D[\Phi] < \infty$. The function defined by

$$\Phi = \begin{cases} 0, & r < \dfrac{a}{2}, \\ -2\cos\theta\,\dfrac{2r-a}{a^2}, & \dfrac{a}{2} \leq r \leq a, \\ 0, & r > a \end{cases}$$

is such an admissible function. Hence there exists a non-negative greatest lower bound d for the values $D[\Phi]$ attained by admissible functions.

[3] This singularity function was introduced by H. Weyl [1].

To prove the existence of the solution is now a simple task. Let $\Phi_1, \Phi_2, \cdots$ be any minimizing sequence. By exactly the same reasoning as in lemma 1.5a, the fundamental relation

$$D[\Phi_n - \Phi_m] \to 0 \tag{2.5}$$

or

$$D[\phi_n - \phi_m] \to 0, \tag{2.5a}$$

for $n, m \to \infty$, is immediately established. We replace the minimizing sequence Φ_n by other piecewise harmonic minimizing sequences obtained from "smoothing processes." Returning to a fixed circle about O of radius $r = r_0 \leq a$, we consider:

1) Smoothing of an admissible function Φ in the disk $r \leq r_0$, i.e. replacing Φ by an admissible function Ψ which is harmonic for $r \leq r_0$ and which, except for an additive constant, coincides with Φ outside the disk. The additive constant is so chosen that $\Psi = 0$ at the origin. Because of Dirichlet's Principle for the circle,

$$D[\Psi] \leq D[\Phi].$$

2) Smoothing of an admissible function Φ for the region G_{r_0} outside the circle $r \leq r_0$.[4] We replace Φ by an admissible function Ω which coincides with Φ for $r \leq r_0$, and for which $\omega = \Omega + S$ is the harmonic function in the subdomain $G_{r_0} : r > r_0$ which has on $r = r_0$ the same values as $\Phi + S$ and which solves the problem

$$D_{G_{r_0}}[\phi] \equiv D_{r_0}[\phi] = \min.$$

with $\phi = \Phi + S$ prescribed on $r = r_0$. The existence of ω has been established in Chapter I, §5.

From $D_{r_0}[\omega] \leq D_{r_0}[\phi]$ we again infer

$$D[\Omega] \leq D[\Phi].$$

For since $\Omega = \Phi$ in $r \leq r_0$, it is sufficient to prove this inequality for the integrals taken over G_{r_0} alone. Now

$$D_{r_0}[\Omega] = D_{r_0}[\omega] + D_{r_0}[S] - 2D_{r_0}[\omega, S]$$

and

$$D_{r_0}[\Phi] = D_{r_0}[\phi] + D_{r_0}[S] - 2D_{r_0}[\phi, S].$$

[4] Note that the discontinuity of Φ at $r = a$ is preserved in this smoothing process.

Subtracting, using the minimum property of ω, and remembering the boundary conditions $\omega - \phi = 0$ on $r = r_0$ and $\partial S/\partial r = 0$ on $r = a$, we find $D_{r_0}[\omega - \phi, S] = 0$ and immediately obtain

$$D_{r_0}[\Omega] - D_{r_0}[\Phi] = D_{r_0}[\omega] - D_{r_0}[\phi] \leq 0.$$

After this preparation we start with an arbitrary minimizing sequence Φ_n and replace this sequence by another, $\Psi_n = \psi_n - S$, obtained from Φ_n by smoothing in the disk $r \leq \beta < a$, where β is a fixed constant. Since $D[\Psi_n] \leq D[\Phi_n]$, the sequence Ψ_n is again a minimizing sequence and

$$D[\Psi_n - \Psi_m] = D[\psi_n - \psi_m] \to 0$$

as $n, m \to \infty$. Restricting the domain of integration to the disk $r \leq \beta$, in which Ψ_n is harmonic, and recalling $\Psi_n = 0$ at the origin, we

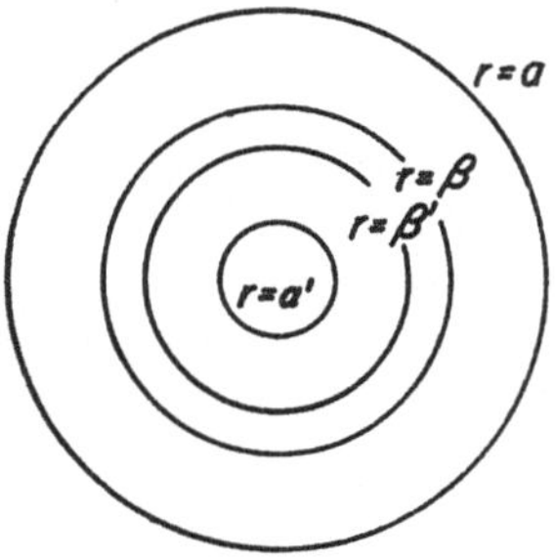

Figure 2.2

infer from lemma 1.3b that the Ψ_n converge uniformly to a harmonic function Ψ in any disk of radius $\beta' < \beta$. Moreover, $D_\beta[\Psi - \Phi_n] \to 0$ by lemma 1.3b, where D_β denotes Dirichlet's integral over the region $r \leq \beta$.

The second step consists in replacing the minimizing sequence Ψ_n by another sequence of functions Ω_n, obtained from Ψ_n by smoothing in the domain $G_{\alpha'}$ exterior to a circle $r = \alpha' \leq \beta'$. Again the functions $\Omega_n = \omega_n - S$ form a minimizing sequence; with $D_{\alpha'}$ denoting Dirichlet's integral over $G_{\alpha'}$,

$$D_{\alpha'}[\Omega_n - \Omega_m] = D_{\alpha'}[\omega_n - \omega_m] \to 0$$

for $n, m \to \infty$, where ω_n is harmonic in $G_{\alpha'}$. Moreover, since $\alpha' \leq \beta'$ the values $\Omega_n(\alpha', \theta) = \Psi_n(\alpha', \theta)$ converge uniformly. Hence it follows (see footnote 5 p. 23) that the sequence ω_n converges uniformly in

$G_{\alpha'}$ to a harmonic function $\omega = \Omega + S$ for which by Lemma 1.3b $D_{\alpha'}[\omega - \omega_n] = D_{\alpha'}[\Omega - \Omega_n] \to 0$; more generally, $D_{\alpha'}[\Omega - \Phi_n] \to 0$ for any minimizing sequence Φ_n which coincides with Ψ_n in the disk $r \leq \alpha'$.

The function U that takes the values of Ψ for $r \leq \alpha'$ and those of Ω for $r > \alpha'$ is therefore admissible and

$$D[U - \Phi_n] = D_{r<\alpha'}[\Psi - \Phi_n] + D_{\alpha'}[\Omega - \Phi_n] \to 0.$$

This implies, as in Chapter I, §5, 3,

$$D[U] = \lim D[\Phi_n] = d.$$

Hence U solves variational problem II; consequently the function u is harmonic throughout G except at the origin, where it has the prescribed singularity. Moreover relation (2.1) holds, and u is uniquely determined, independently of the radius a.

2. *Continuous Dependence of the Solution on the Domain.*[5]

Theorem 2.1: Let G_n be a sequence of subdomains of G tending to G, $U_n = u_n - S$ the solution of variational problem II for G_n (the singularity remaining fixed as n changes). Then U_n and u_n converge in G to functions U and u, respectively. Furthermore, the convergence to zero of $u_n - u = U_n - U$ is uniform in any closed subdomain of G. The function U is the solution of variational problem II for G.

Proof: Convergence of G_n to G means that every point of G is a point of almost all domains G_n. For our purposes it is sufficient to assume monotonic convergence of G_n to G,[6] which means that G_n is a subdomain of G_{n+1}.

We denote admissible functions and greatest lower bounds for G_n by the subscript n. If d is the greatest lower bound in the variational problem for G, and Φ_k a minimizing sequence for G, the functions Φ_k are likewise admissible in the variational problem for a subdomain G_n; hence we have

$$d_n = D_n[U_n] \leq D_n[\Phi_k] \leq D[\Phi_k].$$

[5] The continuous dependence of the solution on the position and direction of the dipole singularity is a fact easily deduced from the compactness properties of harmonic functions.

[6] The continuity theorem without this restriction is easily obtained from that for monotonic approach.

Keeping n fixed and letting k increase to infinity, so that $D[\Phi_k]$ tends to d, we find

$$d_n \leq d.$$

In the same way we find

$$d_n \leq d_m, \qquad m > n$$

since U_m is admissible in the variational problem for G_n. Hence there exists a limit of the sequence:

$$\lim_{n\to\infty} d_n = \delta \leq d.$$

Furthermore, for $m > n$, and with $\zeta = U_n - U_m$ in G_n,

$$d_m \geq D_n[U_m] = D_n[U_n - \zeta] = D_n[U_n] + D_n[\zeta] - 2D_n[U_n, \zeta].$$

Since by the minimum property (2.3) of U_n, $D_n[U_n, \zeta] = 0$,

$$d_m \geq d_n + D_n[\zeta].$$

But $d_m \to \delta$ and $d_n \to \delta$ for $m, n \to \infty$; hence $D_n[\zeta] \to 0$. As a consequence we state: for any fixed closed subdomain G' of G the integral $D_{G'}[\zeta]$ or $D_{G'}[u_n - u_m]$ tends to zero as n and m tend to infinity. Since $U_n = u_n - S = 0$ at the origin, lemma 1.3b implies the convergence of u_n to a harmonic function u in such a way that the convergence of $u - u_n$ and of the first derivatives of $u - u_n$ to zero is uniform in any closed subregion G' of G. Because of the semicontinuity of Dirichlet's integral for harmonic functions (lemma 1.1), we have for $U = u - S$

$$D[U] \leq \lim d_n \leq d,$$

and since U is admissible in the variational problem for G,

$$D[U] = d;$$

thus the theorem is proved.

3. Conformal Mapping of Plane Domains on Slit Domains

In the following articles we shall study the conformal mapping of G by the analytic function $w = u + iv = f(z)$ of the complex variable $z = x + iy$ (u denoting the harmonic function constructed in the preceding section).

1. *Mapping of k-fold Connected Domains.* Let us first consider k-fold connected domains G bounded by k boundary curves or, more generally, connected sets of boundary points $\gamma_1, \gamma_2, \cdots, \gamma_k$:

Theorem 2.2: The analytic function $w = u + iv = f(z)$ maps the k-fold connected domain G conformally on the whole w-plane with k segments parallel to the u-axis excluded. (Some of these "boundary slits" may reduce to points.) The mapping has a single-valued inverse. Choose an arbitrary point z_0 in G and complex constants $\sigma \neq 0$ and λ; the mapping function $f(z)$ is uniquely determined if it is to have the form

$$f(z) = \frac{\sigma}{z - z_0} + \lambda + (z - z_0)R(z),$$

$R(z)$ being regular and analytic in G.

Proof: Since u solves the problem of minimizing $D_{K^*}[\phi]$ for the domain K^* outside the circle κ with the values of ϕ on κ given as those of u, the results of Chapter I, §8 show that $f(z)$ is a single-valued analytic function in G and that the imaginary part v assumes constant boundary values c_ν on the boundary components γ_ν. Let $h = \xi + i\eta$ be any constant with $\eta \neq c_\nu$, $\nu = 1, 2, \cdots, k$. All we have to prove is that the equation $f(z) - h = 0$ has one and only one solution z in G. Let $\gamma_\nu^{(\epsilon)}$ be a sequence of piecewise smooth curves in G tending to γ_ν as $\epsilon \to 0$, and let $\beta_\nu^{(\epsilon)}$ be their images, which are again closed curves (not necessarily simple) in the plane of $w = f(z)$. Then for sufficiently small ϵ, $\beta_\nu^{(\epsilon)}$ is arbitrarily near the line $v = c_\nu$ and hence has positive distance from the point h. Now

$$N - P = \frac{1}{2\pi i} \sum_\nu \int_{\gamma_\nu^{(\epsilon)}} d \log [f(z) - h]$$

is the number of zeros minus the number of poles of $f(z) - h$ in the subdomain G_ϵ of G bounded by the curves $\gamma_\nu^{(\epsilon)}$. Again for sufficiently small ϵ, the point $w = h$ is outside all loops formed by $\beta_\nu^{(\epsilon)}$ and therefore the right-hand side of the equation vanishes, because it represents the total variation of the argument of the vector joining h to a point describing $\beta_\nu^{(\epsilon)}$. Thus $N = P$; since $f(z) - h$ has one pole in G_ϵ, the point O, the function $f(z) - h$ has one zero.

It follows that every point h in the plane of $w = f(z)$ is the image of one and only one point of G, with the possible exception of points h on the straight lines $v = c_\nu$. The image B of G must be k-fold connected, and consequently the boundary of B consists of k slits; these slits lie in a finite domain, because the neighborhood of the origin is mapped on the neighborhood of the point at infinity. Thus the mapping property of $f(z)$ is established.

To prove that the mapping function $f(z)$ is uniquely determined, we assume without loss of generality that $z_0 = 0$, $\lambda = 0$, and $\sigma = 1$, the first two conditions being attainable by translation, the last by rotation and dilatation of the coordinate system. The contention is that no function $f^*(z)$ other than our dipole function $f(z)$ furnishes a conformal mapping of G on a slit domain, if it is required that $f^*(z)$ has, in the neighborhood of the origin, the form $f^*(z) = 1/z + zR^*(z)$, $R^*(z)$ being regular and analytic.

The difference $W = \rho + i\tau = f(z) - f^*(z)$ would be analytic everywhere in G with boundary values of τ constant on every boundary γ_ν. We replace the variable z by $w = f(z)$ and consider the function $W = H(w) = f(z) - f^*(z)$. Obviously $H(w)$ is regular in B, including the point at infinity (the image of O), and thus $W = H(w)$ is a function regular and bounded in the whole slit domain B whose imaginary part τ is constant on each boundary slit β_ν.

We prove uniqueness by the same reasoning concerning the variation of the angle of $H(w) - h$ as used above. Since $H(w)$ is regular everywhere in B, $H(w) - h \neq 0$ for every value of h with the possible exception of values of h on the boundary. This implies $H(w) =$ constant and consequently the uniqueness of the mapping function.

An alternative proof is based on the following general remark: if a single-valued function $H(w) = \rho + i\tau$ is regular, bounded, and not constant in a domain B of the w-plane, the boundary of the subregion of B where τ exceeds a given constant value (actually assumed) must come arbitrarily close to the boundary β of B. For otherwise the boundary of the subregion would include a closed piecewise analytic curve L in B along which $\tau =$ constant, and along which the normal derivative $\partial\tau/\partial n$ of τ could neither vanish identically nor change sign. Therefore the tangential derivative of the conjugate harmonic function ρ would not change sign, and ρ could not be single-valued upon continuation about the closed curve L.

Consequently, in our case, if τ_0 is a value attained by τ in B, τ_0 necessarily coincides with one of the constant boundary values of τ on the slits β_ν, since a curve $\tau = \tau_0$ reaches at least one boundary slit. This is impossible unless τ and hence $W = \rho + i\tau$ is constant in B; thus, since $W = 0$ at $w = \infty$, uniqueness is proved.

Presently we shall extend the scope of the slit theorem by including a much wider class of domains G. One extension, however, should be mentioned here since it is obtained without modification of the reasoning:

The slit domain theorem remains valid for k-fold connected domains G on a Riemann surface with a finite number of sheets. While this statement hardly needs any additional justification, we shall in later sections revert in detail to the general question of the mapping of Riemann domains.

2. *Mapping on Slit Domains for Domains G of Infinite Connectivity.* The solution, in §2, of the variational problem for the dipole function $w = f(z)$ was not restricted to domains G of finite connectivity. The construction of u and the proof that v and $u + iv = f(z)$ are single-valued hold equally well for plane domains G of infinite connectivity. It is of great interest to note that for infinite connectivity of G the dipole function likewise maps G onto a slit domain B, i.e., onto the whole plane of $w = u + iv$ bounded by infinitely many parallel slits; i.e., any connected point set of boundary points in the w-plane consists of a straight segment $v =$ constant.

This generalization is not immediately obtained by the reasoning for k-fold connected domains. It follows, however, by considering G as the limit of a monotonic sequence of domains G_n of finite connectivity. We map G_n on a slit domain B_n by a function $w = f_n(z)$ with the singularity $1/z$ at the fixed point O. According to §2, 2, $f_n(z)$ tends to an analytic function $f(z)$ as $n \to \infty$. If $f_n(z)$ maps G_n on a simple plane domain and if $f_n(z) \to f(z)$, with $f(z) \not\equiv$ constant, $f(z)$ maps G on a simple plane domain B in the w-plane.[7] The con-

[7] For if G' is a suitable closed subdomain of G lying in G_n for sufficiently large n, and if C is the piecewise smooth boundary of G', we may for given h, because of $f(z) \not\equiv$ constant, assume that $f(z) \neq h$ on C. (The condition may be satisfied, if necessary, by a slight deformation of C which enlarges G'.) We have

$$N - 1 = \frac{1}{2\pi i}\int_C d\log\,[f(z) - h] = \lim_{n\to\infty}\frac{1}{2\pi i}\int_C d\log\,[f_n(z) - h],$$

and the last integral is either zero or minus one. Hence the equation $f(z) - h = 0$ has either one or no solution in G'.

dition $f(z) \neq$ constant is ensured by the singularity $1/z$. The domain B obviously has the same connectivity as G.

To show that B is a slit domain we have to prove that any connected point set β^* of boundary points of B consists of points with the same coordinate v.

The image in B of a neighborhood of O is a neighborhood of $w = u + iv = \infty$, which means that the boundary β of B lies within a finite circle of radius a about $w = 0$. If v were not constant on β^*, then β^* would be intersected by a family of lines $v =$ constant $= c$, $c_1 < c < c_2$, see Figure 2.3. We consider the segments of these lines extending from $u = +\infty$ to their first intersection with β^*—

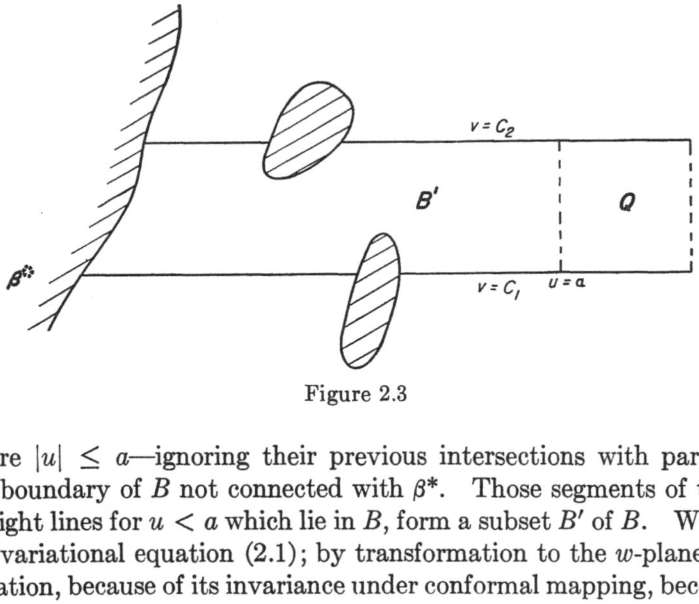

Figure 2.3

where $|u| \leq a$—ignoring their previous intersections with parts of the boundary of B not connected with β^*. Those segments of these straight lines for $u < a$ which lie in B, form a subset B' of B. We use the variational equation (2.1); by transformation to the w-plane this equation, because of its invariance under conformal mapping, becomes simply

$$\iint_B \zeta_u \, du \, dv = 0, \tag{2.1a}$$

valid for any piecewise smooth function $\zeta(u, v)$ that vanishes in a neighborhood of $w = \infty$ and for which $D[\zeta]$ is finite. With Q denoting the square

$$a \leq u \leq a + c_2 - c_1, \qquad c_1 \leq v \leq c_2$$

we define, in particular,

$$\zeta = \begin{cases} (v - c_1)(v - c_2) & \text{in } B', \\ (v - c_1)(v - c_2)\dfrac{a + c_2 - c_1 - u}{c_2 - c_1} & \text{in } Q, \\ 0 & \text{in } B - B' - Q. \end{cases}$$

Then $\zeta_u = 0$ in B', so that (2.1a) reduces to

$$\iint_Q \zeta_u \, du \, dv = 0.$$

But this result is absurd, since ζ_u does not change sign in Q. The contradiction proves our statement concerning β^* and thus establishes the mapping theorem.

For infinite connectivity the statement of *uniqueness* is no longer correct. As first observed by Koebe, the slit domains B obtained from our variational problem have a distinctive property: The boundary β of B can be enclosed in a finite number of curves in the w-plane which contain a total area that can be made arbitrarily small; i.e. *the boundary β forms a point set of content zero.*

The proof is an almost immediate consequence of the variational condition $D[u, \zeta] = 0$ for ζ vanishing in the neighborhood of the origin. If G_ϵ is a domain of finite connectivity—bounded, say, by polygons γ_ϵ—that tends monotonically to G as ϵ tends to zero, and if D_ϵ denotes the Dirichlet integral over G_ϵ, it follows immediately that

$$D_\epsilon[u, \zeta] \to 0$$

as $\epsilon \to 0$. Choosing a function ζ identical with u except in a neighborhood of O and applying Green's formula, we obtain

$$\sum \int_{\gamma_\epsilon} u \frac{\partial u}{\partial n} \, ds = \sum \int_{\gamma_\epsilon} u \frac{\partial v}{\partial s} \, ds \to 0,$$

where s denotes arc length on γ_ϵ. The expression on the left is the total area in the u, v-plane contained by the images β_ϵ of the polygons γ_ϵ; thus the statement concerning the content of β is proved.

It is easy to construct slit domains whose boundary does not have zero content. Such a domain, for example, is obtained if we choose as boundary slits the segments $0 \leq u \leq 1$ for v ranging over all values of a perfect nowhere dense point set of positive measure.

Such slit domains can be mapped onto slit domains as above. Our example shows that in the case of infinite connectivity, uniqueness cannot be expected without additional conditions for B. The subject, however, has not yet been sufficiently explored for further discussion here.

3. *Half-Plane Slit Domains. Moduli.* We return to domains G of finite connectivity k. The corresponding slit domains B depend on $3k$ parameters—the lengths of the k slits and the coordinates of their left endpoints. However, we could have chosen as singularity of $f(z)$ a pole with prescribed residue σ at an arbitrary point z_0 of G. The mapping function would then have the form

$$w = f(z) = \frac{\sigma}{z - z_0} + \lambda + (z - z_0)R(z),$$

$R(z)$ being regular in G. This shows that three complex parameters z_0, σ, λ or 6 real parameters can be fixed arbitrarily in the mapping function, so that in general only $3k - 6$ essential parameters remain.[8]

In such a normalization of slit domains, the $3k - 6$ "moduli" of the domain are not all exhibited as tangible geometrical objects. A more geometrical representation of the moduli can be achieved by a slight modification of our class of normal domains. We replace the slit domains so far considered by "half-plane slit domains." These originate as conformal images of a domain G if we let the singular point O tend to one of the boundary lines while the direction of the dipole at O becomes that of the tangent to the boundary line. Then the image of the boundary curve is an infinite line $v =$ constant, say the u-axis; the image of G is the upper half-plane, except for $k - 1$ finite slits.

We may directly transform any slit domain B into a half-plane slit domain S: By reflection of B in the slit β_k, which is assumed not to shrink to a point, we construct a two-sheeted Riemann surface $\bar{B}$ with $2k - 2$ boundary slits symmetric with respect to β_k, now an interior curve of $\bar{B}$. This domain $\bar{B}$ is mapped[9] onto a plane slit domain $\bar{S}$ by a function that has its singularity on β_k and is symmetric with respect to β_k. Then $\bar{S}$ is symmetric with respect to the real

[8] In the cases $k = 0$ and $k = 1$ we saw that the numbers of essential parameters are 0 and 1, respectively.

[9] See the remark at the end of article 1; also the detailed discussion in §4.

axis, and β_k is mapped onto the real axis in $\bar{S}$; G is obviously mapped on the domain S that constitutes half of $\bar{S}$, as stated.

By a linear transformation: $w = u + iv = az + b$, with real a and b, the left end of one other slit may be given a prescribed position, say $u = 0, v = 1$. Since the point O could also be chosen arbitrarily on β_k, we have proved

Theorem 2.3: A k-fold connected domain G in the z-plane ($k > 1$) may be mapped conformally on a *normalized half-plane slit domain S in the u, v-plane* bounded by $v = 0$ and by $k - 1$ slits in the upper half-plane, one of which has its left endpoint at $u = 0$, $v = 1$. In addition the point at infinity on $v = 0$ can be made to correspond to an arbitrarily chosen boundary point of G.

This mapping on the half-plane slit domain S involves $3k - 6$ parameters; $3k - 5$ parameters describe the geometrical shape of S, but due to the stipulated coordination of a fixed point on the boundary of G to the point $u = \infty$, $v = 0$ the $3k - 5$ geometrical parameters are subject to one condition, which reduces the number of degrees of freedom by one.

The *uniqueness* proof for this mapping is entirely similar to that in article 1.

4. *Boundary Mapping.* The mapping of the open domain automatically implies a coordination of boundary points. Let us first consider bounded plane domains G, or more generally subdomains G' of G, denoting the subdomain and its image again by G and B, instead of G' and B'. If a boundary point Q is the limit of a sequence of points $P_1, P_2, \cdots$ of G, the images P_n^* of the points P_n tend to limiting points on the boundary of B. We associate all these limiting points with Q and consider the corresponding association for the inverse mapping.

If Q is an isolated boundary point of G, there is only one corresponding boundary point of B, i.e. the slit reduces to a point. For the function $f(z)$ is bounded and regular in a neighborhood of Q except possibly *at* Q, and hence has a definite limiting value for z tending to Q.

The following analysis of the behavior of the mapping function $f(z)$ at the boundary depends neither on its minimum property nor on the fact that the boundaries of B are slits, but is based solely on the fact that the Dirichlet integral of u or v is finite:

Theorem 2.4: If $w = f(z) = u + iv$ maps a domain G in the z-plane, part of whose boundary is formed by a Jordan arc C, onto a domain B in the w-plane with a corresponding Jordan boundary arc C^*—e.g. a slit—then the mapping implies a continuous one-to-one coordination of the points of C and C^*.

Proof: The theorem is a consequence of lemma 1.4. We consider polar coordinates r, θ about a point Q on C, with $r\theta = s$, and the integral

$$2 \iint (u_x^2 + u_y^2)\, dx\, dy = \iint (u_x^2 + u_y^2 + v_x^2 + v_y^2)\, dx\, dy = M(a)$$

extended over the portion $r \leq 2a$ of G. This integral represents twice the area of the image of this region; hence it exists and $\epsilon^2(a) \equiv M(a)$ tends to zero for $a \to 0$. We note that

$$u_x^2 + u_y^2 + v_x^2 + v_y^2 \geq u_s^2 + v_s^2 .$$

Under the assumption that Q is not an isolated boundary point, a value ρ can be found so small that the pointset $r \leq \rho$ in G, or a subset of this pointset, forms a simply connected domain G_ρ with Q as a boundary point, bounded by an arc L_ρ of $r = \rho$ and an arc C_ρ of C. Furthermore for $\rho < 2a$ we have

$$\iint_{G_\rho} (u_s^2 + v_s^2)\, dr\, ds \leq \epsilon^2(a).$$

Hence (see the proof of lemma 1.4) there exists, for sufficiently small a, a radius ρ with $a \leq \rho \leq 2a$ and

$$\int_{L_\rho} (u_s^2 + v_s^2)\, ds \leq \frac{\epsilon^2(a)}{a} . \tag{2.6}$$

Now we consider a sequence a_ν of values a tending to zero as $\nu \to \infty$. To the sequence a_ν there corresponds a sequence of radii ρ_ν, tending to zero, for which (2.6) is satisfied, and a sequence of nested subdomains G_{ρ_ν} of G bounded by $L_{\rho_\nu} + C_{\rho_\nu}$. Their images are again nested, simply connected subdomains B_{ρ_ν} of B bounded by the images $L^*_{\rho_\nu}$ of the arcs L_{ρ_ν} and by a part $C^*_{\rho_\nu}$ of C^*. The boundary points on C^* associated with Q necessarily belong to all the arcs $C^*_{\rho_\nu}$.

According to (2.6) and Schwarz' inequality we have, for the length $l_\nu^* = \int_{L_{\rho_\nu}} \sqrt{u_s^2 + v_s^2}\, ds$ of $L_{\rho_\nu}^*$, the estimate

$$l_\nu^{*2} \leq 2\pi\rho_\nu \frac{\epsilon^2(a_\nu)}{a_\nu} \leq 4\pi\epsilon^2(a_\nu).$$

Hence l_ν^* tends to zero as $\rho_\nu \to 0$, and consequently the distance between the endpoints of the Jordan arc $C_{\rho_\nu}^*$ tends to zero. Consequently the diameter of this arc likewise tends to zero; because of the nested character of the domains G_{ρ_ν}, there is one and only one point, common to all the $C_{\rho_\nu}^*$, that corresponds to Q. Since the same proof holds for the inverse mapping also, our theorem is established.

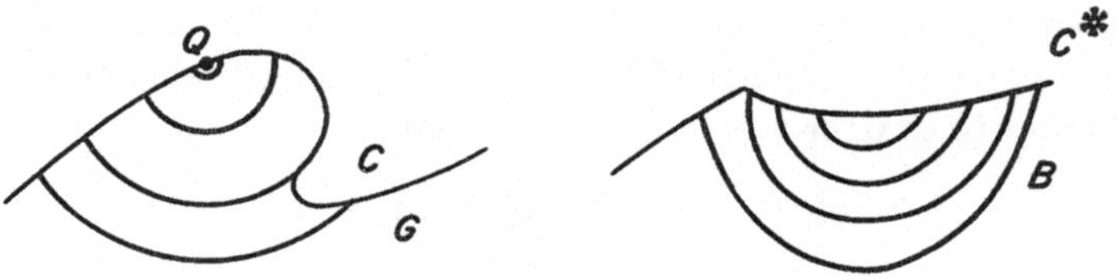

Figure 2.4

It may be pointed out that this reasoning also illuminates the behavior of the mapping at the boundary when the latter is not a Jordan arc. We consider, for example, a domain G whose boundary is not a Jordan curve, while the image B is a slit domain. A boundary point Q^* of B may correspond not to a single boundary point on C, but to a whole pointset Π, a so-called "primend." Then our reasoning shows that all the points of Π are common boundary points of a sequence of nested, simply connected subdomains G_{ρ_ν} of G cut out of G by a sequence of curves whose lengths tend to zero. A further analysis of primends and of their one-to-one correspondence to the points of the slit is, on this basis, rather obvious but will not be pursued here.[10]

4. Riemann Domains

1. *Introduction.* In the construction of the dipole potential u we have so far visualized G as a plane domain. However, wide generalizations offer themselves immediately. The potential u and

[10] See Carathéodory [1], Hurwitz-Courant [1], p. 404, Kerékjártó [1].

the conformal mapping on plane slit domains are obtained in the same manner for the much wider class of "Riemann domains" G, with one modification: if G is not of genus zero, the potential function v conjugate to u is not single-valued and the image slit domain is of a new type, as will be seen in §7.

The construction of u by Dirichlet's Principle was based on the following properties of G:

a) G consists of a finite or denumerable number of non-overlapping "cells" and their boundaries. A cell is defined as a simply connected domain topologically equivalent to a disk by means of a mapping which, except possibly at isolated points, is conformal. (A cell may, therefore, contain a branchpoint of finite order.) In this mapping the boundary of the cell is required to correspond biuniquely to that of the disk.

Instead of visualizing G as the sum of denumerably many cells we may consider it as the limit of a monotonic sequence of subdomains G_n, each of which consists only of a finite number of cells and of the boundaries in common to two adjacent cells.

b) Each point of G is an interior point of a cell in G. In other words: not only can the points interior to cells constituting G—as described in a)—be imbedded in *other* cells all of whose points are in G, but also the boundary points of the constituting cells can be imbedded in such cells. These latter cells, of course, overlap with the cells described in a).

For each cell Dirichlet's Principle, with fixed or partly free boundary values, may be established from the foregoing. Property a) of the domain G serves for the definition of integrals over G, and for the construction of u in each individual cell by smoothing of minimizing sequences. Property b) is essential to establish the potential functions u in adjacent cells as analytic continuations of each other.

No other conditions were needed for solving boundary value problems or for proving the existence of the dipole potential u. As to conformal mapping: for a plane domain G the harmonic function v conjugate to u is, as seen, single-valued, and no other property is needed to prove that G may be mapped on a slit domain. However, for Riemann surfaces G not of genus zero the proof that v is single-valued breaks down. Nevertheless, as we shall see in §7, a parallel-slit theorem can be formulated even for domains G not of genus zero if the concept of slit domain is properly modified.

The properties a) and b) essentially characterize the concept of a "Riemann domain" G. Before formulating a precise abstract definition we consider in detail those cases of Riemann domains which are particularly important for our purposes:
1) *Riemann Surfaces* G spread over the plane of a complex variable $z = x + iy$ possess the properties a) and b). The cells forming G can be chosen as simply connected plane domains—possibly containing the point at infinity—whose boundaries are Jordan curves, or as a finite number of congruent disks connected through a branch point in the center. Obviously such points of G as lie on the boundary of one of the constituting cells can be imbedded in a cell of the same type containing points of several of the cells.
2) *Polyhedral Domains:* The construction of the dipole potential u and the mapping on slit domains also apply to domains on polyhedra. Polyhedral domains G can be decomposed into cells each of which is either a whole face of the polyhedron or part of a face. If x and y are rectangular coordinates on a face the corresponding "cell" is represented by $z = x + iy$ as the *"local complex variable."* A point P of G on an edge, not a vertex, is on the boundary of two adjacent cells Z_1, Z_2. To imbed P in a third cell we rotate the plane of Z_2 about the common edge into the plane of Z_1, so that the two cells together now form a plane cell $Z_1 + Z_2'$; then P is contained in the plane cell $Z_1 + Z_2'$ and we may therefore consider $Z_1 + Z_2$ as a cell[11] on the polyhedron containing P. (The mapping of $Z_1 + Z_2$ on $Z_1 + Z_2'$ is congruent, hence conformal, angles being of course measured in the surface.) Analytically, we choose local coordinates in the cell surrounding P, so that the common edge is the x-axis, the x coordinate of any point in the cell is measured parallel to the edge, and the y coordinate is the perpendicular distance from it.

Dirichlet's integral is now immediately defined as the sum of the integrals over the individual plane cells, each integral referring to the local coordinates. A function of a point P in G is called harmonic if it is harmonic in the local coordinates in each of the cells considered. An analytic function $u + iv$ on G is then defined in the obvious way and maps each of the cells conformally.

As to the vertices V of the polyhedron G, they can also be im-

[11] If part of the common edge is on the boundary of G, $Z_1 + Z_2$ should be connected only along that part of the edge which belongs to G and contains P, in order to ensure the simple connectivity of the cell.

bedded in cells Z consisting of circular sectors with (sufficiently small) radius ρ and center V on all the faces of G converging in V. To verify that these sectors form a cell as defined in a) we subject each of the circular sectors to a conformal mapping of the form $z' = z^\alpha$ where z denotes the local complex variable in the sector, V being the origin, and where the exponent α, the same for all the sectors, is so chosen that the sum of the angles of the resulting circular sectors is 2π; then these sectors may be fitted together to form a plane circular disk Z' of radius $\rho' = \rho^\alpha$ in the z'-plane. This disk Z' is the conformal image of the domain Z on the polyhedron, the conformality being interrupted only at the vertex V. Hence Z is a cell as defined above.

As a consequence we state: *Domains G of genus zero on a polyhedron can be mapped biuniquely onto plane parallel slit domains, the conformality—but not the continuity—of the mapping being interrupted at the vertices alone.*

3) *Curved Surfaces in Space*: The results of the previous sections apply also to curved surfaces G in space as long as they have the property that a suitably small neighborhood of each point of G can be referred to isometric parameters and hence mapped conformally onto a plane domain. The cells Z for such a surface G are simply connected domains conformally equivalent to circular disks in a plane. Functions ϕ of a point P ranging over G may, in a cell Z, be considered as functions $\phi(x, y)$ of the "local variables" (rectangular coordinates x, y, characterizing the image of P in the plane conformal image Z^* of Z). The Dirichlet integral of ϕ over Z is defined as

$$\iint_{Z^*} (\phi_x^2 + \phi_y^2)\, dx\, dy,$$

and is invariant under conformal mapping of Z^*. Dirichlet's integral over G is defined as a sum of such integrals referring to a set of non-overlapping cells that constitute G. Harmonic functions in G are functions which are harmonic in the local variables x, y belonging to the plane conformal images Z^* of the cells Z.[12]

The existence of the dipole potential u then follows as before, and we find: *The mapping theorem remains valid for domains G of genus zero on curved surfaces, provided that the surface permits con-*

[12] See also the general discussion in §5.

formal mapping "in the small" of the neighborhood of each point. In other words: the various local mappings can be replaced by one conformal mapping in the large of the whole domain G onto a plane parallel-slit domain, thus providing a uniform variable $x + iy$ *for the whole surface G.*

4) *Domains defined by Boundary Coordination:* Geometric function theory leads to the consideration of more general "Riemann domains" G such as in the following examples:

(1) G is a parallelogram in the z-plane; however, corresponding points on opposite sides are "identified" by the stipulation that analytic functions in G are to have the same values at corresponding points. Such a "period parallelogram," with corresponding points considered as identical, is a closed domain of genus 1.

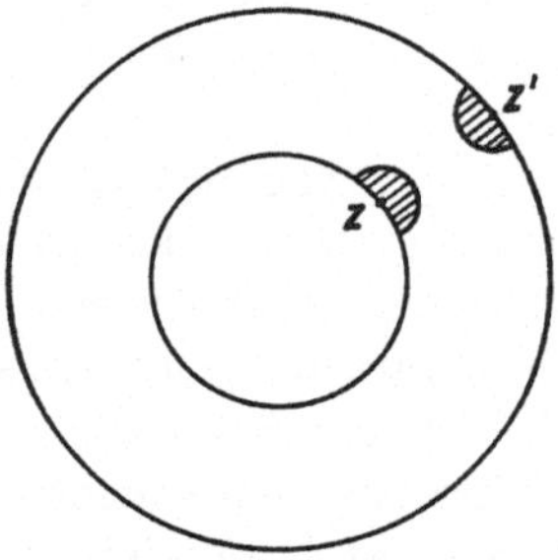

Figure 2.5. Cell formed by boundary coordination.

(2) Another closed domain G of genus 1 obtained from a plane domain $\tilde{G}$ is the ring $1 \leq |z| \leq 2$ in the z-plane, where the boundary points z on $|z| = 1$ are identified with the points z' on $|z| = 2$ by the correspondence $z' = 2z$. This identification means that functions $f(z)$ in G must satisfy the condition $f(2z) = f(z)$.

The coordination of the boundaries in (1) and (2) is effected by a conformal transformation of the plane which applies not only to the domain $\tilde{G}$ but to a whole plane neighborhood of each boundary point of $\tilde{G}$, in such a way that the correspondence of boundary points extends to their neighborhoods. Using this transformation for example (2), we immediately see that a point of G on $|z| = 1$ or $|z| = 2$ can be imbedded in a cell Z as indicated in Figure 2.5. In the z-plane the cell Z appears in two pieces; but these pieces are connected by the boundary identification and together they are conformally equivalent to a plane disk. Thus G is a domain to which our previous construction of the potential u can be applied. The same is true for example (1). (Note, however, that the domains in these two examples are not of genus zero and therefore cannot be mapped onto slit domains of the type considered before.)

2. *The "Sewing Theorem."* Domains G obtained from plane domains $\tilde{G}$ by boundary coordination will occur on different occasions in this book, and lead to remarkable mapping theorems. One important theorem is

Theorem 2.5 (Sewing Theorem): Let a domain $\tilde{G}$ in the x, y-plane be cut by a slit along an analytic arc C that joins the points P and Q. The points A_+ and A_- of the two edges C_+ and C_- of the slit may be coordinated by an analytic transformation $z_+ = t(z_-)$ which leaves

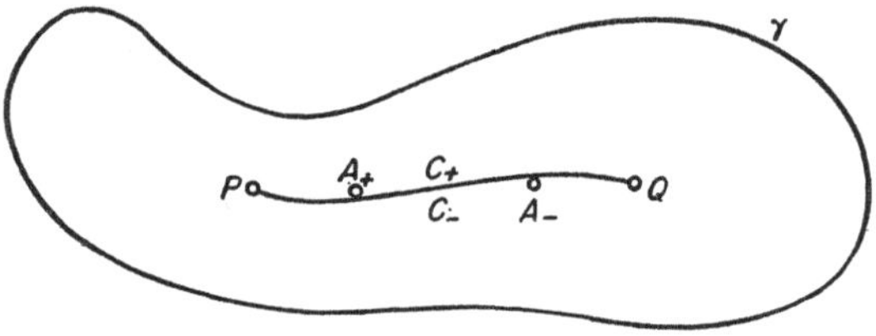

Figure 2.6. Slit PQ with coordination of the two edges.

the endpoints P and Q fixed. By this boundary coordination, a domain G is obtained from $\tilde{G} - C$. Besides the boundary γ of $\tilde{G}$, the domain G has as boundaries the "vertices" P and Q. Then the domain G can be conformally mapped onto a plane domain B, e.g. onto a parallel-slit domain, in such a way that C is mapped on an analytic arc C' and corresponding points A_+ and A_- on the two edges C_+ and C_- of the slit C are mapped onto the same point A' on C'. In other words, the abstractly stipulated identity of the two edges of C can be geometrically reestablished by conformal mapping: the two edges of the cut C can be "sewn together."

Another version of the sewing theorem, simpler because of the absence of vertices, is the following: ,

The plane domains G_+ (consisting of the exterior of an analytic closed curve C_+) and G_- (consisting of the interior of another analytic closed curve C_-) are combined into one region G by an analytic transformation $z_+ = t(z_-)$ that establishes a biunique correspondence between all the points A_+ on C_+ and A_- on C_-. Then the domains G_+ and G_- can be mapped conformally onto two domains G'_+ and G'_-, respectively, in such a way that C_+ and C_- are transformed into the same curve C', that G_+ is transformed into the exterior of C' and G_- into the interior of C' and, furthermore, that corresponding points A_+ and A_- go into the same point A' on C. In other words, by conformal mapping the two separate components

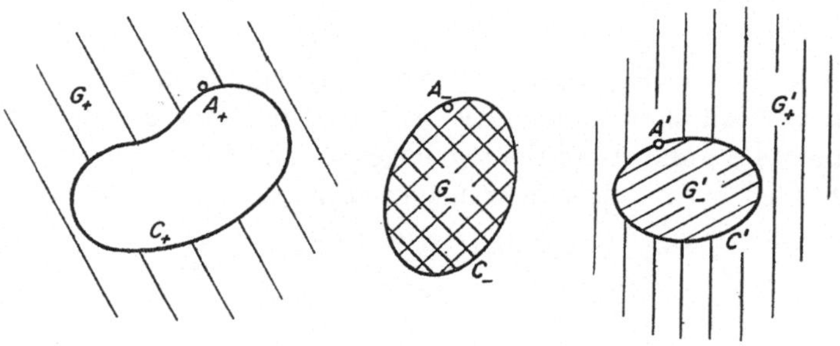

Figure 2.7. Fusing of two regions G_+, G_-.

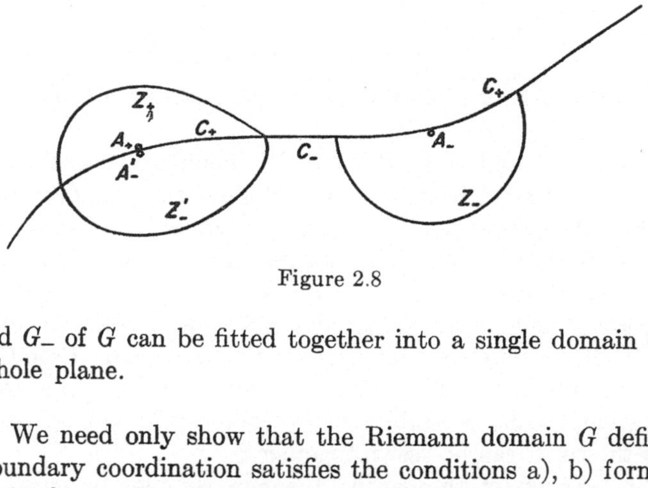

Figure 2.8

G_+ and G_- of G can be fitted together into a single domain G', i.e., the whole plane.

Proof: We need only show that the Riemann domain G defined by our boundary coordination satisfies the conditions a), b) formulated above in §4,1. To this end we ascertain both that the ordinary points of G lie in cells—which is obvious—and that every "point" on the coordinated edges, i.e. the point appearing geometrically as A_+ on C_+ and as A_- on C_-, can be surrounded by a cell which may then be mapped on a plane domain. Such a cell is immediately obtained if we use the function $t(z)$ to map a sufficiently small "half-cell" Z_-, bounded by an arc c_- of C_- and a circular arc with A_- as center, on a domain Z'_-, again represented in the z-plane. (Figure 2.8 refers to the first case of the sewing theorem.)

By this mapping the arc c_- is transformed into the corresponding arc c_+ in such a way that points corresponding in the boundary coordination are brought into coincidence. A function $g(z)$ continuous in G, with $g(z_+) = g(z_-)$, becomes continuous in $Z_+ + Z'_-$ if transplanted from Z_- into Z'_- by the transformation $z = t(z_-)$; here Z_+ denotes a half-cell adjacent to c_+ as indicated in Figure 2.8. These facts characterize $Z_+ + Z_-$ as a cell in G, and thus our theory is immediately applicable in the second version of the theorem, where

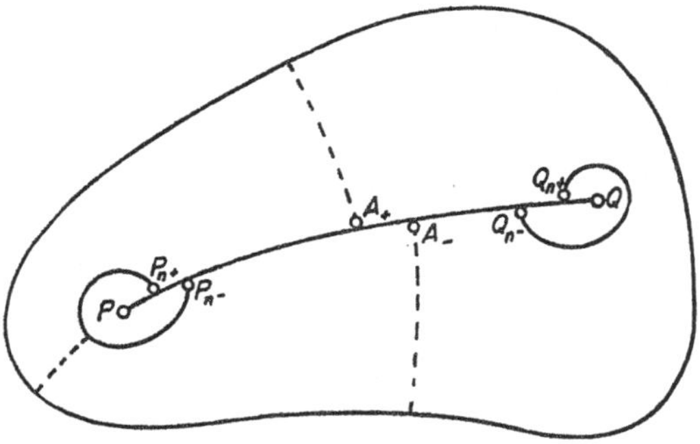

Figure 2.9

C can be covered by a finite number of such cells. In the first form of the sewing theorem, where the endpoints P and Q of the slit C intervene, we consider G as the monotonic limit of domains G_n, which in turn can be decomposed into a finite number of cells, as indicated in Figure 2.9. (G_n replaces G so as to exclude neighborhoods of P and Q; then G_n is dissected into "cells" by the dotted lines and the cut C.)

In the preceding construction the endpoints P and Q of the slit C do not appear imbedded in a cell. Hence, in the mapping of G onto B, it cannot generally be expected that they will again pass into single points rather than into slits, as seen by the following construction (Figure 2.10): We cut the w-plane by an analytic curve S which starts at a point E and winds asymptotically about the segment T: $0 \leq u \leq 1$ of the u-axis. The w-plane with the cut S

may be called B; T is a "primend" of B (see §3, 4). We map B on a slit domain $\tilde{G}$ of the z-plane. The whole boundary slit γ of $\tilde{G}$ corresponds to the cut S together with the primend T, and, as we saw, T is mapped onto a single point of the slit γ. Without loss of generality we may assume that γ is the segment $0 \leq x \leq 1$ of the x-axis and that T corresponds to the point $x = 0$. In this mapping two opposite points A'_+ and A'_- on the two edges of the slit S correspond to different, not opposite, points on the two edges of the slit γ. (For otherwise the mapping would be continuous, hence regular, yielding a conformal mapping of the z-plane with the origin omitted on the w-plane with the slit T omitted; this mapping is impossible).

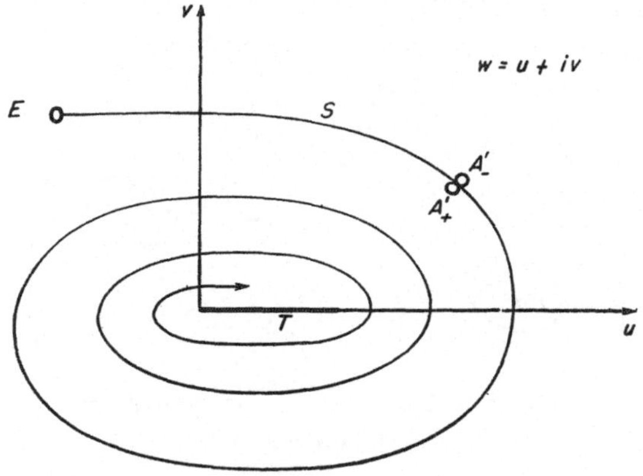

Figure 2.10

We obtain a domain G from $\tilde{G}$ by identifying corresponding points A_+ and A_- on the two edges of the slit γ; then G is mapped onto the whole w-plane except the segment T which corresponds to the single initial point of γ.

A correspondence of P and Q to single points in the mapping is assured by the condition established in the

Corollary to Theorem 2.5: The vertices P and Q go into isolated boundary points of B if the coordination by $z_+ = t(z_-)$ at the boundary satisfies the condition

$$(2.7) \qquad 0 < \alpha < |t'(z_-)| < \frac{1}{\alpha},$$

where α is a positive constant less than 1.

Proof: We take—without loss of generality—the slit C as the straight segment $0 \leq x \leq 1$ of the x-axis. By the function $w = u + iv = f(z)$ the point P: $x = 0$ is mapped on a closed set of boundary points of B, say on a slit p. To show that p is merely an isolated point, we shall construct closed curves in B, surrounding p, of arbitrarily short length.

Consider the integral

$$M(a) = \iint |f'(z)|^2 \, r \, dr \, d\theta,$$

extended over all points of G whose distance r from P is less than $2a$. For sufficiently small a the singularity of $f(z)$ is outside the domain

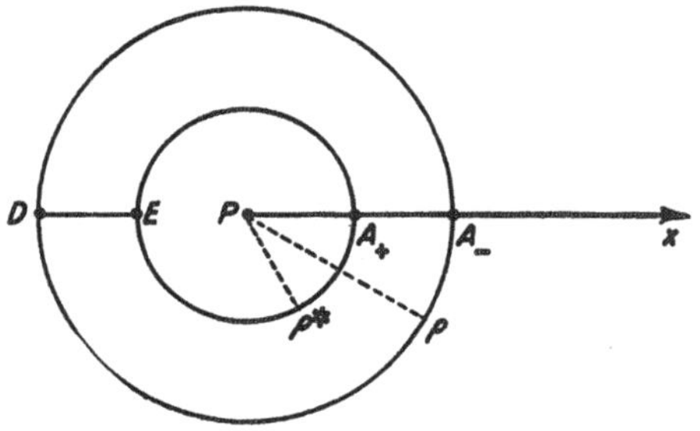

Figure 2.11

of integration; moreover we have $M(a) \to 0$ for $a \to 0$. We shall construct two circles $r = \rho$ and $r = \rho^* = t(\rho)$ about P, with images in the w-plane of arbitrarily small total length for sufficiently small a, which cut the slit C in the two corresponding points A_- and A_+, respectively. (The images of these closed circles in the z-plane are curves closed neither in G nor in B.) Furthermore, we shall see that the length of the image of the radial segment DE (see Figure 2.11) can be made arbitrarily small. Then the curve consisting of the lower semi-circle $A_- D$, the upper semi-circle $A_+ E$, and the segment DE is closed in G; its simple and closed image in B encloses p, and its length can be made arbitrarily small. Hence p is an isolated point.

To carry out the construction we choose a value of a so small that the circle with radius $2a$ about P intersects the slit C and does not contain the singularity of $f(z) = u + iv$. We further note that condition (2.7) and the relations $r^* = t(r)$, $0 = t(0)$ imply, by the mean value theorem, that $\alpha r < r^* < r/\alpha$. Setting $z = r\,e^{i\theta}$ and $z^* = r^*\,e^{i\theta}$ with $r^* = t(r)$ we have

$$\int_0^{a\alpha} dr \int_0^{2\pi} |f'(z)|^2\, r\, d\theta < M(a)$$

and

$$\int_0^{a\alpha} dr^* \int_0^{2\pi} |f'(z^*)|^2\, r^*\, d\theta < M(a).$$

Considering r as independent variable in $r^* = t(r)$ and recalling that $dr^*/dr = t'(r) > \alpha$ we have, *a fortiori*,

$$\alpha \int_0^{a\alpha} dr \int_0^{2\pi} |f'(z^*)|^2\, r^*\, d\theta < M(a).$$

Consequently

$$\int_0^{a\alpha} dr \left\{ \int_0^{2\pi} |f'(z)|^2\, r\, d\theta + \alpha \int_0^{2\pi} |f'(z^*)|^2\, r^*\, d\theta \right\} < 2M(a).$$

Hence there exists for r an intermediate value ρ with $0 < \rho < a$, and for $r^* = t(r)$ a corresponding value $\rho^* = t(\rho)$, such that for $|z| = \rho$ and $|z^*| = \rho^*$ we have simultaneously

$$\rho \int_0^{2\pi} |f'(z)|^2\, d\theta < 2\,\frac{M(a)}{a\alpha}$$

and

$$\alpha\rho^* \int_0^{2\pi} |f'(z^*)|^2\, d\theta < 2\,\frac{M(a)}{a\alpha},$$

(see the similar reasoning in Chapter I, §4, 2.)

The lengths $L(\rho)$ and $L(\rho^*)$ of the images in the w-plane of the circles $r = \rho < a$ and $r = \rho^* < a$ are appraised as before by

$$L^2(\rho) \leq 2\pi\rho\,\frac{2M(a)}{a\alpha} < 4\pi\,\frac{M(a)}{\alpha}$$

and

$$L^2(\rho^*) \leq 2\pi\rho^* \frac{2M(a)}{a\alpha^2} < 4\pi \frac{M(a)}{\alpha^2},$$

both quantities tending to zero with a.

As to the segment DE, if D is at a distance ρ from P, the distance of each point on DE from the slit C is larger than $\eta = \rho\alpha$; for sufficiently small η the circular disk of radius η about any point z on DE lies entirely within G and within a circle of radius $b = \rho(1/\alpha + \alpha)$ about P. Hence by the mean value property of $f'(z)$ and by Schwarz' inequality we have, for each point z on DE,

$$|f'(z)|^2 < \frac{M(b)}{\pi\rho^2\alpha^2},$$

and for the length λ of the image of DE, with $\overline{DE}$ denoting the length of DE,

$$\lambda \leq \overline{DE} \frac{\sqrt{M(b)}}{\sqrt{\pi}\rho\alpha}.$$

Now $\overline{DE} = |\rho - \rho^*| = |\rho - t(\rho)| = \rho|1 - t'(\bar{\rho})|$, where $\bar{\rho}$ is an intermediate value between 0 and ρ. Hence $\overline{DE} < \rho\,(1 + 1/\alpha)$, and

$$\lambda \leq \frac{1 + \alpha}{\alpha^2} \sqrt{\frac{M(b)}{\pi}}.$$

Since b tends to zero with a, for sufficiently small a the lengths $L(\rho)$, $L(\rho^*)$ and λ all become arbitrarily small. Our corollary is proved.

5. *General Riemann Domains. Uniformization*

The domains G discussed in the preceding articles illustrate the abstract concept of a Riemann domain by concrete cases. To arrive at this general concept we augment the two postulates a) and b) of §4 by one small further step which emphasizes the topological core of Riemann's description of analytic functions in the large.

If a Riemann domain G is subjected to a topological deformation resulting in a domain G^*, or more abstractly, if G^* is another two-dimensional manifold in continuous and biunique correspondence with G, then G^* should likewise be considered a Riemann domain. Such a point of view leads to a modification of postulate a): not a conformal, but merely a topological mapping of the cells of G onto plane cells need be stipulated. Thus we arrive at an abstract definition of a Riemann domain G:

1) G is a two-dimensional connected manifold, i.e.

α) G is a "topological neighborhood space."[13]

β) Every point P in G has a neighborhood topologically equivalent to a plane disk.

γ) Any two neighborhoods N_P and N_Q can be connected by a finite chain of neighborhoods $N_P = N_0, N_1, \cdots, N_n = N_Q$ such that N_{k-1} has points in common with N_k for $k = 1, 2, \cdots, n$.

2) With each point P of G there is associated a neighborhood N_P and a particular topological mapping of N_P onto a disk Z_P in the plane of a complex variable $z_P = x + iy$—called a local variable for this neighborhood—such that the following decisive condition is satisfied: If the neighborhoods N_P and N_Q of points P and Q, respectively, have a domain in common, the corresponding regions in the cells Z_P and Z_Q of the local variables z_p and z_q are mapped on each other conformally by the function which takes corresponding points of the cells Z_P and Z_Q into each other.

As a result of these properties of G we can define harmonic functions and analytic functions in G. A function ϕ defined for all points P of G is called harmonic or analytic if it is harmonic or analytic, respectively, in the local variables x, y or $x + iy = z$. Since these local properties are invariant under conformal mapping, the definition, although referring only to local variables, is consistent in the large. Furthermore we call a point set in G a smooth or an analytic arc if its image in the plane of the local variables has the respective properties.

Dirichlet's integral can now be defined by reference to local variables. For this purpose the domain G must be decomposed into a denumerable set of non-overlapping "cells" having the following properties:

a) Each cell is a domain which, together with its boundary, is

[13] See Alexandroff [1], Chapter I. The axioms defining such a space are: G is a collection of points P; to each point P is associated a family of point-sets N_p of G_1, called neighborhoods of P, with the following properties: (1) P is contained in all neighborhoods N_P. (2) The common part of two neighborhoods of P contains a neighborhood of P. (3) If Q lies in N_P then N_P contains a whole neighbornood of Q. (4) Two distinct points P and Q of G possess suitable neighborhoods N_P and N_Q, respectively, which have no points in common. The formulation of these axioms is due to Hausdorff.

contained within a neighborhood—referred to local coordinates z—on a disk Z. It may be assumed that such a cell on Z is a simply connected domain whose boundary consists of piecewise analytic Jordan curves.

b) Each point of G either belongs to a definite cell or is on the boundary of a finite number of cells.

c) Each point of G possesses a neighborhood having points in common with only a finite number of cells.

Such a decomposition of G is called a *triangulation*, since the simply connected cells in the z-plane may be replaced by conformally equivalent triangular domains. The possibility of such a triangulation need not be stipulated, but is a simple consequence of the preceding definitions.

The Dirichlet integral $D[\phi]$ of a function ϕ over a cell is defined as the integral

$$\iint_Z (\phi_x^2 + \phi_y^2)\, dx\, dy,$$

and this definition is invariant under replacement of Z by a conformally equivalent domain. The Dirichlet integral of ϕ over the whole domain G is then the sum of the integrals over the cells of the triangulation. That the Dirichlet integral is independent of the particular triangulation employed is easily seen by standard methods of proof.

With these concepts all the arguments of the preceding sections remain in force; and thus *the existence of the dipole potential u follows for any Riemann domain. The mapping theorem remains valid in the previous form if the conjugate potential v is single-valued in G, i.e. for Riemann domains of genus zero.* Of course, conformality of this mapping always refers to the mapping of the cells Z in the local variables.

The process of mapping G onto an ordinary plane domain is sometimes called "*uniformization*"; i.e. the variety of different local complex variables in these respective cells can be replaced by one and the same complex variable w for the whole domain G.

It must be mentioned that the *orientability of Riemann domains G* is a consequence of the preceding general definition: Consider a cell N in G and the corresponding cell Z in the z-plane, assigning to the latter an orientation, say the rotation in the negative sense with respect to the z-plane. Then consider a chain of overlapping cells $N = N_0, N_1, \cdots, N_n$ and the corresponding cells $Z = Z_0, Z_1, \cdots, Z_n$ in the z-plane. Since two adjacent cells Z_k and Z_{k+1} are related by a conformal correspondence of the images of points in the overlapping regions on G, the orientation of Z is transferred to all cells Z_k and thus a definite orientation is assigned to the whole domain G.

Hence we are not allowed to consider non-orientable surfaces, such as a Moebius strip, as Riemann domains. On the other hand, we want to include them in our theory of conformal mapping. We resolve the difficulty by considering the "orientable covering surface" G of the non-orientable surface $\tilde{G}$; i.e. we count each cell twice, once with each orientation. Intuitively we distinguish the two sides of the non-orientable surface $\tilde{G}$ as if they were different or separated by a thin wall. The covering surface is then a Riemann domain in the proper sense.

To carry out the construction of the *dipole potential for a non-orientable surface* it suffices to stipulate that the functions ϕ and consequently the potential u depend on the position of P on $\tilde{G}$ alone and not on the orientation of $\tilde{G}$ in the neighborhood of P. As we shall see, however, the value of the conjugate potential depends on the orientation.

6. *Riemann Domains Defined by Non-Overlapping Cells*

For our purposes Riemann domains may preferably be visualized in an intuitive way[14] as represented by a denumerable number of simply connected plane cells each of which is bounded by a finite number of analytic arcs. Each point in such a cell represents a point of G, and points in different cells represent different points of G. Points on an analytic boundary arc c also represent points in G, provided that c and another boundary arc c' of a cell are point-wise identified by an analytic transformation of the complex variable $z' = t(z)$ that transforms a neighborhood of a point of c in the z-plane into a neighborhood of a point of c' in the z'-plane. By defining

[14] The main deviation from the preceding discussion is that in defining G we avoid the use of *overlapping* cells and establish the connection of *adjacent* cells by analytic transformation of the analytic boundary arcs of the cells alone.

the point z on c and the point z' on c' as representing the same point of G we establish G as a Riemann domain. For we have seen in §4, 2 that such points on the arcs c or c' can again be imbedded in a cell consisting, except for a conformal transformation, of a part of Z adjacent to c and a part of Z' adjacent to c'. As to the endpoints of these coordinated arcs, the "vertices," they must be considered as boundary points of G. However, the corollary to theorem 2.5 can be generalized to the following theorem, whose proof is entirely similar to that given for the corollary:

Theorem 2.6: Suppose that the cells $Z_1, Z_2, \cdots, Z_r, Z_{r+1} = Z_1$ converge at a vertex V. Without loss of generality we may assume

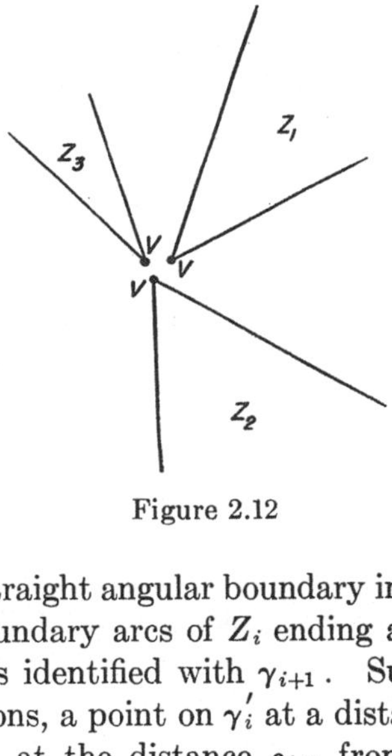

Figure 2.12

that each cell has a straight angular boundary in a neighborhood of V. The two straight boundary arcs of Z_i ending at V may be called γ_i and γ_i', so that γ_i' is identified with γ_{i+1}. Suppose that, by these boundary coordinations, a point on γ_i' at a distance ρ_i^* from V passes into a point on γ_{i+1} at the distance ρ_{i+1} from V in the next cell. Then the condition

$$0 < \alpha < \frac{d\rho_2}{d\rho_1^*}, \frac{d\rho_3}{d\rho_2^*}, \cdots, \frac{d\rho_{r+1}}{d\rho_r^*} < \frac{1}{\alpha}$$

and consequently

$$0 < \alpha < \frac{\rho_2}{\rho_1^*}, \frac{\rho_3}{\rho_2^*}, \cdots, \frac{\rho_{r+1}}{\rho_r^*} < \frac{1}{\alpha},$$

valid in a neighborhood of V for a fixed positive $\alpha < 1$, are sufficient to ensure that V in G can be imbedded in a cell. The point V is therefore considered an interior point of G under this condition, while in general the mapping is not conformal at V.

7. *Conformal Mapping of Domains Not of Genus Zero*

1. *Introduction.* If a Riemann domain G is of genus zero, i.e. if it is dissected into separate parts by any closed curve in G,[15] then, as seen before, the harmonic function v conjugate to the dipole potential u is single-valued in G and the function

$$w = u + iv$$

maps G onto a plane parallel-slit domain. If, however, G is not of genus zero, while u by definition is single-valued in G, the conjugate potential $v = v(P)$ need not return to its original value as the point P describes a non-dissecting circuit. Of course, the function $w = u + iv$ can no longer provide a conformal mapping of G onto a simple plane domain, which would necessarily be of genus zero. Still we shall see that the analytic function $w = u + iv$ maps the domain G onto a domain consisting of the whole w-plane with the exception of finite boundary slits parallel to the u-axis; in addition the plane is cut by pairs of parallel slits $v =$ constant which extend from $u = -\infty$ to a finite value of u and whose edges are coordinated in such a manner that the image slit domain B in the w-plane has the prescribed topological structure.

2. *Description of Slit Domains Not of Genus Zero.* To define such plane slit domains we first consider domains without boundary slits. There are two types of "interior" slit pairs: 1) a pair of slits $v =$ constant in the u,v-plane extends from $u = -\infty$ to the same finite value of u; we coordinate or identify the four edges as in Figure 2.13, so that corresponding points have the same u-coordinate and corresponding edges are marked with the same figures (*pair of the first type*). 2) A similar pair of slits with the sole difference that the coordination is as in Figure 2.14 (*pair of the second type*). These boundary coordinations imply that paths meeting an edge of a slit

[15] In recent German literature on conformal mapping such domains are called "schlicht-artig," since they can be mapped on simple plane or "schlicht" domains.

have to be continued from the corresponding slit as indicated in Figures 2.13 and 2.14.

A domain with a pair of slits of the second type is non-orientable. This is immediately seen from Figure 2.17, which shows that the boundary coordination between the slits reverses the orientation of an adjacent angle.

For a pair of slits of the first type, however, the orientation is not changed by the coordination of the edges of the slit. Plane regions which contain only slit pairs of the first type are orientable.

We may cut the plane by k pairs of slits, with the edges coordinated as in the diagrams. Such a plane with coordinated slits need not have the topological structure of a closed surface. Although, by the identification of edges, any point on one of the infinite slits is obviously an interior point of our slit domain B, this will be true of the point at

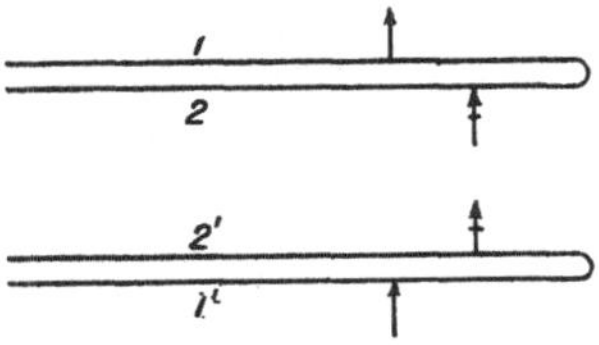

Figure 2.13. Slit pair of first kind.

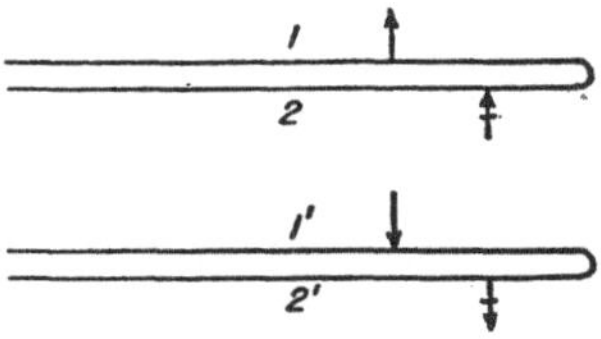

Fig. 2.14. Slit pair of second kind.

infinity only if the slits are properly arranged. If the slit domain B is to correspond topologically to a Riemann domain G, with the point at infinity corresponding to an interior point O, then a closed curve on G enclosing O will have to correspond to a single closed line in the slit domain which separates this domain into two parts. Such a line in the slit domain is made up of segments separated visually in the w-plane but connected by the stipulated coordination. The *condition that the point at infinity in the slit domain is an interior point* can be simply expressed by the following equivalent statement:

A vertical line u = constant = c that intersects all the slits is fully described if, on meeting an edge of a slit, one continues at the corresponding edge in the same or the opposite direction accc**r**ding as the pair is of the first or the second type, and proceeds in this manner until the path is closed. This condition implies a certain interlocking of the slits. (The condition is not satisfied, for example, by a plane with just one pair of slits of the first type.[16] See Figure 2.13.)

[16] The domain corresponds topologically to a plane with two points removed.

In order to have an admissible domain, at least one more pair of slits is necessary. If the second pair is again of the first type we obtain an orientable surface, the torus (Figure 2.15); if the second pair of slits is of the second type, we obtain a non-orientable surface, the non-orientable torus or "Klein bottle" (Figure 2.16).[17] It should be remarked that, in the second example, the orientation of the surface is preserved along the closed path consisting of the slits of the second type and reversed along the slits of the first type.

On the other hand, a plane with just one pair of slits of the second type represents an admissible surface, the projective plane or cross cap, which may also be visualized as a disk where diametrically opposite points are identified (Figure 2.17).

A closed surface is characterized topologically by its characteristic number and by whether it is orientable. Orientability, as discussed above, depends on the absence of pairs of slits of the second type. The characteristic number is related to the total number of slits.

Any closed surface G can be considered topologically as a polyhedron having a certain number F of faces, E of edges, and V of vertices. Then the characteristic number L is defined as

$$L = 2 - F + E - V.$$

If G is orientable, of finite genus p, i.e., can be described visually as a sphere with p "handles" attached,

$$L = 2p.$$

If G is non-orientable, i.e. is visually described by p "handles" and q "cross caps,"

$$L = 2p + q$$

For a closed surface which is represented by a plane slit domain B with s pairs of slits, we have $F = 1$, $E = 2s$ (corresponding edges being identified), and $V = s + 1$, the point at infinity counting as a single vertex. Hence $s = L$. Thus, *if the characteristic number of our slit domain is to be L, the domain must have exactly L pairs of infinite slits*, e.g. $2p$ pairs of the first type if it is orientable and of genus p.

The class of slit domains B is now extended by including the

[17] That Figures 2.15, 2.16, and 2.17 correspond to the intuitive models of the torus, Klein bottle, and projective plane, respectively, will be seen on page 90 by means of the streamline diagrams (Figures 2.33, 2.34, 2.35).

limiting cases obtained if different pairs coincide entirely or in part. (As illustrations examine Figures 2.18, 2.19, 2.20, where corresponding edges are indicated by the same number.)

Similar situations arise when pairs of slits of the second type are present.

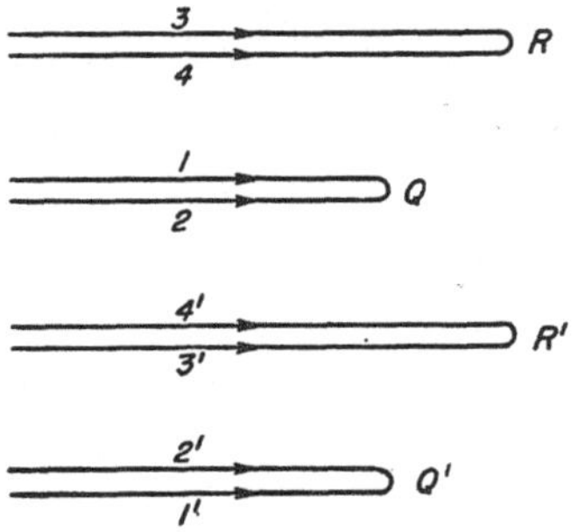

Fig. 2.15. Torus in slit representation.

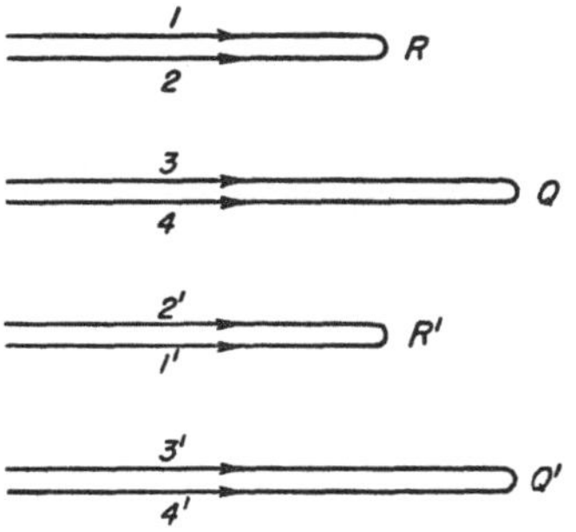

Figure 2.16. Klein's bottle in slit representation.

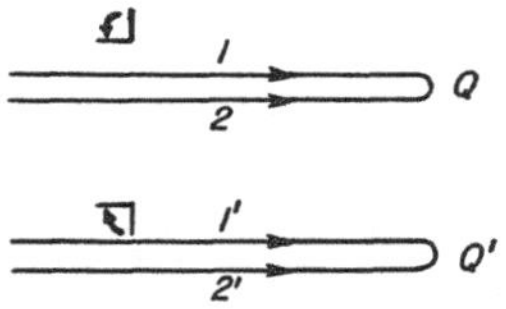

Figure 2.17. Slit pair of second kind showing reversal of orientation and representing projective plane.

So far our slit domains B were considered closed; we now introduce r boundary slits. A boundary slit may be either a finite slit $v =$ constant as in the preceding sections—not coinciding with one of the infinite slits—or it may coincide with a part of an infinite slit not containing the endpoint, as illustrated by Figure 2.21. In the latter case the coordination of the two edges of an infinite slit is interrupted along a finite segment (α and β in the diagram). Otherwise a boundary slit may at the same time form the end part of a pair of corresponding infinite slits; in this case the endpoints of the coordinated pair need not have the same u-coordinate, and the coordination of the ends is interrupted, as in Figure 2.22. Here the segments α, β, γ, δ form the boundary slit under consideration. In a similar way this

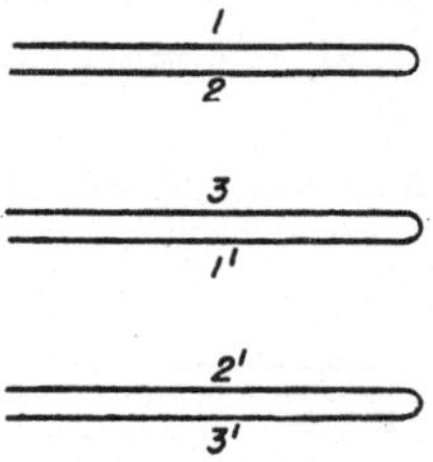

Fig. 2.18

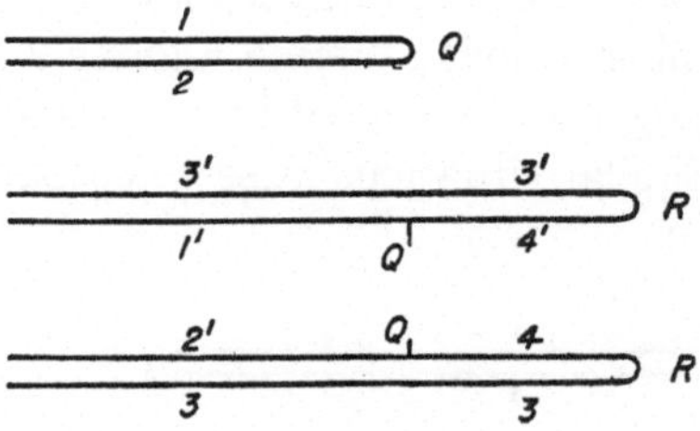

Fig. 2.19

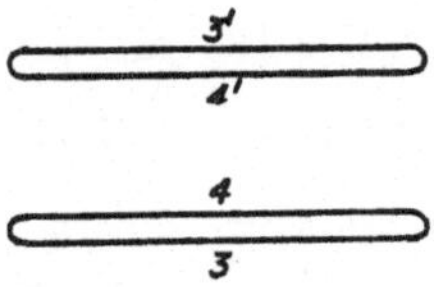

Fig. 2.20

Figures 2.18, 2.19, and 2.20. Limiting cases of closed slit domains.

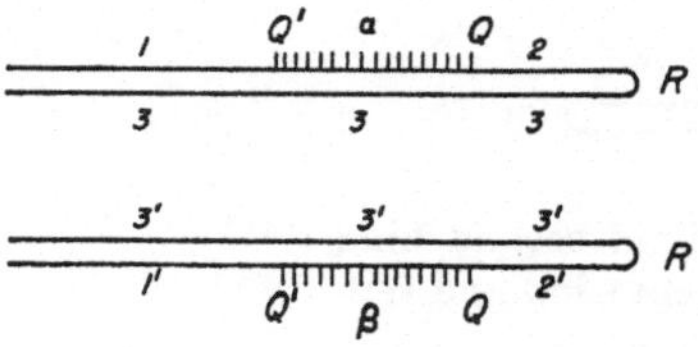

Fig. 2.21

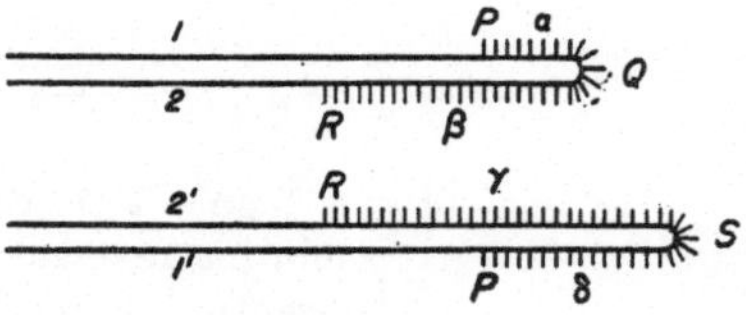

Fig. 2.22

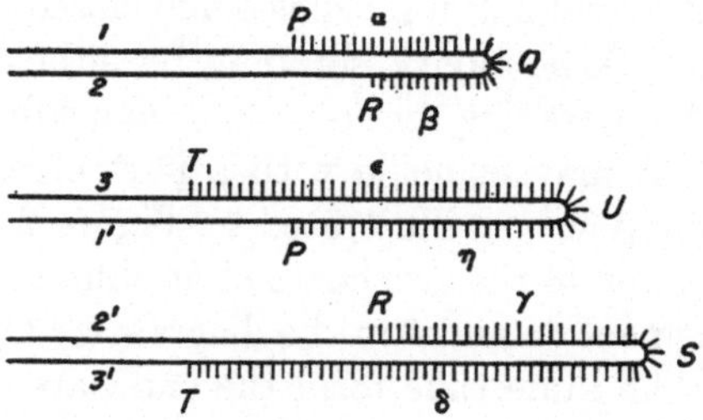

Fig. 2.23

Figures 2.21, 2.22, and 2.23. Slits carrying segments of boundary points

may happen if the pair of infinite slits is of the second type or if, as in Figure 2.23, the coordination embraces more than two slits.

Again, as before for genus zero, we shall find it convenient to consider *half-plane slit domains* lying in the half-plane $v > 0$, with $v = 0$ corresponding to a boundary line; otherwise nothing need be changed in our preceding description. For example, a Moebius strip is represented by Figure 2.24; the limiting case where one of the infinite slits falls into $v = 0$ is represented in Figure 2.25, with the boundary indicated by shading.

For such half-plane domains we may stipulate a normalization similar to that used for $L = 0$, namely place the point at infinity on

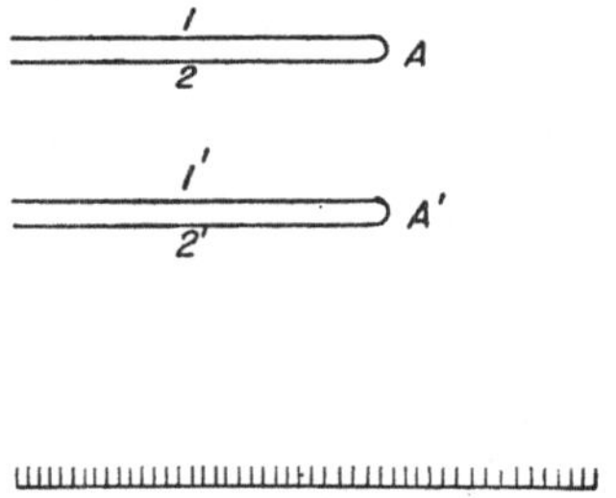

Fig. 2.24

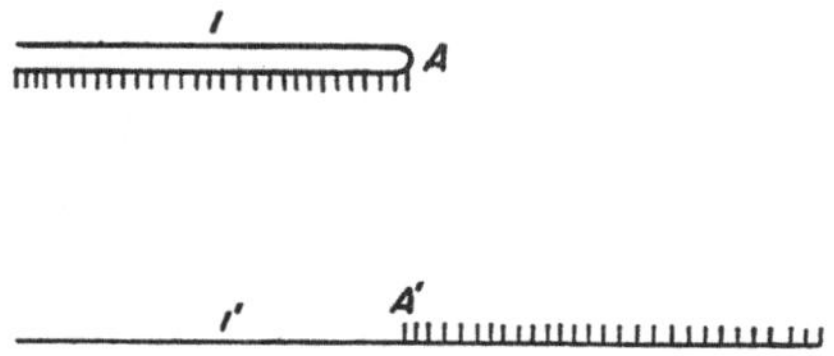

Fig. 2.25

Figures 2.24 and 2.25. Moebius strip in slit representation.

the boundary slit and fix the end of one interior slit at $u = 0$, $v = 1$.

3. *The Mapping Theorem.* We now consider the class of all slit domains B with r such boundary slits and with characteristic number L, and state the main

Theorem 2.7: Every Riemann domain G, as well as every non-orientable domain with a Riemann covering surface (see the end of Section 5), with characteristic number L and with r boundaries can be mapped conformally on a slit domain B. The mapping on a normalized half-plane slit domain is uniquely determined if the point at infinity is to correspond to a fixed point O on the boundary of G.

Proof: The mapping is again provided by the analytic function $u + iv = f(x + iy)$ (with the singularity at the origin O) which was constructed in §2 and §3. As observed, for $L > 0$ the function $f(z)$

can no longer be single-valued in G. Thus while u, by the nature of its definition, remains single-valued in G, the imaginary part v has additive periods along non-bounding closed curves. We proceed to examine the streamlines $v =$ constant of the function $f(z) = u + iv$.

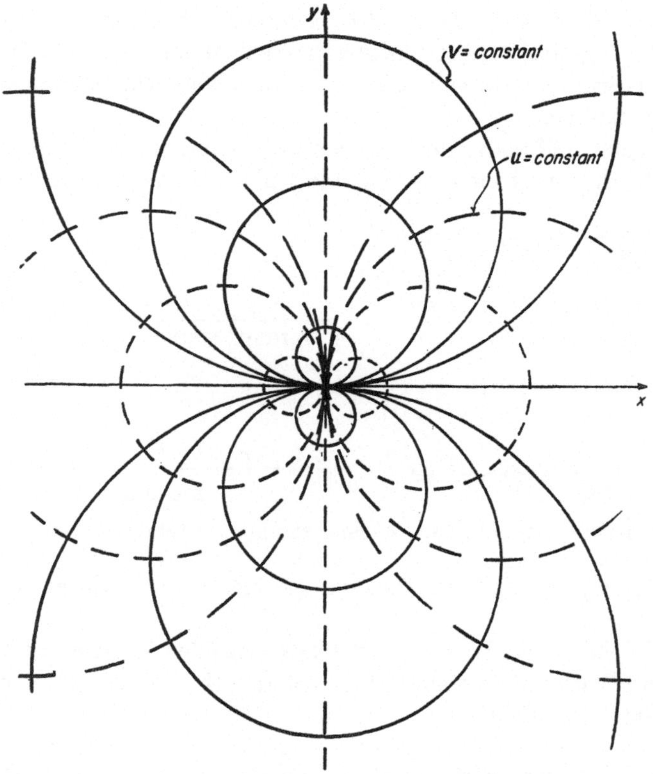

Figure 2.26. Streamlines and equipotential lines through dipole

First we recall a fact from the elements of the theory of complex functions: in the neighborhood of the pole O of a function $f(z)$ the equipotential lines $u =$ constant and the streamlines $v =$ constant form, approximately, two orthogonal families of circles, the first family tangent to the y-axis and the second tangent to the x-axis if the residue is real. For constants c of sufficiently large absolute value, the curve $u = c$ becomes an arbitrarily small closed (nearly circular) curve through O, and the same holds for the curve $v = c$.

Secondly the streamlines $v = c$ are analytic curves in G along which the values of u chang monotonically. Exceptions are provided only by the "crossing points," i.e. the zeros of $f'(z)$. The streamlines at a crossing point Q of the first order are schematically shown in Figure 2.27. Four branches issue from Q, at equal angles: the arrows indicate the direction of decreasing u. At a crossing point of order m (i.e. an m-fold zero of $f'(z)$), we have, similarly, $2m + 2$ branches of streamlines through Q forming equal angles, as indicated for $m = 2$ in Figure 2.28. Finally we recall that each boundary line γ of G can be considered as part of a streamline $v =$ constant. To study the behavior of streamlines in the neighborhood of a boundary curve γ we may visualize γ as analytic, e.g. a circle. For, according to the results obtained previously, we can map a doubly connected strip Z of G, bounded by γ and by another curve, conformally on a plane domain in such a manner that γ becomes a slit or a circle, and that the boundary coordination is biunique. Thus the topological behavior of the mapping by $u + iv$ in the vicinity of γ is completely exhibited by the behavior in the cell Z. Then the function $u + iv$, v being constant on γ, can be analytically extended beyond γ and therefore may be considered regular on γ.

Since u is single-valued along γ, we have, if s is the arc length and n the direction normal to γ,

$$\int_\gamma \frac{\partial v}{\partial n}\, ds = 0,$$

a relation which implies that $\partial v/\partial n$ has an even number $2\mu + 2$ of zeros on γ. Since $\partial v/\partial s$ vanishes on γ, these zeros are crossing points. If $\mu = 0$ we have the situation treated in §3. This case is illustrated in Figure 2.29, where a streamline meets the boundary γ at a point Q and splits into two branches which flow in different senses about γ, unite at another point Q' of γ, and there leave γ for the interior of G. In this case $\mu = 0$ we say that γ represents a simple element and is not a crossing element for G. If, however, $\mu > 0$ we shall say that γ represents a crossing element of order μ for the flow. The cases $\mu = 1$ and $\mu = 2$ are shown by Figures 2.30 and 2.31, with the direction of decreasing values of u again indicated in the diagrams. Finally we observe that each streamline, if followed in the direction of decreasing values of u, even through crossing elements, must ultimately lead to the pole O and arrive there from the side $x < 0$, on the

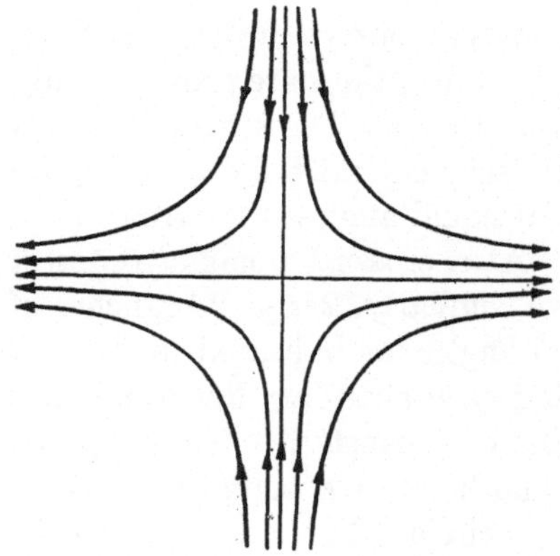

Fig. 2.27. Crossing point of first order.

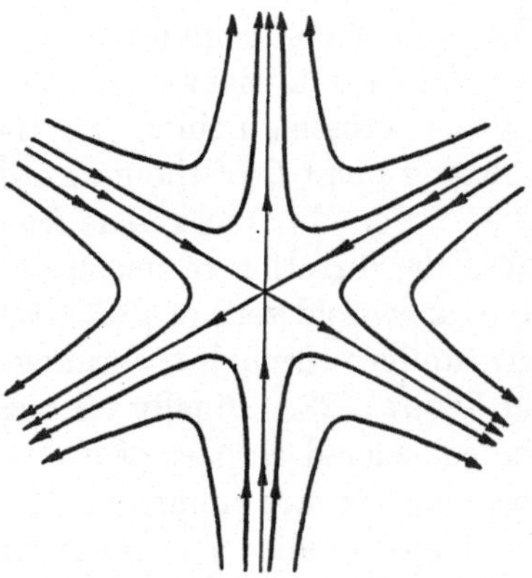

Figure 2.28. Crossing point of second order.

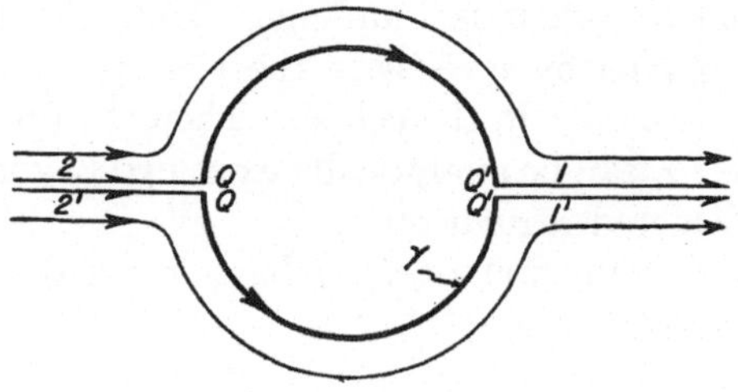

Figure 2.29. Boundary curve representing a simple element

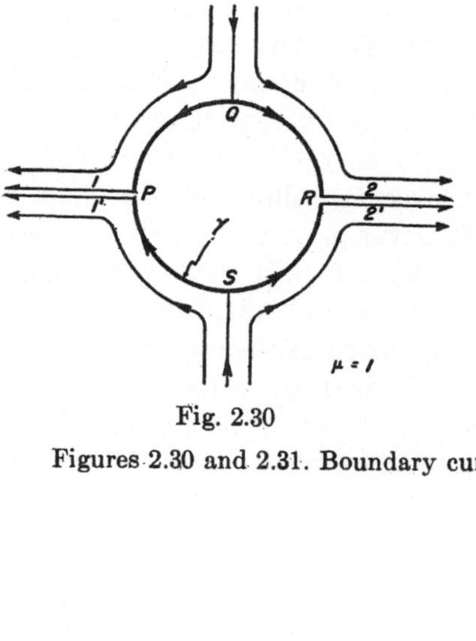

Fig. 2.30

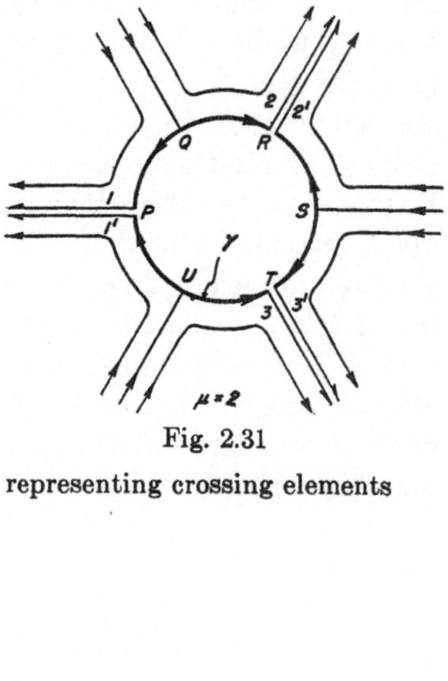

Fig. 2.31

Figures 2.30 and 2.31. Boundary curves representing crossing elements

left of the y-axis, where the curves $u = c$ with large negative values of c are situated. These streamlines cannot be closed since u is single-valued; they can end only at O because at all other points of G the function $f(z)$ is regular.

Our mapping theorem is easily obtained if we dissect G along streamlines into a region G^* so that the function v becomes single-valued in G^*. For this purpose we consider all the crossing points Q and crossing boundary elements γ and cut G along those branches of the streamlines v = constant, emanating from Q or γ, along which u decreases. The cuts end at the pole O, entering it there from the side $x < 0$. On the other hand, from each point P of G^*, by following a streamline in the direction of increasing—instead of decreasing—u, we can reach O from the side $x > 0$ where no cuts occur; hence G^* remains a connected domain, and through each point P of G^* there exists a single streamline v = constant along which u varies monotonically from $-\infty$ to $+\infty$. Therefore this streamline is mapped in a one-to-one way onto a straight line v = constant in the u, v-plane, and the function $f(z)$ is single-valued in G^*. For, as we saw, the streamline through any point P of G^* leads to O from $x > 0$ in an unambiguous way, since it cannot pass through a crossing element; but in the neighborhood of O, for $x > 0$, the values of v are, save for an arbitrary additive constant, uniquely determined for G^*.

Moreover the curves $v = c_1$ and $v = c_2$, for $c_1 \neq c_2$, have no common point in G^* except O, because near O for $x > 0$ they are different curves and cannot intersect in the domain G^* in which there are no crossing elements. Hence G^* is mapped in a one-to-one way onto a domain B in the u, v-plane. The boundaries of this domain B are segments v = constant. Inasmuch as these boundary lines correspond to the cuts through crossing elements, they are parallel slits as described above, and the identification of opposite points on different edges of a cut on G is effected by coordinating such points as have equal values of u on the slits corresponding to the different edges of the cut.

To complete the analysis, we first consider the typical "general" case of a simple crossing point Q on a streamline which passes through no other crossing element. The cut through this point Q, as in Figure 2.32, is marked by a heavy line and corresponding edges of the same part of the cut are identified by the same figures. The thin line indicates those branches of the streamline through Q which lead to

$u \to +\infty$. Typical instances of simple crossing points are furnished by the torus and the projective plane. The torus, represented as a rectangle with opposite sides identified, is mapped on a slit domain by the elliptic ζ-function of Weierstrass whose streamlines are sche-

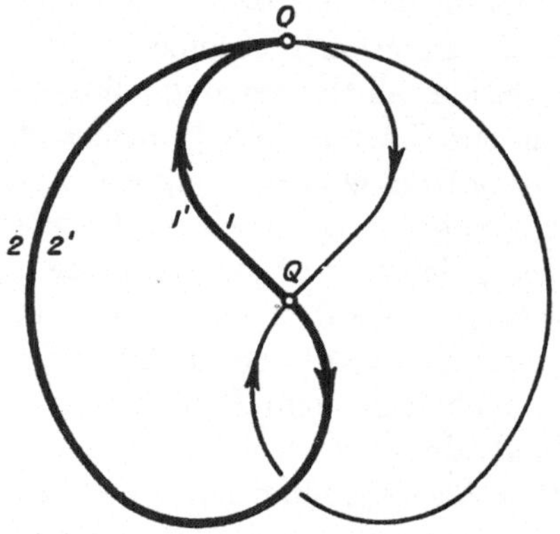

Figure 2.32. Streamline through critical point in projective plane.

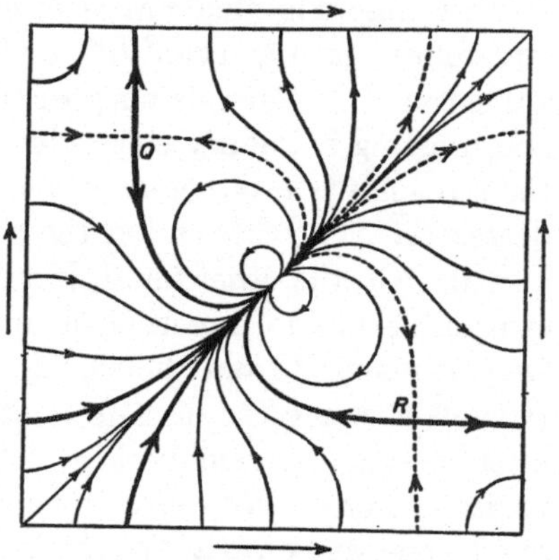

Figure 2.33. Dipole flow pattern on torus.

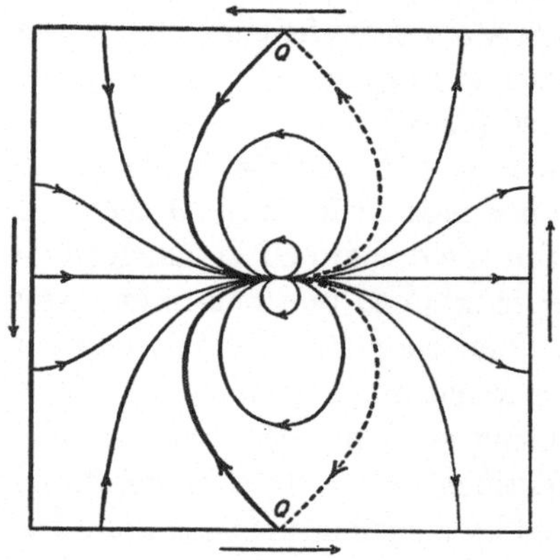

Figure. 2.34. Dipole flow pattern on projective plane.

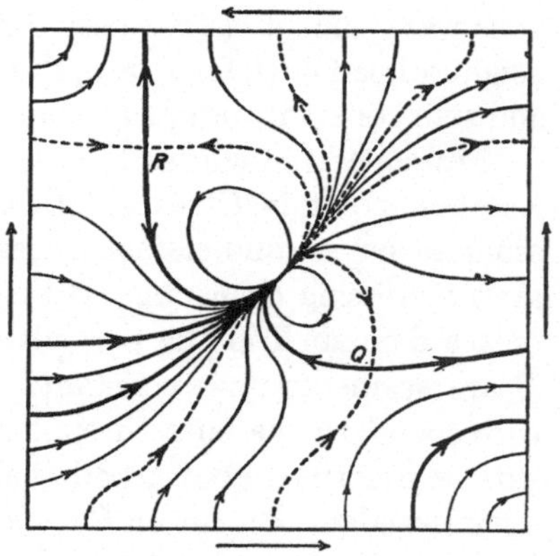

Figure 2.35. Dipole flow pattern on Klein bottle.

matically drawn in Figure 2.33. The cuts issuing from the two crossing points Q and R are shown by heavy lines (cf. Figure 2.37). The projective plane, represented as a square with opposite sides identified reversely, is mapped by a function whose streamlines are schematically indicated in Figure 2.34 (cf. Figure 2.17). We also show the Klein bottle in Figure 2.35 (cf. Figure 2.16).

If a crossing point Q has higher multiplicity—i.e. if it can be considered as the limiting case of several simple crossing points approaching one another—then the situation remains essentially the same and

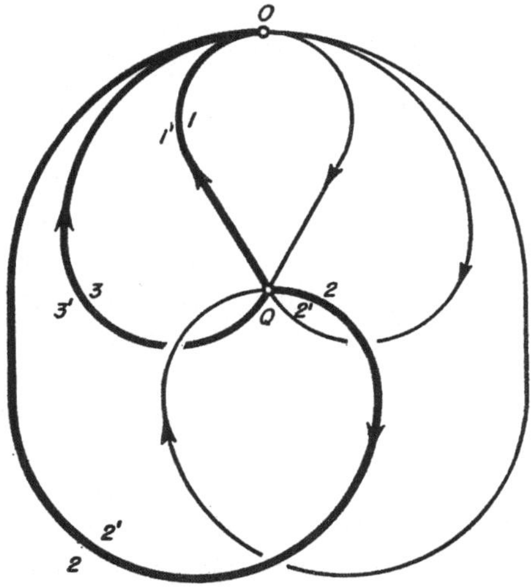

Figure 2.36. Streamline through critical point of second order on torus.

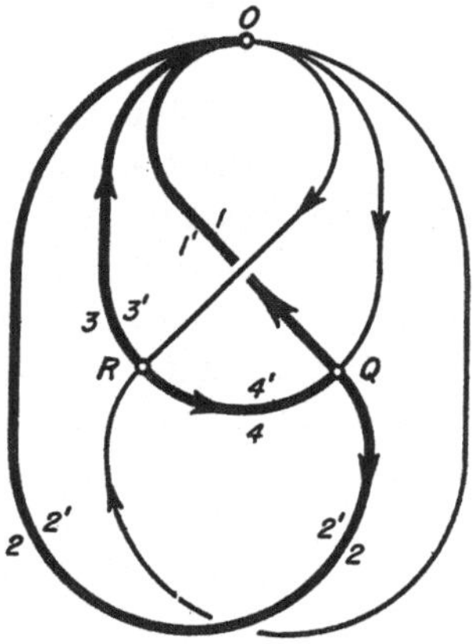

Figure 2.37. Critical streamlines on torus in general position.

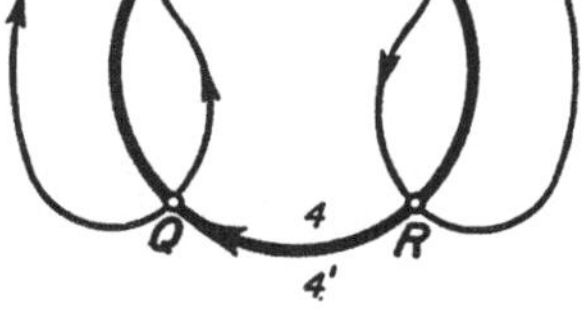

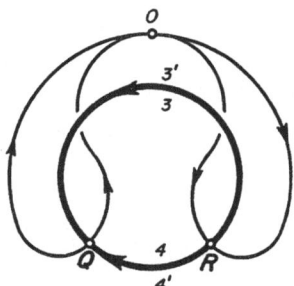

Figure 2.38. Critical streamlines on torus in symmetric position.

is illustrated in sufficient generality by Figure 2.36, representing a crossing point of order 2. The image in the u, v-plane is given by Figure 2.18.

If a branch of a streamline coming from the crossing point R, with decreasing values of u, meets another crossing point Q, both of the first order, we obtain, in general, the situation illustrated in Figure 2.37. The image in the u, v-plane is shown in Figure 2.19. If, in addition, the other part of the cut through R likewise passes through the same crossing point Q, we obtain the diagram in Figure 2.38. Here a single closed circuit, which does not pass through O, replaces the two independent circuits belonging to R and Q. The image of Figure 2.38 corresponds to the case of two pairs of cuts which partly coincide in such a way that the discontinuity of the image disappears at the infinite ends of the slits (Figure 2.20). It is immediately clear how a repetition or accumulation of such possibilities is expressed by a coalescing of infinite slits in our plane slit domain.

Finally we have to account for the mapping of the boundaries γ. If a boundary curve γ is part of a streamline not serving as one of our cuts, its image is simply a finite cut $v =$ constant separated from the infinite cuts. If γ is on one of the cuts through a crossing point but is not itself a crossing element, the situation in B is represented in Figure 2.21 and the image is given by Figure 2.29 (see pp. 84 and 88). If, thirdly, γ is a crossing element of order μ, e.g. $\mu = 1$ or 2, the situation in B is given by Figures 2.22 and 2.23 and the image of γ is given by Figures 2.30 and 2.31 (see pp. 84 and 88).

Thus the statement of our theorem is proved.

4. *Remarks. Half-Plane Slit Domains.* While the preceding theorem can easily be extended to the case where there are infinitely many boundary elements γ, it is not obvious that it can be generalized to the case of an infinite characteristic number L (e.g. an orientable domain G of infinite genus). It may further be remarked that, by theorem 2.2, the mapping function $f(z)$ is uniquely determined up to an arbitrary additive complex constant if the singular point O (situated, for example, at $z = z_0$), and the singularity $\sigma/(z - z_0)$ at that point, i.e. the complex constant σ, are prescribed. Thus there is, for a given domain G, a family of conformally equivalent slit domains which depends on three arbitrary complex—or six real—constants.

In applications it is often desirable to characterize a definite "normalized" representative slit domain conformally equivalent to a domain G. This is easily possible in the case of non-closed domains G with at least one boundary element γ. Instead of mapping γ on a finite slit, we can map it on the whole real axis $v = 0$ and G on a *half-plane slit domain* $v > 0$. The procedure is quite similar to that for $L = 0$ in §3, 3. Without loss of generality we may suppose that, by previous mapping of the adjacent part of G, γ is already a slit $y = 0$ through the origin. Then we let the singularity O tend to this boundary in such a way that the streamlines in O become tangent to $y = 0$. Alternatively, by reflecting G on the boundary slit γ we may form a domain $\bar{G}$ and then map $G + \bar{G}$ on a slit domain by a function $f(z)$ with the singularity $1/z$ at the origin on γ. The image of G is now a slit domain B' similar to that above but consisting only of the upper half-plane. The boundary slit γ is mapped on the real axis $v = 0$, while all the other infinite slits and boundary slits remain as before. Now we may, by a dilatation and a translation in the u-direction, introduce a suitable normalization. For example, the half-plane slit domain which is the conformal image of G may be uniquely characterized by fixing the vertex of a boundary slit or of any infinite slit at $u = 0, v = 1$.

CHAPTER III

Plateau's Problem

1. Introduction

Intimately connected with Dirichlet's Principle and conformal mapping is Plateau's problem which has long challenged mathematicians by the contrast between its simplicity of statement and difficulty of solution: to find the surface G of least area spanned in a given closed Jordan curve γ. If the surface G is represented in x,y,z-space by a function $z(x,y)$ with continuous derivatives, the area A is given by

$$A = \iint_B (1 + z_x^2 + z_y^2)^{1/2}\, dx\, dy,$$

where B is the domain in the x,y-plane bounded by the projection β of γ. The surface G is obtained as solution of the boundary value problem for Euler's (non-linear) differential equation

$$z_{xx}(1 + z_y^2) - 2z_{xy}z_xz_y + z_{yy}(1 + z_x^2) = 0.$$

This approach, although it has been pursued with a remarkable measure of success,[1] has proved to be both cumbersome and essentially inadequate. Surfaces G are excluded if they cannot be represented by a function $z(x,y)$, while the geometrical minimum problem, formulated for arbitrary closed curves γ, in no way permits such a restriction. Instead we consider surfaces represented parametrically by a vector $\mathfrak{x}(u,v)$ with components x,y,z, or x_1, x_2, x_3, given as functions of two parameters u,v which range over a domain B of the u,v-plane bounded by a curve β. The area is expressed by the integral

$$A(\mathfrak{x}) = \iint_B \sqrt{eg - f^2}\, du\, dv = \iint_B W\, du\, dv, \tag{3.1}$$

[1] See Radó [1].

where, in vector notation,

$$e = \mathfrak{x}_u^2 = \sum_{\nu=1}^{3}\left(\frac{\partial x_\nu}{\partial u}\right)^2, \qquad g = \mathfrak{x}_v^2 = \sum_{\nu=1}^{3}\left(\frac{\partial x_\nu}{\partial v}\right)^2,$$

$$f = \mathfrak{x}_u\,\mathfrak{x}_v = \sum_{\nu=1}^{3}\frac{\partial x_\nu}{\partial u}\frac{\partial x_\nu}{\partial v}, \qquad W^2 = eg - f^2.$$

Euler's conditions for the variational integral (3.1) are a system of differential equations

$$\frac{\partial}{\partial u}\frac{\partial W}{\partial x_{\nu_u}} + \frac{\partial}{\partial v}\frac{\partial W}{\partial x_{\nu_v}} = 0, \qquad \nu = 1, 2, 3, \tag{3.2}$$

which expresses the fact that the mean curvature of the surface $\mathfrak{x}(u,v)$ is zero. Surfaces of vanishing mean curvature are called minimal surfaces even though they may not actually furnish a minimum of area under given conditions.

Although the mathematical problem of proving the existence of a surface $\mathfrak{x}(u,v)$, that solves the preceding differential equations and is bounded by a prescribed curve γ, has long defied mathematical analysis, an experimental solution is easily obtained by a simple physical device. Plateau,[2] a Belgian physicist, studied the problem by dipping an arbitrarily shaped wire frame representing γ into a soap solution. The soap film forming within the wire frame attains a position of stable equilibrium, which corresponds to a relative minimum of area, and thus produces a minimal surface spanned in γ.[3] Plateau's experiments led to the name "Plateau's problem" for the general question of minimal surfaces bounded by prescribed contours. It is appropriate to distinguish from this general problem the more specific one of finding a surface that furnishes the smallest area at least relative to neighboring surfaces or, physically speaking, the problem of stable equilibria of the soap film. As we shall see in Chapter VI, there are contours bounding unstable minimal surfaces, whose areas do not furnish relative minima.

During the 19th century Plateau's problem was solved for many specific contours. Progress was made in the general theory by Riemann, Schwarz, Weierstrass, and others mainly on the basis of one idea: The geometrical meaning of the problem makes it obvious

[2] See Plateau [1], [2].

[3] We neglect the influence of gravity.

that the system of differential equations for x_1, x_2, x_3 is invariant under arbitrary transformations of the parameters u,v. Taking advantage of the freedom of choice of these parameters one can simplify the nonlinear differential equations, reducing them to the linear harmonic equation $\Delta\mathfrak{x} = 0$.[4] This reduction is effected if we assume the possibility of introducing *isometric parameters* u,v on the surfaces G, i.e. parameters which are characterized by the equations

$$e - g = 0, \qquad f = 0, \tag{3.3}$$

or, equivalently, by the fact that the mapping of G onto the domain B in the u,v-plane is conformal. Then $W = e = g$, and the differential equations (3.2) immediately become

$$\Delta\mathfrak{x} = \frac{\partial^2 \mathfrak{x}}{\partial u^2} + \frac{\partial^2 \mathfrak{x}}{\partial v^2} = 0, \quad \text{or} \quad \Delta x_\nu = 0, \qquad \nu = 1, 2, 3. \tag{3.4}$$

Surfaces $\mathfrak{x}(u,v)$ for which $\Delta\mathfrak{x} = 0$ are called *harmonic surfaces.*

Harmonic functions $x_\nu(u,v)$ may be considered as real parts $\mathcal{Re}[f_\nu(w)]$ of analytic functions

$$f_\nu(w) = x_\nu + i\tilde{x}_\nu$$

of the complex variable $w = u + iv$ (where $\tilde{x}_\nu(u,v)$ is conjugate harmonic to $x_\nu(u,v)$); consequently the Cauchy-Riemann equations imply

$$\phi(w) = \sum_{\nu=1}^{3} f_\nu'(w)^2 = \sum_{\nu=1}^{3} \left(\frac{\partial x_\nu}{\partial u} - i\frac{\partial x_\nu}{\partial v}\right)^2 = e - g - 2if.$$

It follows that, for any harmonic vector $\mathfrak{x}(u, v)$, the expression

$$\phi(w) = (e - g) - 2if$$

is an analytic function of the complex variable $w = u + iv$. This analytic function vanishes identically for harmonic vectors which represent minimal surfaces. We call the equation

$$\phi(w) = 0$$

the characteristic equation for minimal surfaces. Discarding in the

[4] This linearization corresponds to that of the differential equations of geodesics by the introduction of arc length as parameter; it is a special case of a more general fact concerning quasilinear partial differential equations in two independent variables (see Courant [2] Chapter III, §2, 1).

preceding considerations whatever is motivation rather than proof, we formulate as basic definition:

A minimal surface G is a surface represented by a harmonic vector $\mathfrak{x}(u, v)$ *for which*

$$\phi(w) = e - g - 2if = 0. \tag{3.5}$$

As domain B for the variables u,v or for $w = u + iv$ we may choose the disk $u^2 + v^2 \leq 1$. Plateau's problem is to solve the equation $\Delta \mathfrak{x} = 0$ for B under the additional conditions that $\phi(w) = 0$ and that $\mathfrak{x}$ maps the boundary $u^2 + v^2 = 1$ onto the prescribed contour γ. After this simplification the nonlinear character of the problem remains only in the boundary condition and in condition (3.5).

While relations (3.3) have the appearance of side conditions in the form of nonlinear partial differential equations of the first order, they are really of a much less restrictive character. The equivalent form (3.5) of these conditions shows that the characteristic relation $\phi(w) = 0$ is identically true in B, if we merely know, for example, that the boundary values of the real part $e - g$ are zero and that the imaginary part $-2f$ vanishes at one point of B.

A theory of minimal surfaces based on the preceding or similar definitions has led to famous results concerning minimal surfaces spanned in specific contours of simple geometric shapes, such as quadrilaterals and other polygons. More recently Garnier has attacked Plateau's problem for a general contour γ by using classical methods for polygonal boundaries. Complete and satisfactory success was however achieved only in 1930 and 1931, independently, by T. Radó and J. Douglas.[5] Their success was due to a combination of the classical approach with that of the modern calculus of variations.[6]

[5] See Douglas [1], Radó [1], and Radó [2] Chapter V.

[6] Quite a different line of approach to the problem of least area, by direct methods of the calculus of variations, was attempted by Lebesgue in a classical paper [1]. More recently McShane [1] has pursued this line under remarkably general conditions for the surfaces in competition. However, one should keep in mind the fact that it is not sufficient to give an existence proof by widening the concept of solution, e.g. by giving up conditions of differentiability. To obtain the relevant information that the actual solution belongs to the narrower class of "reasonable" functions one has to overcome practically the same difficulties and to use similar methods as would occur in an attack under more restrictive conditions.

The harmonic character of the vector $\mathfrak{x}(u,v)$ representing a minimal surface makes it natural to seek a connection between Plateau's problem and Dirichlet's Principle as applied to the integral

$$(3.6)\quad D[\mathfrak{x}] = \frac{1}{2}\iint_B (\mathfrak{x}_u^2 + \mathfrak{x}_v^2)\, du\, dv = \frac{1}{2}\iint_B (e + g)\, du\, dv.$$

Such a connection is moreover suggested by the following consideration. The inequality

$$\frac{e+g}{2} \geq \sqrt{eg},$$

where the equality sign holds only for the case $e = g$, implies

$$\frac{e+g}{2} \geq \sqrt{eg - f^2},$$

and

$$\frac{e+g}{2} = \sqrt{eg - f^2}$$

for $e - g = f = 0$ alone. Hence for all surfaces $\mathfrak{x}$

$$(3.7)\qquad D[\mathfrak{x}] \geq A(\mathfrak{x}),$$

and

$$(3.7\text{a})\qquad D[\mathfrak{x}] = A(\mathfrak{x})$$

if and only if (3.3) is satisfied, i.e. if the surface, without necessarily being harmonic, is represented by isometric parameters u,v.

If it is again assumed that such an isometric parametrization is possible for all surfaces under consideration, the equality (3.7a) is always attainable for suitably chosen parameters u,v. For a fixed surface, $A(\mathfrak{x})$ is independent of the choice of parameters. The integral $D[\mathfrak{x}]$, on the other hand, is not; as implied by the preceding inequalities, it becomes as small as possible for isometric parameters.

Combining the preceding facts we conclude: To minimize the area $A(\mathfrak{x})$ among all areas to be compared it is sufficient to minimize Dirichlet's integral (3.6) under the prescribed boundary conditions. The vector which minimizes Dirichlet's integral automatically solves the problem of least area; it is not only a harmonic vector, but also satisfies (3.3) or, as is equivalent for harmonic vectors, the relation $\phi(w) = 0$.

The possibility of introducing isometric parameters u,v for all surfaces to be compared remains unproved in this argument. Instead of supplying the proof we find it more convenient to consider the preceding discussion merely as a motivation. As the point of departure for a rigorous treatment of the problem of least area, we formulate independently

Variational Problem III: To minimize $D[\mathfrak{x}]$ under the given conditions at the boundary.

We shall show

1) that the problem possesses a solution $\mathfrak{x}$,
2) that the solution actually represents a minimal surface, and
3) that the minimal surface is a surface of least area.

Before carrying out this program we emphasize that its scope extends far beyond the original problem of a surface of least area spanned in a closed contour. Douglas has observed that the problem might equally well be stated and solved for any number m of coordinates $x_1, x_2, \cdots, x_m$, and that in the case $m = 2$ its solution is tantamount to a proof of Riemann's theorem on conformal mapping of simply connected domains on the interior of a circle. Moreover, Douglas has envisaged and attacked[7] the much more general problem of minimal surfaces of prescribed topological structure bounded by k given contours.

Douglas restricts the vectors admissible in variational problem III to harmonic vectors, so that Dirichlet's integral $D[\mathfrak{x}]$ can be expressed in terms of the boundary values of $\mathfrak{x}$ alone. As a consequence, the variational problem is reduced to a problem for functions of one variable. However, as the author and, independently, L. Tonnelli have observed, a considerable simplification may be gained by admitting a wider class of vectors; moreover, this approach makes possible solutions of the general problem of Douglas for higher topological structure (see Chapter V), of free boundary problems, and of problems concerning unstable minimal surfaces (see Chapter VI).

In the following chapters these general theories with some of their ramifications will be discussed. The present chapter will be restricted to Plateau's problem for simply connected minimal surfaces with prescribed boundary γ. We shall not only give existence

[7] See Douglas [5].

proofs, but also analyze questions such as that of the dependence of the solution on the given data, i.e. on the prescribed boundary.

2. Formulation and Solution of Basic Variational Problems

1. *Notations.* The following notations will be used consistently. The domain of the parameters u, v in the complex plane of $w = u + iv$ will be called B, its boundary β; in this chapter B will generally be the unit circle $|w| < 1$. The minimal surface will be called G and its given Jordan boundary γ. We shall call "admissible" all vectors $\mathfrak{x}$ which are continuous in $B + \beta$ and piecewise smooth in B, and which define a continuous mapping of β on γ, monotonic in the following sense: as a boundary point describes β monotonically, the image point describes γ monotonically. Furthermore Dirichlet's integral for any admissible vector is assumed to be finite. The totality of these vectors forms a set called the *function space* $\mathfrak{S}$. The subspace of $\mathfrak{S}$ consisting of all harmonic vectors in $\mathfrak{S}$ will be called $\mathfrak{P}$.

Our basic variational problem is *to find an admissible vector* $\mathfrak{x}(u, v)$ *for which the Dirichlet integral* $D[\mathfrak{x}]$ *attains its least value* d. Generally we shall indiscriminately speak of vectors $\mathfrak{x}(u, v)$ or surfaces G.

That the variational problem makes sense, i.e. that there actually exist admissible vectors, is not *a priori* evident; as a matter of fact, it is not difficult to construct contours γ for which $D[\mathfrak{x}]$ is always infinite if the vector $\mathfrak{x}$ satisfies the conditions of continuity and the boundary condition imposed. Therefore we expressly state as an assumption that admissible vectors exist. This assumption, however, cannot be considered essentially restrictive; for example, it is automatically satisfied for all rectifiable contours, see §8, 3.

2. *Fundamental Lemma. Solution of Minimum Problem.* The construction of the solution of the variational problem, in other words the existence proof for the solution, is based on an important lemma, in statement and proof very similar to lemma 1.4. Instead of functions ϕ we discuss a whole class of vectors with uniformly bounded Dirichlet integral:

Lemma 3.1.: In a domain B of the u,v-plane consider the class of piecewise smooth vectors $\mathfrak{x}$ for which Dirichlet's integral is uniformly bounded by the constant M:

$$D_B[\mathfrak{x}] \leq M.$$

About an arbitrarily fixed point O we draw circles of radius r. Denote by C_r an arc or a set of arcs of such a circle contained in B, by s arc length on C_r. Then for every positive $\delta < 1$ there exists a value ρ, depending on $\mathfrak{x}$, with $\delta \leq \rho \leq \sqrt{\delta}$, such that

$$\int_{C_\rho} \mathfrak{x}_s^2 \, ds \leq \frac{\epsilon(\delta)}{\rho} \tag{3.8}$$

with

$$\epsilon(\delta) = \frac{4M}{\log 1/\delta} \tag{3.9}$$

tending to zero for $\delta \to 0$. Furthermore, the square of the length L_ρ of the image C'_ρ of C_ρ in x,y,z-space has the bound

$$L_\rho^2 \leq 2\pi \, \epsilon(\delta), \tag{3.10}$$

i.e. the oscillation of $\mathfrak{x}(u,v)$ on C_ρ is at most $\sqrt{2\pi\epsilon(\delta)}$.

Proof: If the vector $\mathfrak{x}$ has continuous first derivatives,

$$\int_\delta^{\sqrt{\delta}} dr \int_{C_r} \mathfrak{x}_s^2 \, ds \leq D[\mathfrak{x}] \leq 2M.$$

To obtain (3.8), write the left member in the form

$$\int_\delta^{\sqrt{\delta}} \frac{dr}{r} \cdot p(r)$$

with

$$p(r) = r \int_{C_r} \mathfrak{x}_s^2 \, ds$$

and apply the mean value theorem of the integral calculus. If $\mathfrak{x}$ is only piecewise smooth, the reasoning has to be slightly modified, exactly as in Chapter I. Relation (3.10) follows by Schwarz' inequality, since

$$L_\rho = \int_{C_\rho} \sqrt{\mathfrak{x}_s^2} \, ds.$$

To solve the variational problem, we start, as usual, with a minimizing sequence $\mathfrak{x}_1, \mathfrak{x}_2, \cdots$ of admissible vectors, i.e. with a sequence for which $D[\mathfrak{x}_n] \to d$; d now denotes the greatest lower bound for $D[\mathfrak{x}]$. If we replace each of the vectors $\mathfrak{x}_n$ by an admissible harmonic vector with the same boundary values on β, the new vectors—again denoted by $\mathfrak{x}_n$—*a fortiori* form a minimizing sequence by Dirichlet's Principle. Our intention is to construct the solution $\mathfrak{x}$ as limit of a uniformly convergent subsequence of the vectors $\mathfrak{x}_n$.

The possibility of such a procedure depends on a further preparatory step. Because of the invariance of $D[\mathfrak{x}]$ under conformal mapping, we may transform the unit circle B in the complex w-plane into itself by an arbitrary linear transformation that takes a point P into a point Q and $\mathfrak{x}(P)$ into $\mathfrak{y}(Q)$ with $D[\mathfrak{x}] = D[\mathfrak{y}]$. A sequence $\mathfrak{y}_n$ thus obtained is again a minimizing sequence.[8] This freedom is utilized to "normalize" the minimizing sequence by imposing the "three point condition": Admissible vectors $\mathfrak{x}$ are required to map three fixed points P_1, P_2, P_3 of β onto three fixed points P_1', P_2', P_3' of γ. The three point condition can be satisfied by substituting for each admissible vector another with the same Dirichlet integral. The new vectors are obtained by the linear transformation of the unit circle into itself that takes three points Q_1, Q_2, Q_3 into P_1, P_2, P_3, respectively.

Now we state

Lemma 3.2: The boundary values of the admissible vectors $\mathfrak{x}$, satisfying the inequality $D[\mathfrak{x}] \leq M$ and the three point condition, are equicontinuous. Consequently these boundary values form a compact set on β; the harmonic admissible vectors form a compact set in the region $B + \beta$.

Proof: We apply the fundamental lemma 3.1, as well as a basic property of Jordan curves γ. For any positive σ, there exists a positive quantity $\tau(\sigma)$ having the following property: if Q, Q' are a pair of points on γ at a distance not exceeding τ, the diameter of one of the two arcs of γ determined by Q and Q' does not exceed σ. For given σ, we choose δ, $0 < \delta < 1$, so that the quantity $2\pi\epsilon(\delta) = 8\pi M/\log(1/\delta)$ of lemma 3.1 satisfies the inequality $2\pi\epsilon(\delta) \leq \tau^2$. Thus, for any point

[8] By letting the sequence of linear transformations which take $\mathfrak{x}_n$ into $\mathfrak{y}_n$ converge to a degenerate one, we could prevent the convergence of a minimizing sequence to the solution; hence the necessity for a normalization as in the following paragraph. Incidentally, other methods of normalization which achieve the same purpose can be given and are preferable in other cases, see Chapters IV, V, VI.

P on β, there exists a quantity ρ, $\delta \leq \rho \leq \sqrt{\delta}$, such that the oscillation of $\mathfrak{x}$ on the circle C_ρ about P is not greater than $\sqrt{2\pi\epsilon(\delta)} \leq \tau$. By the three point condition, for δ sufficiently small, the larger arc of β cut out by C_ρ contains at least two of the fixed points P_1, P_2, P_3; hence the image of the smaller arc of β is the "smaller arc" of γ. For points R and R' on β at a distance not exceeding δ, therefore, $|\mathfrak{x}(R) - \mathfrak{x}(R')| < \sigma$. This inequality expresses the equicontinuity of the boundary values of $\mathfrak{x}$, since δ depends only on the quantity σ. This equicontinuity together with the boundedness of the admissible vectors $\mathfrak{x}$ implies the compactness by Arzelà's theorem.

We apply this lemma to a normalized minimizing sequence of harmonic vectors $\mathfrak{x}_n$. Consequently we can choose a subsequence of the minimizing sequence with uniformly convergent boundary values, again denoted by $\mathfrak{x}_n$. Since the $\mathfrak{x}_n$ were assumed to be harmonic, we infer immediately that they converge uniformly in $B + \beta$ to an admissible vector $\mathfrak{x}$ harmonic in B.

By the lower semicontinuity of Dirichlet's integral (lemma 1.1),

$$D[\mathfrak{x}] \leq \text{lim inf.}\ D[\mathfrak{x}_n] = d.$$

Since d is the greatest lower bound for $D[\mathfrak{x}]$, it follows that

$$D[\mathfrak{x}] = d.$$

The harmonic vector $\mathfrak{x}$ solves our variational problem.

3. *Remarks. Semicontinuity.* The preceding reasoning actually yields a somewhat more general result. Instead of the fixed curve γ we may consider a sequence of continuous curves γ_n converging to γ "smoothly," or in the "Fréchet sense," i.e. in such a manner that if two points P_n, Q_n on γ_n tend to the points P, Q of γ, then the whole arc P_nQ_n of γ_n tends to one of the two arcs PQ of γ. We need not even require the γ_n to be simple curves, but may permit them to have multiple points and corresponding small loops, as long as these disappear in the limit. We assume the existence of admissible vectors with finite Dirichlet integral for all γ_n. Now consider a sequence of vectors $\mathfrak{x}_n$ mapping β on γ_n and having uniformly bounded Dirichlet integrals. Such vectors $\mathfrak{x}_n$ can be normalized by the *three point condition*: to three fixed points P, Q, R on β correspond three points P_n', Q_n', R_n' on γ_n which tend to three prescribed fixed points P', Q', R' on γ. Then the boundary values of these vectors $\mathfrak{x}_n$ form a compact set of functions, as is seen by reasoning similar to that in

article 2. As before, we may replace $\mathfrak{x}_n$ by a sequence of harmonic vectors; we choose a uniformly convergent subsequence and again denote it by $\mathfrak{x}_n$.

Denoting the limit vector by $\mathfrak{x}$, we have once more

$$D[\mathfrak{x}] \leq \text{lim inf. } D[\mathfrak{x}_n].$$

In particular, we may take for $\mathfrak{x}_n$ a solution of the variational problem for γ_n (if γ_n is a Jordan curve). The sequence $\mathfrak{x}_n$ converges to a vector $\mathfrak{x}$ admissible in the problem for γ, therefore $d \leq D[\mathfrak{x}] \leq$ lim inf. $D[\mathfrak{x}_n]$. This argument proves

Theorem 3.1: If d_n is the greatest lower bound of $D[\mathfrak{x}]$ under the condition that $\mathfrak{x}$ maps the boundary β on γ_n, then

$$d \leq \text{lim inf. } d_n .$$

In other words: the minimum d of the variational problem is a lower semicontinuous function of the curve γ, if convergence is understood in the sense of Fréchet.

3. *Proof by Conformal Mapping That Solution is a Minimal Surface*

That the solution $\mathfrak{x}$ of problem III is a minimal surface, i.e. that it satisfies the condition $\phi(w) = 0$, can be proved in a simple way by using the results of Chapter II. For the construction of the solution $\mathfrak{x}$ of the variational problem the parameter domain B was fixed as a circle and the three point condition was imposed. However, since Dirichlet's integral is invariant under any conformal mapping, our solution will furnish the minimum even compared with a wider class of vectors $\mathfrak{z}$; specifically we show that this is the case for certain discontinuous vectors $\mathfrak{z}$.[9] We cut the domain B along a circle κ enclosing the disk K, in which r, θ may be polar coordinates. With an arbitrary analytic periodic function $\lambda(\theta)$ and a parameter ϵ define the function $\psi = \theta + \epsilon\lambda(\theta)$. For sufficiently small ϵ the correspondence between θ and ψ is analytic and biunique. We further introduce the vector

[9] Their introduction is motivated by the observations that the expression for the area $A(\mathfrak{x})$ yields the same value for discontinuous parametric representations as well and that we are aiming at cases where $A(\mathfrak{x})$ and $D[\mathfrak{x}]$ are identical.

$\mathfrak{z}(u,v)$ defined in B by

$$\mathfrak{z}(r,\theta) = \begin{cases} \mathfrak{x}(r,\psi) & \text{in } K, \\ \mathfrak{x}(u,v) & \text{in } B - K; \end{cases}$$

$\mathfrak{z}(r,\theta)$ is discontinuous on κ but maps B again on the surface G in space. We state

$$D[\mathfrak{z}] \geq D[\mathfrak{x}]$$

or what is equivalent

$$D_K[\mathfrak{z}] \geq D_K[\mathfrak{x}].$$

In other words, $\mathfrak{x}$ also yields a minimum with respect to these dis-

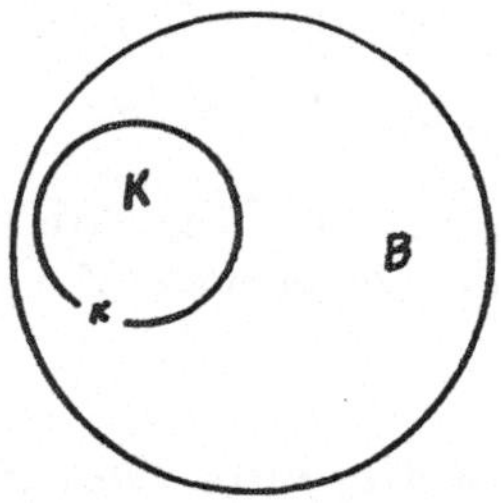

Figure 3.1.

continuous vectors $\mathfrak{z}$. As a consequence, since $\mathfrak{x}$ and $\mathfrak{z}$ have continuous derivatives (of all orders)[10] with respect to u,v,ϵ in $K + \kappa$

$$\text{(3.11)} \qquad \frac{d}{d\epsilon} D_K[\mathfrak{z}] = \iint_K (\mathfrak{z}_u \mathfrak{z}_{u\epsilon} + \mathfrak{z}_v \mathfrak{z}_{v\epsilon})\, du\, dv = 0$$

for $\epsilon = 0$. The proof is immediate if we apply the sewing theorem to the domain B^* obtained from B by cutting along κ and identifying the point θ on the outer edge of the cut with the point $\psi = \theta + \epsilon\lambda$ on the inner edge. The Riemann domain B^* may again be mapped conformally on the unit circle B. This mapping takes the vector $\mathfrak{z}$, which is continuous in B^*, into a vector $\mathfrak{z}'$ in B that satisfies all conditions of admissibility in the original variational problem. Since the invariance of Dirichlet's integral implies $D[\mathfrak{z}] = D[\mathfrak{z}']$ and since $D[\mathfrak{z}'] \geq D[\mathfrak{x}]$ the statement is proved.

[10] A similar reasoning could not be applied to the whole domain B since these derivatives need not exist at the boundary β.

By Green's formula (3.11) is equivalent to

$$-\iint_K \mathfrak{z}_\epsilon \Delta\mathfrak{z}\, du\, dv + \int_\kappa \mathfrak{z}_\epsilon \mathfrak{z}_r\, r\, d\theta = 0,$$

for $\epsilon = 0$; all expressions under the integral signs are continuous in the coordinates and the parameter. For $\epsilon = 0$ we have

$$\mathfrak{z}_\epsilon = \lambda \mathfrak{x}_\theta, \; \mathfrak{z}_r = \mathfrak{x}_r, \quad \text{and} \quad \Delta\mathfrak{z} = 0$$

so that

$$\int_\kappa \lambda \mathfrak{x}_r \mathfrak{x}_\theta\, r\, d\theta = 0.$$

The arbitrary character of λ implies that

$$\mathfrak{x}_r \mathfrak{x}_\theta = 0,$$

or

$$\mathfrak{x}_r \mathfrak{x}_s = 0,$$

everywhere on κ, s being arc length on κ. Through any point P in B we can draw such a circle κ in a given direction; hence for any point P and any two orthogonal directions indicated by s and n we have $\mathfrak{x}_s \mathfrak{x}_n = 0$. In particular, if κ is parallel to the u-axis at P, we have $\mathfrak{x}_s \mathfrak{x}_n = \mathfrak{x}_u \mathfrak{x}_v = f = 0$, and if κ is parallel to the line $u - v = 0$, we obtain $(\mathfrak{x}_u - \mathfrak{x}_v)(\mathfrak{x}_u + \mathfrak{x}_v) = e - g = 0$. The relation $\phi(w) = e - g - 2if = 0$ is thus proved, and our solution is identified as a minimal surface.

4. *First Variation of Dirichlet's Integral*

The proof that $\phi(w) = 0$ can be obtained, if desired, without use of the sewing theorem or of conformal mapping altogether. The following method employs a simple expression for the first variation of $D[\mathfrak{x}]$. Since this expression will be of basic importance in more general problems later we shall discuss it in somewhat greater detail than necessary for the present purpose; in particular we shall not necessarily assume the domain B to be fixed or simply connected.

1. *Variation in General Space of Admissible Functions.* The variations of the vector $\mathfrak{x}$ which we shall consider are of two types:

a) Variation of the parametric representation while the geometric surface is kept fixed;
b) Variation of the surface.[11]

As always in the calculus of variations, it is quite useless to consider the most general variations compatible with the conditions of admissibility.

To introduce variations of type a) we transform the domain B into itself or into another admissible domain B' by a biunique transformation of the form

$$
\begin{aligned}
u &= u' + \epsilon\Lambda \\
v &= v' + \epsilon\mathrm{M}
\end{aligned}
\tag{3.12}
$$

or, in complex notation, with $w = u + iv$, $w' = u' + iv'$,

$$w = w' + \epsilon(\Lambda + i\mathrm{M}). \tag{3.12a}$$

The parameter ϵ will take on a suitably small value; the quantities Λ and M are continuous functions of u,v,ϵ in B and of u',v',ϵ in B', with piecewise continuous first derivatives. We further suppose that the absolute values of the first derivatives have a common bound b. Finally, using the notation

$$
\begin{aligned}
\Lambda(u',v',0) &= \lambda(u',v'), \\
\mathrm{M}(u',v',0) &= \mu(u',v'),
\end{aligned}
\tag{3.13}
$$

we require the inequalities

$$
\begin{aligned}
|\Lambda_{u'} - \lambda_{u'}| &< \epsilon b, \qquad |\Lambda_{v'} - \lambda_{v'}| < \epsilon b, \\
|\mathrm{M}_{u'} - \mu_{u'}| &< \epsilon b, \qquad |\mathrm{M}_{v'} - \mu_{v'}| < \epsilon b
\end{aligned}
\tag{3.14}
$$

to hold; similar inequalities are assumed to hold for the derivatives with respect to u,v as independent variables. These inequalities imply

$$|\Lambda_u - \lambda_{u'}| < 3\epsilon b, \quad \text{etc.,} \tag{3.14a}$$

and

$$|\Lambda_{u'} - \lambda_u| < 3\epsilon b, \quad \text{etc.} \tag{3.14b}$$

[11] For present purposes only variations of type a) are required. Variations of type b), affecting the surface as well, are easily expressed, see (3.19) and (3.20). We shall make use of them in Chapter VI.

From these assumptions we deduce, by a short calculation, the inequalities

$$(3.15)\qquad \begin{aligned} &\left| \frac{\partial(u', v')}{\partial(u, v)} - (1 - \epsilon\lambda_u - \epsilon\mu_v) \right| \leq 2\epsilon^2(b^2 + b), \\ &\left| \frac{\partial(u, v)}{\partial(u', v')} - (1 + \epsilon\lambda_{u'} + \epsilon\mu_{v'}) \right| \leq 2\epsilon^2(b^2 + b) \end{aligned}$$

for the Jacobian of the transformation.

We introduce the variation a) of the vector $\mathfrak{x}$, which replaces $\mathfrak{x}$ in B by the vector

$$\mathfrak{z}(u', v') = \mathfrak{x}(u, v)$$

in B'. Denoting by D^* the Dirichlet integral taken over B' with respect to u' and v', we have

$$D^*[\mathfrak{z}] = \frac{1}{2} \iint_B \{[\mathfrak{x}_u(1 + \epsilon\Lambda_{u'}) + \mathfrak{x}_v\,\epsilon\mathrm{M}_{u'}]^2 + [\mathfrak{x}_u\,\epsilon\Lambda_{v'} + \mathfrak{x}_v(1 + \epsilon\mathrm{M}_{v'})]^2\} \frac{\partial(u', v')}{\partial(u, v)}\, du\, dv.$$

By a simple calculation using (3.13), (3.14), (3.14a), (3.14b), (3.15), and the inequality

$$\iint_B |\, \mathfrak{x}_u \mathfrak{x}_v \,|\, du\, dv \leq D[\mathfrak{x}],$$

we obtain the basic variational formula

$$(3.16)\qquad D^*[\mathfrak{z}] = D[\mathfrak{x}] + \tfrac{1}{2}\epsilon \mathrm{V}(\mathfrak{x}, \lambda, \mu) + \epsilon^2 \mathrm{R}.$$

Here the "*first variation* V of $D[\mathfrak{x}]$" is defined by

$$(3.17)\qquad V(\mathfrak{x}, \lambda, \mu) = \iint_B [p(\lambda_u - \mu_v) - q(\lambda_v + \mu_u)]\, du\, dv,$$

with the abbreviations

$$p = \mathfrak{x}_u^2 - \mathfrak{x}_v^2, \qquad q = -2\mathfrak{x}_u\mathfrak{x}_v\,.$$

The remainder term R has the bound

$$|\, R \,| < c\, D[\mathfrak{x}]$$

with a constant c depending only on b, not on $\mathfrak{x}$. The existence of the (improper) integral V and hence that of $D^*[\mathfrak{z}]$ is assured because

$$(3.18) \quad | p(\lambda_u - \mu_v) - q(\lambda_v + \mu_u) | \leqslant 2b(| p | + | q |) \leqslant 4b(\mathfrak{x}_u^2 + \mathfrak{x}_v^2).$$

Although in general only the variation a) of the parametric representation of a fixed surface will be needed, we note briefly that variations affecting the surface can be expressed by a vector $\mathfrak{t}(u',v')$ vanishing at the boundary of B'. We replace $\mathfrak{x}$ in B by

$$(3.19) \qquad \mathfrak{z}(u',v') = \mathfrak{x}(u,v) + \epsilon\mathfrak{t}(u',v',\epsilon).$$

Supposing $D^*[\mathfrak{t}]$ to be uniformly bounded in ϵ, we find

$$(3.20) \qquad D^*[\mathfrak{z}] = D[\mathfrak{x}] + \tfrac{1}{2}\epsilon V + \epsilon D[\mathfrak{x}, \mathfrak{t}] + \epsilon^2 T,$$

where $\mathfrak{t}$ stands for $\mathfrak{t}(u,v,0)$ and the remainder T remains uniformly bounded if the integrals $D[\mathfrak{x}]$ and $D[\mathfrak{t}]$ are bounded.

The expression

$$\lim_{\epsilon \to 0} \frac{D^*[\mathfrak{z}] - D[\mathfrak{x}]}{\epsilon} = D[\mathfrak{x}, \mathfrak{t}] + \tfrac{1}{2} V$$

is called *the first variation of* $D[\mathfrak{x}]$ for the variation (3.19) of the vector $\mathfrak{x}$. Since $\mathfrak{t} = 0$ on γ, the term $D[\mathfrak{x}, \mathfrak{t}]$ vanishes for harmonic vectors $\mathfrak{x}$, as shown in Chapter I, and the first variation of $D[\mathfrak{x}]$ has in this case the value $V/2$ even if $\mathfrak{t}$ is not identically zero.

A vector $\mathfrak{x}$ is called *stationary* if the first variation vanishes for all admissible variations of the form (3.19). Since the deformation λ,μ of the parameter domain and the vector $\mathfrak{t}$ do not depend on each other, the stationary character of $\mathfrak{x}$ is equivalent to the separate conditions $D[\mathfrak{x}, \mathfrak{t}] = 0$ and $V = 0$. The first relation is equivalent to $\Delta\mathfrak{x} = 0$. For the proof of $\phi(w) = 0$ we may therefore, assuming $\mathfrak{x}$ to be harmonic,[12] concentrate on evaluating the condition $V = 0$, thus restricting our attention to variations of the boundary representation of γ by the boundary values of $\mathfrak{x}$ on β.

2. *First Variation in Space of Harmonic Vectors.* Before proceeding to prove that $\phi(w) = 0$ we interject a remark of independent

[12] If $\mathfrak{x}$ is not necessarily harmonic and if we demand $\mathfrak{t} = 0$, then $V = 0$ expresses the stationary character of the parameter representation of a given surface. The following arguments then yield, not the harmonic character of $\mathfrak{x}$, but the relation $p = q = 0$, which expresses the fact that the stationary parameter representation is isometric.

interest. Douglas' treatment of Plateau's problem was based on the consideration of harmonic vectors $\mathfrak{x}$ alone. We shall show that this restriction has no effect on the form V of the first variation; more precisely, if $\mathfrak{x}$ is a harmonic vector in B and if $\mathfrak{x}'$ is the harmonic vector in B' with the boundary values $\mathfrak{x}'(u',v') = \mathfrak{x}(u,v)$ defined by transformation (3.12) (but in general differing from $\mathfrak{z}(u',v')$ in B'), then

$$\lim_{\epsilon\to 0} \frac{D^*[\mathfrak{x}'] - D[\mathfrak{x}]}{\epsilon} = \tfrac{1}{2} V(\mathfrak{x}, \lambda, \mu).$$

In other words, in the space $\mathfrak{P}$ of harmonic vectors we obtain for the first variation of $D[\mathfrak{x}]$ the same expression as before defined in the wider space $\mathfrak{S}$, provided that the initial vector $\mathfrak{x}$ is harmonic.[13] For the proof we observe that $D^*[\mathfrak{x}'] \leq D^*[\mathfrak{z}]$ since $\mathfrak{x}'$ and $\mathfrak{z}$ coincide at the boundary β' of B'; hence

$$D^*[\mathfrak{x}'] \leq D[\mathfrak{x}] + \tfrac{1}{2}\epsilon V + \epsilon^2 R. \tag{3.21}$$

However, we could equally well start from $\mathfrak{x}'$ in B' and transform $\mathfrak{x}'(u',v')$ into $\mathfrak{y}(u,v) = \mathfrak{x}'(u',v')$ by transformation (3.12) interpreted inversely. Then we obtain, from inequalities (3.14), (3.14a), (3.14b), (3.15),

$$D[\mathfrak{x}] \leq D[\mathfrak{y}] \leq D^*[\mathfrak{x}'] - \tfrac{1}{2}\epsilon V + \epsilon^2 R'. \tag{3.22}$$

The remainder $|R'|$ is bounded if $D[\mathfrak{x}]$ is bounded, consequently also if $D^*[\mathfrak{x}']$ is bounded. The two inequalities together can be written as

$$\tfrac{1}{2}V - \epsilon R' \leq \frac{D^*[\mathfrak{x}'] - D[\mathfrak{x}]}{\epsilon} \leq \tfrac{1}{2}V + \epsilon R$$

for ϵ tending to zero. This relation proves our assertion.

Remarks: In accordance with the invariance of $D[\mathfrak{x}]$ formula (3.17) for V shows that $V = 0$ if $\lambda + i\mu$ is analytic in B so that $w' = w - \epsilon(\lambda + i\mu)$ is a conformal transformation of B into B' and $\lambda_u - \mu_v = \lambda_v + \mu_u = 0$. Likewise, if $\lambda + i\mu$ is an analytic function of w in a part B_1 of B and if $B_2 = B - B_1$, then

$$V = \iint_{B_2} [p(\lambda_u - \mu_v) - q(\lambda_v + \mu_u)]\, du\, dv, \tag{3.23}$$

[13] Note that it is not obvious whether the change from $\mathfrak{x}$ to $\mathfrak{x}'$ can be expressed in the form (3.12).

since the integrand vanishes in B_1 by the Cauchy-Riemann differential equations.

3. *Proof That Stationary Vectors Represent Minimal Surfaces.* The expression V for the first variation of $D[\mathfrak{x}]$ leads to very simple proofs of the condition $\phi(w) = 0$ for the solution $\mathfrak{x}$ of our variational problem. Obviously if $\mathfrak{x}$ is a minimal surface, V vanishes identically for all admissible functions λ, μ. Conversely, we show that the stationary character of $\mathfrak{x}$, i.e. the relation $V = 0$ together with the harmonic character of $\mathfrak{x}$, implies $p = q = 0$ or $\varphi(w) = 0$. As a matter of fact it is sufficient to assume that $V = 0$ for a suitable family of variations λ, μ. (It should be noted once more that $p = e - g = 0$ and $q = f = 0$, or the isometric character of the coordinates, follow from $V = 0$ alone; the surface $\mathfrak{x}$ need not be harmonic, nor need it actually furnish a minimum of Dirichlet's integral.)

1) *Proof using Riemann's mapping theorem:* If we presuppose the mapping theorem for simply connected domains, $\mathfrak{x}$ is seen to furnish the solution of the variational problem even if the parameter domains are permitted to vary into others equivalent conformally to the interior of a circle—say domains B' obtained from a circular disk B by a transformation

$$u' = u + \epsilon\lambda(u,v), \qquad v' = v + \epsilon\mu(u,v),$$

where λ,μ are arbitrary continuous functions with piecewise continuous first derivatives in $B + \beta$ and ϵ is so small that the correspondence between $B + \beta$ and $B' + \beta'$ is biunique. The differential equations $\lambda_u - \mu_v = H$, $\lambda_v + \mu_u = K$ can be solved[14] for arbitrary functions H and K piecewise continuous in $B + \beta$. The relation

$$\iint_B (pH - qK)\, du\, dv = 0$$

for arbitrary piecewise continuous H and K then immediately implies the condition $p = q = 0$ by the basic lemma of the variational calculus.

2) *Elementary variant of the proof:* For a harmonic vector $\mathfrak{x}$, we can conclude from $V = 0$ that $\phi(w) = p + iq = 0$ by making use only

[14] Suppose, e.g., $H = 0$ and hence $\lambda = w_v$, $\mu = w_u$ with a function w for which $w_{uu} + w_{vv} = K$. Any particular solution w of this equation will furnish λ and μ as required.

of the conformal mapping of the interior of an ellipse onto that of a circle. We set $\lambda = u$, $\mu = 0$; obviously B' is an ellipse and $V = 0$ becomes

$$\iint_B p\, du\, dv = 0.$$

By the mean value theorem of potential theory it follows that $p = 0$ at the center of the disk B. Setting $\mu = 0$, $\lambda = v$ we find, in the same way, that $q = 0$ at the center; hence $\phi(w) = 0$ for $w = 0$. To prove $\phi(w) = 0$ for any other value $w = a$ in B, we first transform the disk B into itself by the linear transformation $w' = (w - a)/(\bar{a}w - 1)$, so that $w = a$ goes into $w' = 0$. The preceding reasoning then applies to the transformed vector $\mathfrak{x}'(u',v') = \mathfrak{x}(u,v)$ for which $p' + iq' = \phi'(w') = \mathfrak{x}_{u'}'^2 - \mathfrak{x}_{v'}'^2 - 2i\mathfrak{x}_{u'}'\mathfrak{x}_{v'}' = (dw/dw')^2\phi(w)$. Since $D^*[\mathfrak{x}'] = D[\mathfrak{x}]$ the vector $\mathfrak{x}'$ likewise makes $D^*[\mathfrak{x}']$ stationary, and hence $p' + iq' = 0$ for $w' = 0$. But $(dw/dw')^2 \neq 0$ in B; we conclude that $\phi(w) = 0$ for $w = a$, and the proof is complete.

3) *Proof without use of conformal mapping:* The fact that $V = 0$ implies $\varphi(w) = 0$ for harmonic vectors can be proved without reference to any mapping theorem. If γ is taken to be a plane curve we obtain thereby a proof of Riemann's mapping theorem as a by-product. Such a version of the proof deserves our particular interest since it is preparatory to the investigation of conformal mapping of multiply connected domains which we shall carry out in Chapter V on a similar basis.

For the proof we restrict Λ, M in such a way that B' is again the unit circle: We consider an arbitrary real function $\alpha(r,\theta)$ of the polar coordinates r,θ with piecewise continuous derivatives in B, and define Λ, M by the relation

$$\Lambda + i\,\mathrm{M} = -w\,\frac{e^{i\epsilon\alpha} - 1}{\epsilon}. \tag{3.24}$$

The transformation inverse to (3.12a) takes the form $w' = we^{i\epsilon\alpha}$, and for sufficiently small ϵ maps the circle β onto itself in a one-to-one way.

Let B_r denote the interior of the circle β_r of radius r concentric with β. The relation $V = 0$ is equivalent to

$$\iint_{B_r} [p(\lambda_u - \mu_v) - q(\lambda_v + \mu_u)]\, du\, dv \to 0$$

as $r \to 1$. Integration by parts yields

$$(3.25) \qquad \int_{\beta_r} [\lambda(p\, dv + q\, du) + \mu(p\, du - q\, dv)] \to 0,$$

the domain integral vanishing since $\mathfrak{x}$ is harmonic and p and q, respectively the real and imaginary parts of the function $\phi(w)$, satisfy the Cauchy-Riemann equations. With $\mathscr{I}m$ denoting the imaginary part of a complex quantity, we may write (3.25) in the more convenient form

$$(3.26) \qquad \mathscr{I}m \left[\int_{\beta_r} (\lambda + i\mu)\phi(w)\, dw \right] \to 0$$

as $r \to 1$. For the particular variation (3.24) we have

$$\lambda + i\mu = -iw\alpha(r, \theta).$$

Since $dw = iw\, d\theta$, relation (3.26) becomes

$$(3.27) \qquad \mathscr{I}m \left[\int_0^{2\pi} \alpha(r, \theta)\, w^2 \phi(w)\, d\theta \right] \to 0$$

as $r \to 1$.

The function $w^2\phi(w)$ is regular and analytic in B, so that $\mathscr{I}m[w^2\phi(w)] = H(r, \theta)$ is harmonic in B. We consider a fixed point Q in B with polar coordinates ρ, ψ and construct the Poisson kernel

$$K(r, \theta; Q) = \frac{1}{2\pi} \frac{r^2 - \rho^2}{r^2 - 2r\rho \cos(\theta - \psi) + \rho^2}.$$

With a value $r_1 > \rho$ we define an admissible variation $\alpha(r, \theta)$ by the relation

$$\alpha(r, \theta) = K(r, \theta; Q)F(r, \theta).$$

The function $F(r, \theta)$ is given by

$$F(r, \theta) = \begin{cases} 1, & r > r_1, \\ \dfrac{2r - \rho - r_1}{r_1 - \rho}, & r_1 \geq r \geq \dfrac{\rho + r_1}{2}, \\ 0, & \dfrac{\rho + r_1}{2} > r. \end{cases}$$

With the variation α, relation (3.27) becomes, for $r \to 1$,

$$\int_0^{2\pi} K(r, \theta; Q)H(r, \theta)\, d\theta \to 0;$$

since the left side is the value of the harmonic function H at the point Q, we have $H(\rho,\psi) \to 0$, and therefore $H(\rho,\psi) = 0$ for any point Q in B.

Thus the regular analytic function $w^2\phi(w)$, having a vanishing imaginary part, is real and constant in B. At $w = 0$, $w^2\phi(w)$ is zero since ϕ is regular in B; hence $w^2\phi(w) = 0$ everywhere, and $\phi(w) = 0$ as stated. As a consequence we may now assert: *Stationary surfaces are minimal surfaces.*

5. *Additional Remarks*

In this section we supplement the preceding solution of Plateau's problem by establishing some significant properties of the solution.

1. *Biunique Correspondence of Boundary Points.* In our formulation of the variational problem in §2, 1, admissible vectors were required to map β on γ in a continuous and monotonic[15] way. This requirement, as we now show, implies that the boundary correspondence is biunique.

Since the mapping is monotonic we have only to prove that a whole arc c on β can never correspond to the same point, e.g. the point $\mathfrak{x} = 0$ on γ. Otherwise by a conformal mapping we could transform the unit circle B to the half-plane $v > 0$, and c to a segment on $v = 0$, again denoting them by B and c, respectively. Since $\mathfrak{x} = 0$ on c, the harmonic vector $\mathfrak{x}$ can be analytically extended by reflection in c into the lower half-plane; on c itself we have $\mathfrak{x}_u = 0$, hence by $\mathfrak{x}_u^2 - \mathfrak{x}_v^2 = 0$ also $\mathfrak{x}_v = 0$. This argument leads to the absurd conclusion that the harmonic vector $\mathfrak{x}$ is identically constant. Hence c cannot be a segment, but must reduce to a single point, as was to be proved..

2. *Relative Minima.* The solution of the variational problem in §2, 2 proceeds unchanged if the vector $\mathfrak{x}$ is subject to suitable inequality conditions. We may for instance, in the case of harmonic admissible vectors, prescribe that $| \mathfrak{x}(u,v) - \mathfrak{a}(u,v) | \leq b$ (b constant),

[15] "Monotonic" is always understood in the weak sense only.

$\mathfrak{a}(u,v)$ being a prescribed admissible harmonic vector. If we admit from the outset only harmonic vectors $\mathfrak{x}$ with $|\mathfrak{x} - \mathfrak{a}| \leq b$, the existence of a solution can be proved exactly as in §2. Suppose that the vector $\mathfrak{x}$ solving this minimum problem is in the interior of the "sphere" $|\mathfrak{x} - \mathfrak{a}| \leq b$; our variational reasoning in §3 or §4 establishes $\mathfrak{x}$ as a minimal surface, which furnishes a *relative* minimum of Dirichlet's integral. If, however, $|\mathfrak{x} - \mathfrak{a}| = b$, i.e. if $\mathfrak{x}$ is on the boundary of the "sphere" at least for some points of $B + \beta$, our variational reasoning is not applicable and $\mathfrak{x}$ need not be a minimal surface throughout.

3. *Proof that Solution of Variational Problem Solves Problem of Least Area.* The minimal surface $\mathfrak{x}$ which minimizes Dirichlet's integral possesses the least possible area. For the proof of this fact, which was really the initial goal, we cannot dispense with theorems on conformal mapping (cf. following article). The simplest argument is based on the results of Chapter II, §4, namely, that a simply connected polyhedral domain bounded by a Jordan curve can be mapped in a biunique and—except at the vertices—conformal way on the interior of the unit circle.[16]

Let G be any surface admissible in problem III for the contour γ, A its area. Then there exists a sequence of polyhedra π_n, whose boundaries γ_n tend smoothly to γ, for which

$$\lim_{n\to\infty} A(\pi_n) = A.$$

The possibility of a conformal mapping of π_n on the unit circle B implies that there is an isometric parameter representation $\mathfrak{y}_n(u,v)$ of π_n, with the unit circle as parameter domain. Hence, by formula (3.7a),

$$D[\mathfrak{y}_n] = A(\pi_n).$$

Let $\mathfrak{x}_n$ be the solution of problem III for the polygonal contour γ_n that bounds π_n, $\mathfrak{x}$ the solution for the boundary γ. Then

$$D[\mathfrak{x}_n] \leq D[\mathfrak{y}_n],$$

[16] Approximation by polyhedra and their mapping on the interior of the unit circle is the principal tool used by Radó in his solution of Plateau's problem.

and theorem 3.1 on the lower semicontinuous dependence of the minimum d on the boundary gives

$$d = D[\mathfrak{x}] \leq \lim_{n\to\infty} \inf. D[\mathfrak{x}_n] \leq \lim_{n\to\infty} A(\pi_n) = A,$$

or

$$d \leq A.$$

On the other hand d is the area of the minimal surface spanned in γ. Hence d is the smallest possible value for the area A of an admissible surface bounded by γ: the solution of problem III likewise solves the problem of least area.

4. *Role of Conformal Mapping in Solution of Plateau's Problem.* The solution of Plateau's problem requires no knowledge of potential theory beyond Poisson's solution of the boundary value problem for harmonic functions in the circle. Since the conformal equivalence of the unit circle B with other plane domains bounded by Jordan curves is a consequence, we might have chosen for B, instead of a circle, such a more general domain. Apparently this remark removes the objection that the method refers to a special type of parameter domain B. However, the original geometrical nature of the problem of spanning surfaces of least area in a given contour requires even greater independence of specific parametric representations than the Riemann mapping theorem provides in the form mentioned and proved previously.

Up to now we have obtained Riemann's mapping theorem merely for *plane domains* bounded by Jordan curves. The preceding method does not yield a proof of the mapping theorem for more general Riemann domains B. Consider, for example, the domain constructed from the vertical strip $-1 \leq u \leq 1$, $-\infty < v < \infty$ by removing the disk $u^2 + v^2 < 1/4$, and identifying the boundary points of the strip with equal values of v on the two vertical lines $u = -1$ and $u = 1$. *A priori* it is conceivable that we would obtain for Dirichlet's integral a minimum different from d, and therefore a minimal surface different from the one constructed in §2. In order to exclude this possibility we must know beforehand that such Riemann domains may be mapped on the unit circle.

The indispensability of mapping theorems for a more thorough analysis of the situation is also shown by the following remark. The condition that a surface is a *minimal surface* represents a purely *local property*; it is equivalent to the relation $\phi(w) = 0$ for a local parametric representation alone. Our attempt at a solution, however, starts with the assumption that a *uniform representation* $\mathfrak{x}(u,v)$ is given for the *whole* surface. We must realize that, for the definition of area, a surface may be dissected into parts; each part could be given a local parameter representation, and all these representations would be independent of one another. It is obvious that the possibility of a *uniformization* must be ascertained before we can expect to solve both Plateau's problem and the problem of least area. The fact that a set of properly coordinated local conformal representations can be made uniform by one single representation valid for the whole surface is essentially the mapping theorem for Riemann domains proved in Chapter II.

As a conclusion we state: The method based on the study of the first variation is valuable in itself, particularly because it yields theorems on conformal mapping—a point of view which will become even more important for problems of higher connectivity in Chapters IV and V. On the other hand, methods based on mapping theorems are more pertinent if the minimum problem refers to the geometrical quantity, area, rather than to the substitute Dirichlet integral, which coincides with area only for isometric representations.

6. Unsolved Problems

1. *Analytic Extension of Minimal Surfaces.* The existence proofs given for Plateau's problem leave unanswered many pertinent questions, for example, that of the analytic extension of minimal surfaces. For a plane curve γ we have simply a conformal mapping of B onto the plane domain bounded by γ; then we know from the elementary theory of analytic functions that the mapping function can be analytically extended beyond the boundary, provided γ is an analytic curve. A corresponding theorem for minimal surfaces in n-dimensional space, $n \geq 3$, has not been proved. In other words the truth of the following statement, plausible as it is, remains an open question:

A minimal surface can be analytically extended (as a minimal surface) beyond any analytic arc of the boundary curve.

The difficulty of the problem will be appreciated if one notes that the analytic boundary γ may conceivably be represented by a vector whose components are non-analytic functions of the arc length on β.

In the case of *straight* segments of the boundary γ the analytic extension of the minimal vector can be easily effected by the *principle of reflection*. In fact we prove

Theorem 3.2: If the boundary γ of a minimal surface contains a straight segment α, the minimal surface can be extended analytically as a minimal surface beyond α.

Proof: Again it is convenient to use the upper half-plane instead of the unit circle as a parameter domain. For simplicity we restrict ourselves to three-dimensional x,y,z-space for the vector $\mathfrak{x}$. Let c be the open segment of the boundary line $v = 0$ which corresponds to α. Without loss of generality we may further assume that α is a segment of the z-axis, i.e. that $x = y = 0$ along α. By reflection x and y can be extended harmonically beyond c and are regular in a neighborhood S of any closed subsegment c^* of c; in this neighborhood $x(u,-v) = -x(u,v)$ and $y(u,-v) = -y(u,v)$. The derivatives of x and y are regular in S and x_u, y_u tend uniformly to zero as the ordinate v of the point P with coordinates u,v in S tends to zero and P tends towards c^*. From $f = 0$, $e = g$, i.e. from

$$x_u x_v + y_u y_v + z_u z_v = 0, \qquad x_u^2 + y_u^2 + z_u^2 = x_v^2 + y_v^2 + z_v^2$$

in B, we infer, for P tending to c^*, that $z_u z_v \to 0$ uniformly; hence z_v tends uniformly to 0, since $x_u \to 0$, $y_u \to 0$, and $z_u^2 \geq z_v^2 - x_u^2 - y_u^2$. It follows that z_v has boundary values zero on c^*; the harmonic function $z(u,v)$ can be analytically extended by reflection beyond c^* in such a way that $z(u,-v) = z(u,v)$.

2. *Uniqueness. Boundaries Spanning Infinitely Many Minimal Surfaces.* In strictly linear boundary value problems of elliptic differential equations the proof of the uniqueness of the solution usually does not present a serious difficulty and is of an essentially simpler character than the existence proof. For Plateau's problem the situation is quite different. Little insight of a general character has been obtained concerning the number of solutions of Plateau's problem for a given contour. The following example suggested by soap film experiments (see Figures 3.2, 3.3) demonstrates the existence

of two or more different solutions, both furnishing relative minima with different values for the area d.

We first recall, from elementary variational calculus, a fact concerning the surface of revolution of least area between two parallel circles of equal radius, placed, for example, perpendicularly to the z-axis and with centers on the z-axis at a distance a. It is well known[17] that for certain values of a the catenoid, i.e. the surface of revolution of a catenary, gives a strong relative minimum with respect to all neighboring surfaces of revolution. On the other hand, a can be so chosen that the area d of the catenoid is not an absolute minimum, but exceeds by more than a positive quantity 2α the total area d_0 of the two disks within the boundary circles. We make the further observation that the catenoid furnishes minimum area even if we admit to competition surfaces that are not surfaces of

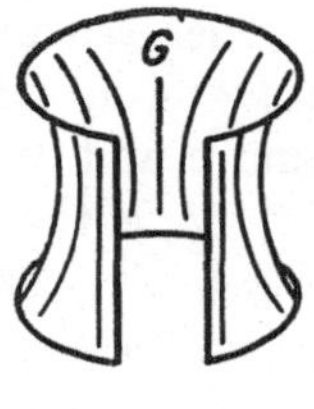

Fig. 3.2

Fig. 3.3

Figure 3.2 and 3.3. Two different minimal surfaces spanning the same contour.

revolution. This can be seen by the classical symmetrization process; the area of every admissible surface is decreased by replacing each cross section z = constant by the circle with center on the z-axis having the same area as the cross section. The catenoid is a doubly connected surface; by cutting out of it a small strip s bounded by two generating catenaries, cf. Figure 3.2, we produce a simply connected surface G.[18] This simply connected surface is bounded by one piecewise analytic Jordan curve γ, and its area may be assumed larger than $d_0 + \alpha$ if we make the strip s sufficiently narrow. The new surface G certainly furnishes a relative minimum for the curve γ, since it is part of a larger minimal surface furnishing a minimum. On the other

[17] Cf. Bliss [1], p. 122.

[18] This example was suggested by N. Wiener to J. Douglas, see Douglas [1], p. 269.

hand, if it is suitably chosen the strip s together with the two disks at the end forms a surface G' simply connected and bounded by γ, with area less than $d_0 + \alpha$. Hence, our previous theory proves the existence of a minimal surface G^* spanned by γ whose area is less than $d_0 + \alpha$. The surface G^* is a solution of Plateau's problem different from G.

This contour γ can also be used to construct *rectifiable Jordan curves spanned by infinitely*—even *non-denumerably—many minimal surfaces*. The construction is based on the following lemma, for which a complete proof can be given by the methods developed in Chapter VI.

Lemma 3.3: Let γ_1 and γ_2 be two Jordan curves bounding minimal surfaces $\mathfrak{x}_1$ and $\mathfrak{x}_2$, respectively. Construct a new Jordan curve γ_ϵ

Figure 3.4. Scheme of composition of two contours.

by connecting γ_1 and γ_2 with a thin strip η consisting of two almost parallel lines at a distance not exceeding ϵ, and by omitting corresponding small arcs of γ_1 and γ_2. Then there exists a minimal surface $\mathfrak{x}_\epsilon$ spanned in γ_ϵ which converges to $\mathfrak{x}_1$ in γ_1 and to $\mathfrak{x}_2$ in γ_2, as ϵ tends to zero.

We apply lemma 3.3 to two contours γ_1 and γ_2 geometrically similar to γ (but not necessarily of the same size). With these two Jordan curves we construct a new contour $\gamma' = \gamma_\epsilon$ by the method described in the preceding lemma, see Figure 3.4. Then γ' is spanned by at least four stable minimal surfaces because a solution for the new contour γ' can have the appearance of G or G^* (see Figs. 3.2, 3.3) in each part γ_1 and γ_2. Now we construct still another simple contour by linking successively not two but infinitely many contours $\gamma_1, \gamma_2, \cdots$ similar to γ by narrow "bridges." If we take care to decrease suitably the size of consecutive parts as well as the lengths and widths of connecting bridges, e.g. in a geometric progression, we obtain a rectifiable contour Σ spanned by an infinite number of minimal surfaces each of which has, in every link γ_n of Σ, the appearance either of G or of G^*. If, for a given minimal surface of this type in Σ, we associate with γ_n the number 0 or 1 according as the surface resembles G or G^* in γ_n, we see that every such minimal surface in Σ may be associated with a representation in the binary system of a number between 0 and 1, in a uniquely determined way, and that to each such representation corresponds a minimal surface in Σ. Hence *the number of possible simply connected minimal surfaces in the rectifiable contour Σ has the cardinal number of the continuum.* Let T denote the minimal surface, bounded by Σ, of the type of G in all parts γ_1, γ_2, $\cdots$ of Σ; let T_n denote the minimal surface of the type of G in all parts of Σ except in γ_n. It is plausible that $T_n \to T$ as n tends to infinity, so that we obtain an example of a minimal surface which is a limit of a sequence of other minimal surfaces in the same contour. (Similar examples were suggested by Paul Lévy.)

In Chapter VI we shall prove that a contour which may be spanned by two stable minimal surfaces admits at least one unstable minimal surface. Anticipating this result, we may conclude that in addition to G and G^* there exists an unstable minimal surface G^{**} in the contour γ. As a matter of fact, it is plausible that there exist non-denumerably many unstable minimal surfaces bounded by Σ.

The somewhat paradoxical phenomena of this example seem to confirm the feeling that reasonable geometrical problems may become unreasonable if the data are not properly restricted, e.g. if such abstractions as general rectifiable curves or curves with highly irregular points are permitted to occur.

These heuristic considerations have still to be substantiated by a complete proof. Nor have the following questions been fully answered: For what contours can uniqueness of the solution be established or a bound for the number of solutions given? Are there cases in which *"blocks"* of non-isolated minimal surfaces exist, a block being defined as a *continuum* of minimal surfaces with the same boundary and consequently the same value of the area? For these questions not even examples have been found to indicate plausible answers.

3. *Branch Points of Minimal Surfaces.* If a parametric representation of a minimal surface includes points for which $e = g = f = 0$, the mapping may cease to be conformal at these points.

Such points are called branch points, in analogy with the case of analytic functions for $m = 2$. Although our preceding solution of Plateau's problem does not exclude the possibility of obtaining a minimal surface with branch points, there are few explicit examples of such minimal surfaces known. See Rado [2] and Courant [12].

The following illustration shows what one may expect in such a case. Introduce polar coordinates r, θ in the plane and consider the rays $\theta = 0$, $\theta = 2\pi/3$, $\theta = 4\pi/3$. Let P and Q, $\overline{OP} \neq \overline{OQ}$, be two points on $\theta = 0$ and $\theta = 2\pi/3$ respectively; connect P and Q by a curve η in space which lies above the plane and construct a minimal surface bounded by OP, OQ, and η. This minimal surface lies entirely above the plane. According to article 1, it can be extended by reflection in OQ. The extension lies entirely below the plane; in particular, the point P is reflected into a point P' on $\theta = 4\pi/3$ with $\overline{OP} = \overline{OP'}$. The extension can itself be reflected in OP' so that a new piece $OP'Q'$ is added, Q' being a point on $\theta = 0$ with $\overline{OQ'} = \overline{OQ}$. In general, these three pieces will obviously not be connected analytically along $\theta = 0$. Only after three more reflections, in $\theta = 0$, $\theta = 2\pi/3$, and $\theta = 4\pi/3$, do we obtain a minimal surface G containing a full neighborhood of the origin. The boundary of G is a Jordan curve; the surface has a branch point at O, and $\theta = 0$, $\theta = 2\pi/3$, $\theta = 4\pi/3$ are lines of self-intersection.

We could of course have started with an angle $2\pi/5$ or $2\pi/(2n + 1)$; then the same construction would yield other minimal surfaces with simple branch points. In all these examples the geometrically remarkable feature is that they have more than one line of self-intersection, the branch lines forming equal angles with each other. The questions now arise: Is the behavior of the branch lines in our example typical? What can be stated about branch points of higher order? More generally, we may consider the problem of finding examples and of characterizing classes of curves γ for which the problem of least Dirichlet integral will necessarily lead to minimal surfaces with branch points.

7. *First Variation and Method of Descent*[19]

In itself the first variation $V(\mathfrak{x}, \lambda, \mu)$ introduced in §4, 1— not merely the relation $V = 0$ — is a useful tool for a more penetrating study of minimal surfaces. In an ordinary Euclidean $\mathfrak{x}$-space one

[19] See Courant [10].

can obtain stationary points of a function $f(\mathfrak{x})$ by starting at an arbitrary point $\mathfrak{x} = \mathfrak{x}_0$ and proceeding from it along the line of steepest descent for the function, e.g. along the direction opposite from that given by the gradient of the function. If the function is bounded, this process necessarily leads to a point $\mathfrak{x}$ where the gradient vanishes, i.e. to a stationary point, not necessarily a minimum. A similar idea can be applied to function space, as was first suggested by Hadamard.[20] Initially certain difficulties inherent in the variational calculus for several independent variables seemed to present serious obstacles. Remarkably enough, for Plateau's problem and the related questions of conformal mapping the idea of steepest descent can be successfully applied, with certain modifications. In the first place, it is not necessary to perform the variations along the lines of steepest descent; we shall be satisfied with what we call "*safe descent.*" Secondly, we shall make use of the fact that, by Dirichlet's Principle, we can always reduce our functions to harmonic functions which depend on the boundary values alone.

The method of safe descent for Plateau's problem simply consists in the following procedure. We start with an arbitrary admissible harmonic vector $\mathfrak{x}_0$. If the variation $V(\mathfrak{x}_0,\lambda,\mu)$ does not vanish identically in λ and μ, we can vary the vector $\mathfrak{x}_0$ into another vector $\mathfrak{z}$ for which, by formula (3.16), $D[\mathfrak{z}] < D[\mathfrak{x}_0]$. Replacing the vector $\mathfrak{z}$ by the harmonic vector $\mathfrak{s}$ with the same boundary values we obtain a variation of $\mathfrak{x}_0$ into $\mathfrak{s}$, which decreases the value of Dirichlet's integral. The iteration of this process of descent converges to a minimal surface $\mathfrak{y}$, if we make the descent safe by the following specifications:

For a variation of the type (3.24) and with α chosen as at the end of §4, 3, V reduces to the expression

$$V = \mathcal{I}m[w^2\phi(w)] = H(\rho,\psi),$$

see page 115, where $w = \rho e^{i\psi}$ can be chosen arbitrarily within the circle $|w| < 1$. The vector $\mathfrak{z}$ obtained from $\mathfrak{x}_0$ by this variation is not necessarily normalized in the sense of §2, 2. But since $D[\mathfrak{z}]$ is unchanged under a linear transformation, we can normalize $\mathfrak{z}$ without destroying the validity of formula (3.16). The normalized vector $\mathfrak{z}$ is next replaced by the harmonic vector $\mathfrak{s}$ with the same boundary values.

Consider for the vector $\mathfrak{s}$ the values of $\mathcal{I}m[w^2\phi(w)]$ as a function

[20] See Hadamard [1].

of w in a closed, fixed subdomain of B, e.g. in the circle $|w| \leq 1/2$. Let Q be the point in the circle where $\delta_0 = \mathcal{I}m[w^2\phi(w)]$ has the largest absolute value. If we choose this to be the point Q in the definition of α, we have

$$V = \delta_0 ,$$

and therefore, by (3.16),

$$D[\mathfrak{z}] < D[\mathfrak{x}_0] + \frac{1}{2}\epsilon\delta_0 + \epsilon^2 cD[\mathfrak{x}_0].$$

Let M be an upper bound for $D[\mathfrak{x}]$, and set

$$\epsilon = -\frac{\delta_0}{4Mc};$$

then

$$D[\mathfrak{z}] < D[\mathfrak{x}_0] - \frac{\delta_0^2}{16Mc}.$$

Repeat the same procedure with $\mathfrak{x}_1 = \mathfrak{z}$ instead of $\mathfrak{x}_0$. Continuing in this fashion we obtain a sequence of harmonic vectors $\mathfrak{x}_0, \mathfrak{x}_1, \cdots$ and a sequence of numbers $\delta_0, \delta_1, \cdots$ satisfying the inequality

$$D[\mathfrak{x}_{\nu+1}] < D[\mathfrak{x}_\nu] - \frac{\delta_\nu^2}{16Mc}.$$

It follows that for $\nu \to \infty$ the values $D[\mathfrak{x}_\nu]$ converge to a limit, hence that

$$\delta_\nu \to 0. \tag{3.28}$$

The harmonic vectors $\mathfrak{x}_\nu$ form a compact set, consequently contain a subsequence converging uniformly in $B + \beta$ to a harmonic vector $\mathfrak{x}$. This vector represents a minimal surface; for the functions $\phi_\nu(w)$ corresponding to the convergent subsequence of the $\mathfrak{x}_\nu$ converge uniformly in $B + \beta$ to the function $\phi(w)$ belonging to $\mathfrak{x}$. By relation (3.28),

$$\mathcal{I}m[w^2\phi_\nu(w)] \to 0$$

for $|w| \leq 1/2$; hence $\phi(w) = 0$ in the circle $|w| \leq 1/2$, therefore in the whole circle B. Our method of descent leads to a minimal surface spanned in γ.

It should be emphasized that we have merely constructed a minimal surface, without proving that this minimal surface is "stable" or gives a relative minimum. It may well be that the procedure of safe descent leads to a stationary vector $\mathfrak{x}$ representing an unstable minimal surface. In this case there are vectors $\mathfrak{y}$ in any neighborhood of $\mathfrak{x}$ with $D[\mathfrak{y}] < D[\mathfrak{x}]$, and hence the process of descent can be continued, leading to still another minimal surface of smaller area, etc.

8. Dependence of Area on Boundary

In §2, 3 we proved a theorem concerning the lower semicontinuous dependence on the curve γ of the absolute minimum d of Dirichlet's integral. Under suitable restrictions in the definition of convergence of a sequence γ_n to γ the absolute minimum can be shown to be a *continuous* functional of the boundary curve.

1. *Continuity Theorem for Absolute Minima.* We begin with a relatively simple theorem concerning the continuous dependence on the boundary contour of the absolute minimum d of Dirichlet's integral.

Theorem 3.3: Let

$$x_i' = x_i + \xi_i(x_1, x_2, \cdots, x_m) \qquad i = 1, 2, \cdots, m \tag{3.29}$$

be a transformation of the $\mathfrak{x}$-space into an $\mathfrak{x}'$-space. The functions ξ_i are assumed continuous with piecewise continuous derivatives, and are required to satisfy the inequalities

$$|\xi_i| \leq \epsilon, \qquad \left|\frac{\partial \xi_i}{\partial x_k}\right| \leq \epsilon.$$

The transformation takes γ into a curve γ', and the minimizing vector $\mathfrak{x}$ belonging to γ into a vector $\mathfrak{x}'$, continuous in $B + \beta$ and with piecewise continuous derivatives in B, which maps B on a surface bounded by γ'. Let the greatest lower bounds of Dirichlet's integral for surfaces bounded by γ and γ' be d and d', respectively. Then

$$d' \leq (1 + \delta)d;$$

the quantity $\delta = (1 + m\epsilon)^2 - 1$ tends to zero with ϵ.

Proof: By Schwarz' inequality

$$D[\mathfrak{x}' - \mathfrak{x}] = \sum_{i=1}^{m} \frac{1}{2} \iint_B \left[\left(\sum_{k=1}^{m} \frac{\partial \xi_i}{\partial x_k} \frac{\partial x_k}{\partial u} \right)^2 + \left(\sum_{k=1}^{m} \frac{\partial \xi_i}{\partial x_k} \frac{\partial x_k}{\partial v} \right)^2 \right] du\, dv$$
$$\leq m\epsilon^2 D[\mathfrak{x}].$$

Application of the triangle inequality gives

$$(\sqrt{D[\mathfrak{x}']} - \sqrt{D[\mathfrak{x}]})^2 \leq m^2 \epsilon^2 D[\mathfrak{x}];$$

since $D[\mathfrak{x}] = d$, $D[\mathfrak{x}'] \geq d'$, we have $d' \leq (1 + m\epsilon)^2 d$. The theorem is proved.

For sufficiently small ϵ the transformation (3.29) has an inverse. With this inverse transformation the relationship between d and d' can be considered in reverse order; consequently, ϵ can be chosen so small that the minimum values d and d' for γ and γ' differ by an arbitrarily small quantity.

Examples for such *continuous dependence of the minimum on the curve* γ are given by piecewise smooth curves γ approximated by inscribed polygons γ'. If the approximation is fine enough, the existence of a transformation of $\mathfrak{x}$ into $\mathfrak{x}'$ which takes γ into γ' and satisfies the conditions of the preceding theorem is evident.

In article 5 we shall prove a theorem, less obvious than the preceding one, referring to arbitrary minimal surfaces and independent of their minimum character.

2. *Lengths of Images of Concentric Circles.* We consider a harmonic surface $\mathfrak{x}(u,v) = \mathfrak{x}(r,\theta)$ in the unit circle $B: r < 1$. Suppose that the boundary β is mapped onto a rectifiable curve γ; this implies that the components of the vector $\mathfrak{x}(1,\theta)$ are functions of bounded variation. Denote by $L(r)$ the length of the image of the circle of radius r concentric with the unit circle, by $L = L(1)$ the length of γ:

Theorem 3.4: $L(r)$ increases monotonically in the interval $0 \leq r \leq 1$ and

$$L = L(1) = \lim_{r \to 1} L(r).$$

Proof: It is sufficient to prove

$$L(r_1) \geq L(r_0), \qquad r_0 < r_1 \leq 1,$$

since the limit equation for $r \to 1$ is an immediate consequence of the lower semicontinuity of length for the curves $\mathfrak{x}(r, \theta)$ in r. We carry out the proof for $r_1 = 1$. For $r_1 < 1$, the same argument holds with the simplification that $\mathfrak{x}(r_1, \theta)$ is analytic; this is no longer true for $r_1 = 1$.

The vector $\mathfrak{x}(r_0, \theta)$ can be expressed in terms of the boundary values

$$\mathfrak{s}(\theta) = \mathfrak{x}(1, \theta)$$

by the integral

$$\mathfrak{x}(r_0, \theta) = \int_0^{2\pi} K(r_0, \alpha - \theta)\mathfrak{s}(\alpha)\, d\alpha,$$

where $K(r_0, \alpha - \theta)$ is the Poisson kernel

$$K(r_0, \alpha - \theta) = \frac{1}{2\pi}\frac{1 - r_0^2}{1 - 2r_0 \cos(\alpha - \theta) + r_0^2}.$$

Consider

$$\mathfrak{x}_\theta = \int_0^{2\pi} K_\theta\, \mathfrak{s}(\alpha)\, d\alpha = -\int_0^{2\pi} K_\alpha\, \mathfrak{s}(\alpha)\, d\alpha.\text{[21]}$$

We integrate the last expression by parts, as may be done if integration is understood in the sense of Stieltjes, since $\mathfrak{s}(\alpha)$ is of bounded variation. Hence

$$\mathfrak{x}_\theta = \int_0^{2\pi} K(r_0, \alpha - \theta)\, d\mathfrak{s}(\alpha)$$

and, since $K(r_0, \alpha - \theta) > 0$,

$$|\mathfrak{x}_\theta| \le \int_0^{2\pi} K(r_0, \alpha - \theta)\, |d\mathfrak{s}(\alpha)|.$$

Substituting this integral in the expression $L(r_0) = \int_0^{2\pi} |\mathfrak{x}_\theta|\, d\theta$ for the length $L(r_0)$ and using the relation

$$\int_0^{2\pi} K(r_0, \alpha - \theta)\, d\theta = 1$$

we obtain

[21] The substitution of $-K_\alpha$ for K_θ is the decisive step in the proof.

$$L(r_0) \leq \int_0^{2\pi} d\theta \int_0^{2\pi} K(r_0, \alpha - \theta) \mid d\mathfrak{s}(\alpha) \mid = \int_0^{2\pi} \mid d\mathfrak{s}(\alpha) \mid = L(1),$$

as asserted in the theorem.

3. *Isoperimetric Inequality for Minimal Surfaces.*[22]

Theorem 3.5: Let L be the length of the boundary curve γ of any simply connected minimal surface (not necessarily a surface of absolute minimum area), A the area of the minimal surface. The inequality

$$L^2 - 4\pi A \geq 0$$

always holds, and the equality sign is true only for plane circular disks. In other words, the area of a minimal surface is never larger than that of the disk with the same boundary length.

Proof: First we confine ourselves to the part of the minimal surface $\mathfrak{x}(r, \theta)$ for which $r \leq \rho < 1$, thus gaining the advantage of analytic boundary values on $r = \rho$; theorem 3.4 will then permit passage to the limit as $\rho \to 1$. For convenience we again denote by β and γ the boundaries of the circle and of the minimal surface respectively, by L the length of γ, and by A the area

$$A = \frac{1}{2} \iint_{r \leq \rho} (\mathfrak{x}_u^2 + \mathfrak{x}_v^2)\, du\, dv.$$

Let s be arc length on β, σ the corresponding arc length on γ, and $t = 2\pi\sigma/L$; express differentiation with respect to t by a dot. Along the boundary we consider $\mathfrak{x}$ as a function of σ or t and have

$$\mid \mathfrak{x}_\sigma \mid = 1; \qquad \mid \dot{\mathfrak{x}} \mid = \frac{L}{2\pi}. \tag{3.30}$$

L and A remain invariant if the coordinate system in the $\mathfrak{x}$-space is subjected to a translation. Hence we may assume that the origin is at the center of gravity of $\mathfrak{x}(t)$, i.e. that

$$\int_0^{2\pi} \mathfrak{x}\, dt = 0. \tag{3.31}$$

[22] See Carleman [1] and [3], also Radó [2], Chapter III, §24 and §25, and Radó [3].

Since $\Delta\mathfrak{x} = 0$, by Green's formula

$$A = \frac{1}{2}\int_0^{2\pi\rho} \mathfrak{x}\mathfrak{x}_r \, ds,$$

and hence

$$A \leq \frac{1}{2}\int_0^{2\pi\rho} |\,\mathfrak{x}\,|\,|\,\mathfrak{x}_r\,| \, ds.$$

Furthermore, since $\mathfrak{x}$ is a minimal surface, $|\,\mathfrak{x}_r\,| = |\,\mathfrak{x}_s\,|$; introducing t instead of s, we have, therefore

$$A \leq \frac{1}{2}\int_0^{2\pi} |\,\mathfrak{x}\,|\,|\,\dot{\mathfrak{x}}\,| \, dt.$$

Relations (3.30) imply

$$\frac{L^2}{2\pi} = \int_0^{2\pi} \dot{\mathfrak{x}}^2 \, dt.$$

Hence

$$\begin{aligned} \frac{L^2}{2\pi} - 2A &\geq \int_0^{2\pi} [\dot{\mathfrak{x}}^2 - |\,\mathfrak{x}\,|\,|\,\dot{\mathfrak{x}}\,|] \, dt \\ &= \frac{1}{2}\int_0^{2\pi} [(|\,\dot{\mathfrak{x}}\,| - |\,\mathfrak{x}\,|)^2 + \dot{\mathfrak{x}}^2 - \mathfrak{x}^2] \, dt \end{aligned} \tag{3.32}$$

and, *a fortiori*,

$$\frac{L^2}{2\pi} - 2A \geq \frac{1}{2}\int_0^{2\pi} (\dot{\mathfrak{x}}^2 - \mathfrak{x}^2) \, dt. \tag{3.33}$$

The inequality

$$\int_0^{2\pi} (\dot{\mathfrak{x}}^2 - \mathfrak{x}^2) \, dt \geq 0 \tag{3.34}$$

for periodic $\mathfrak{x}$ with continuous derivatives subject to condition (3.31) is proved in a well-known manner: with $\mathfrak{a}_\nu$ and $\mathfrak{b}_\nu$ denoting the Fourier coefficients of $\mathfrak{x}(t)$, (3.31) implies that $\mathfrak{a}_0 = 0$ and the integral in (3.33) takes the form

$$\frac{\pi}{2}\sum_{\nu=1}^{\infty} (\nu^2 - 1)(\mathfrak{a}_\nu^2 + \mathfrak{b}_\nu^2),$$

an expression which is positive unless $\mathfrak{x}$ has the form

$$\mathfrak{x} = \mathfrak{a}_1 \cos t + \mathfrak{b}_1 \sin t,$$

i.e. unless $\mathfrak{x}$ describes a circle. Since, for a circle, $L^2 - 4\pi A = 0$, our statement is completely proved. It may also be noted that inequality (3.32), by (3.33) and (3.34), implies the same statement: $L^2 - 4\pi A = 0$ only if $|\dot{\mathfrak{x}}| = |\mathfrak{x}|$, i.e. if $\mathfrak{x}(s)$ describes a circle. Finally, as stated before, the passage to the limit for $\rho \to 1$ together with theorem 3.4 establishes the isoperimetric inequality for the whole minimal surface.

We infer that if the boundary curve is rectifiable, Plateau's problem certainly has a solution. For if we approximate γ and its length by polygons γ_n, the minimal surfaces spanning the polygons will, by the isoperimetric inequality, have uniformly bounded Dirichlet integrals; therefore theorem 3.1 implies the existence of an admissible vector.

4. *Continuous Variation of Area of Minimal Surfaces.* For continuous variations of the lengths of rectifiable boundaries, theorem 3.1 on semicontinuous dependence of the area on the boundary can be replaced by a more precise theorem concerning continuous dependence:

Theorem 3.6:[23] Let $\mathfrak{x}_n(u,v) = \mathfrak{x}_n(r,\theta)$ be a sequence of minimal surfaces in $B: r \leq 1$ mapping the circle $\beta: r = 1$ on rectifiable curves γ_n of bounded lengths L_n. If, for $r \leq 1$, the surface $\mathfrak{x}_n$ tends to a minimal surface $\mathfrak{x}$ whose boundary γ has the length

$$L = \lim_{n \to \infty} L_n,$$

then the areas A_n of the minimal surfaces $\mathfrak{x}_n$ tend to the area A of $\mathfrak{x}$.

Proof: As a preliminary remark we observe: By the isoperimetric inequality all the areas A_n are uniformly bounded, say $A_n < M$, and therefore by lemma 3.2 the convergence of $\mathfrak{x}_n$ to $\mathfrak{x}$ is uniform in B.

The proof will be completed if for any prescribed positive ϵ we can find an h so small that

$$D_{S_h}[\mathfrak{x}_n] < \epsilon$$

for all sufficiently large n, where S_h is the ring $1 - h \leq r \leq 1$. For in the inner circle $r \leq 1 - h$ the Dirichlet integral of $\mathfrak{x}_n$ converges

[23] This theorem, which has important applications (see Chapter VI), was discovered by Morse and Tompkins [4] and given a more general formulation by Shiffman [8].

to that of $\mathfrak{x}$, since here the derivatives of $\mathfrak{x}_n$ converge uniformly to those of $\mathfrak{x}$.

An estimate for $D_{S_h}[\mathfrak{x}_n]$ will be obtained by subdividing the ring into sufficiently small simply connected cells which depend on $\mathfrak{x}_n$, and by estimating Dirichlet's integral for each cell, using the isoperimetric inequality. We denote by $L(r_0,\alpha,\beta)$ the length of the arc defined on the minimal surface $\mathfrak{x}(r,\theta)$ by $r = r_0$, $\alpha \leq \theta \leq \beta$, and by $L_n(r_0,\alpha,\beta)$ the corresponding arc length on $\mathfrak{x}_n(r,\theta)$. The expression $L(1,\alpha,\alpha + \eta)$, being a continuous function of α, has a maximum $\delta = \delta(\eta)$ and a minimum $\sigma = \sigma(\eta)$, i.e.

$$\sigma \leq L(1,\alpha,\alpha + \eta) \leq \delta, \qquad \text{for all } \alpha. \tag{3.35}$$

If $\eta \to 0$, δ and σ tend to 0 as well. As long as $\eta > 0$, $\sigma \neq 0$; otherwise an arc of β would be mapped on one point of γ, and this mapping has been proved impossible, see §5, 1.

The division of S_h into cells depends on a preliminary step. The inequality $D[\mathfrak{x}_n] < M$ implies, for $h < 1/2$,

$$I_h(\mathfrak{x}_n) \equiv \int_0^{2\pi}\int_{1-h}^{1} \mathfrak{x}_{n_r}^2 \, dr \, d\theta < 4M.$$

For $\eta = 2\pi/N$ (N an integer), we can write

$$I_h(\mathfrak{x}_n) = \int_0^{\eta}\int_{1-h}^{1} \sum_{\nu=0}^{N-1} \mathfrak{x}_{n_r}^2(r, \theta + \nu\eta) \, dr \, d\theta = \int_0^{\eta}\int_{1-h}^{1} p_n(r, \theta) \, dr \, d\theta,$$

with $p_n(r,\theta) = \sum_{\nu=0}^{N-1} \mathfrak{x}_{n_r}^2(r,\theta + \nu\eta)$. By the same reasoning which led to lemma 1.4, the inequality $I_h(\mathfrak{x}_n) < 4M$ implies, for each $\mathfrak{x}_n$ (with fixed η and h), the existence of a set of equidistant angles α_n, $\alpha_n + \eta$, $\cdots$, $\alpha_n + \nu\eta$, $\cdots$, $\alpha_n + (N - 1)\eta$ such that along the corresponding radii

$$\int_{1-h}^{1} p_n \, dr < \frac{4M}{\eta}.$$

Therefore *a fortiori* on each of these radii

$$\int_{1-h}^{1} \mathfrak{x}_{n_r}^2 \, dr < \frac{4M}{\eta}. \tag{3.36}$$

Let $R_\nu^{(n)}$ be the length of the curve on the minimal surface defined by $\mathfrak{x}_n(r, \theta)$, $\theta = \alpha_n + \nu\eta$, $1 - h \leq r \leq 1$. Since

$$(R_\nu^{(n)})^2 = \left(\int_{1-h}^{1} |\mathfrak{x}_{n_r}| \, dr\right)^2 \leq h \int_{1-h}^{1} \mathfrak{x}_{n_r}^2 \, dr,$$

we have, by (3.36),

$$R_\nu^{(n)} \leq \sqrt{h} \sqrt{\frac{4M}{\eta}}. \tag{3.37}$$

We divide the ring S_h into N cells by the radii $\theta = \alpha_n + \nu\eta = \alpha_\nu$, $\nu = 0, 1, \cdots, N - 1$. With an arbitrarily small positive number $\kappa < 1$, we fix the quantities η, h. First we choose η so small that $\delta < \kappa$. Let $\delta_\upsilon = L(1, \alpha_\upsilon, \alpha_{\upsilon+1})$. By (3.35) we have

$$\sigma \leq \delta_\nu \leq \kappa. \tag{3.38}$$

Then we choose for h a value so small that

$$R_\nu^{(n)} < \sigma, \tag{3.39}$$

as is possible by (3.37); moreover so small that also

$$| L(1 - h, \alpha, \alpha + \eta) - L(1, \alpha, \alpha + \eta) | < \sigma, \tag{3.40}$$

as can be done on the basis of theorem 3.4 and the lower semicontinuity of length. Finally we choose n so large that

$$| L_n(1 - h, \alpha, \alpha + \eta) - L(1 - h, \alpha, \alpha + \eta) | < \sigma \tag{3.41}$$

and

$$| L_n(1, \alpha, \alpha + \eta) - L(1, \alpha, \alpha + \eta) | < \sigma.\text{[24]} \tag{3.42}$$

[24] The assumption $L_n \to L$ together with the well-known property of lower semicontinuity of length implies that $L_n(1, \alpha, \alpha + \eta) \to L(1, \alpha, \alpha + \eta)$ uniformly in the end-points α, $\alpha + \eta$. Suppose, on the contrary, that the sequence of subscripts n contained a subsequence for which, with some positive ϵ,

$$| L_n(1, \alpha_n, \alpha_n + \eta_n) - L(1, \alpha_n, \alpha_n + \eta_n) | > \epsilon.$$

We could determine a further subsequence for which $\alpha_n \to \alpha$, $\alpha_n + \eta_n \to \alpha + \eta$. Since for this final subsequence

$$L_n(1, \alpha_n, \alpha_n + \eta_n) \to L(1, \alpha, \alpha + \eta),$$

the above inequality would imply

$$\liminf. L_n(1, \alpha_n, \alpha_n + \eta_n) - L(1, \alpha, \alpha + \eta) \geq \epsilon$$

For relation (3.42) we use the assumption $L_n \to L$, while (3.41) is a consequence of the uniform convergence of $\mathfrak{x}_{n\theta}$ to $\mathfrak{x}_\theta$, for $r = 1 - h$.

Consider the ν-th cell in S_h on the minimal surface $\mathfrak{x}_n$, and let $\Lambda_\nu^{(n)}$ be the length of the image of its boundary curve. We find, by inequalities (3.38)–(3.42),

$$\Lambda_\nu^{(n)} < \sigma + \sigma + (\delta_\nu + \sigma) + (\delta_\nu + 2\sigma) < 7\delta_\nu .$$

Applying the isoperimetric inequality (theorem 3.5) to the ν-th cell we have, for the area $A_\nu^{(n)}$ of the minimal surface $\mathfrak{x}_n$ corresponding to that cell,

$$A_\nu^{(n)} \leq \frac{(\Lambda_\nu^{(n)})^2}{4\pi}$$

and therefore

$$D_{S_h}[\mathfrak{x}_n] = \sum_{\nu=0}^{N-1} A_\nu^{(n)} \leq \sum_{\nu=0}^{N-1} \frac{(\Lambda_\nu^{(n)})^2}{4\pi} < \frac{49}{4\pi} \sum_{\nu=0}^{N-1} \delta_\nu^2 < \frac{49\kappa}{4\pi} \sum_{\nu=0}^{N-1} \delta_\nu = \frac{49\kappa L}{4\pi}.$$

Hence, for all sufficiently great n, $D_{S_h}[\mathfrak{x}_n]$ will be less than ϵ, if we choose $\kappa = 4\pi\epsilon/49L$; our theorem is proved.

5. *Continuous Variation of Area of Harmonic Surfaces.* Morse and Tompkins have proved a more general form of the preceding theorem, in which the surfaces under consideration are assumed to be only harmonic and not necessarily minimal surfaces:

Theorem 3.7: Let $\mathfrak{x}^{(n)}$ be a sequence of harmonic surfaces having rectifiable boundary curves and suppose that $\mathfrak{x}^{(n)}$ tends uniformly to a surface $\mathfrak{x}$ with rectifiable boundary curve, in such a way that the length L_n of the boundary of $\mathfrak{x}^{(n)}$ converges to the length L of the boundary of $\mathfrak{x}$. Then the area A_n of $\mathfrak{x}^{(n)}$ tends to the area A of $\mathfrak{x}$.

---

the absolute value sign being superfluous because of the lower semicontinuity of length. On the other hand, with K_n denoting the arc of γ_n complementary to α_n, $\alpha_n + \eta_n$,

$$\liminf. L_n(K_n) - L(K) \geq 0$$

by the lower semicontinuity of length, where K is the arc of γ complementary to α, $\alpha + \eta$. By addition, these two inequalities yield

$$\liminf. L_n - L \geq \epsilon,$$

contrary to $L_n \to L$.

Proof: Morse and Tompkins proceed by direct calculation of the integral defining the area. A modification of their proof follows.

We first show, by this method, that an inequality, implying

$$A \leq \tfrac{1}{4} L^2 \tag{3.43}$$

(a substitute for the isoperimetric inequality) holds for all harmonic surfaces $\mathfrak{x}$ with rectifiable boundary curves.

Let $\mathfrak{x}$ again be defined in the unit circle B of the u,v-plane. We then have, in the usual vector notation,

$$A = \iint_B |\mathfrak{x}_u \times \mathfrak{x}_v| \, du \, dv = \iint_B |\mathfrak{x}_r \times \mathfrak{x}_\theta| \, dr \, d\theta.$$

The harmonic vector $\mathfrak{x}$ is expressed in terms of its boundary values $\mathfrak{x}(\theta)$ by Poisson's integral:

$$\mathfrak{x} = \frac{1}{2\pi} \int_0^{2\pi} \mathfrak{x}(\phi) \frac{1 - r^2}{\Omega(r, \theta; \phi)} \, d\phi,$$

where $\Omega(r, \theta; \phi) = 1 - 2r \cos(\phi - \theta) + r^2$.
Similarly as in article 2, we have

$$\mathfrak{x}_\theta = \frac{1}{2\pi} \int_0^{2\pi} \mathfrak{x}(\phi) \frac{\partial}{\partial\theta} \left[\frac{1 - r^2}{\Omega(r, \theta; \phi)} \right] d\phi$$

$$= -\frac{1}{2\pi} \int_0^{2\pi} \mathfrak{x}(\phi) \frac{\partial}{\partial\phi} \left[\frac{1 - r^2}{\Omega(r, \theta; \phi)} \right] d\phi,$$

and, using integration by parts,

$$\mathfrak{x}_\theta = \frac{1}{2\pi} \int_0^{2\pi} \frac{1 - r^2}{\Omega(r, \theta; \phi)} \, d\mathfrak{x}(\phi). \tag{3.44}$$

The Stieltjes integral in (3.44) exists, because $\mathfrak{x}(\phi)$ is rectifiable and therefore of bounded variation.

The harmonic vector $r\mathfrak{x}_r$ conjugate to $\mathfrak{x}_\theta$ is expressed in terms of the boundary values $\mathfrak{x}(\theta)$ by

$$r\mathfrak{x}_r = \frac{1}{2\pi} \int_0^{2\pi} \frac{2r \sin(\phi - \theta)}{\Omega(r, \theta; \phi)} \, d\mathfrak{x}(\phi). \tag{3.45}$$

From (3.44) and (3.45) we obtain

$$(3.46)\qquad \mathfrak{x}_r \times \mathfrak{x}_\theta = \frac{1}{4\pi^2}\int_0^{2\pi}\int_0^{2\pi} \frac{2(1-r^2)\sin(\phi-\theta)}{\Omega(r,\theta;\phi)\Omega(r,\theta;\psi)}\,d\mathfrak{x}(\phi)\times d\mathfrak{x}(\psi).$$

The same product may be expressed by (3.46) with ϕ and ψ interchanged. Adding the two expressions and using the fact that

$$d\mathfrak{x}(\psi)\times d\mathfrak{x}(\phi) = -d\mathfrak{x}(\phi)\times d\mathfrak{x}(\psi),$$

we obtain

$$\mathfrak{x}_r \times \mathfrak{x}_\theta = \frac{1}{4\pi^2}\int_0^{2\pi}\int_0^{2\pi} \frac{[1-r^2][\sin(\phi-\theta)-\sin(\psi-\theta)]}{\Omega(r,\theta;\phi)\Omega(r,\theta;\psi)}\,d\mathfrak{x}(\phi)\times d\mathfrak{x}(\psi).$$

Since $\sin(\phi-\theta)-\sin(\psi-\theta) = 2\cos[\frac{1}{2}(\phi+\psi)-\theta]\sin\frac{1}{2}(\phi-\psi)$, and $|d\mathfrak{x}(\phi)\times d\mathfrak{x}(\psi)| \le |d\mathfrak{x}(\phi)|\cdot|d\mathfrak{x}(\psi)|$, the last formula yields the relation

$$|\mathfrak{x}_r\times\mathfrak{x}_\theta| \le \frac{1}{4\pi^2}\int_0^{2\pi}\int_0^{2\pi}\frac{2(1-r^2)\,|\sin\frac{1}{2}(\phi-\psi)|\cdot|d\mathfrak{x}(\phi)|\cdot|d\mathfrak{x}(\psi)|}{\Omega(r,\theta;\phi)\Omega(r,\theta;\psi)}.$$

From this inequality we obtain an estimate for the area A. To that end let $A(\sigma,\rho)$ be the area of the part of $\mathfrak{x}$ which corresponds to the ring $\sigma \le r \le \rho$. Then

$$A(\sigma,\rho) = \int_\sigma^\rho\int_0^{2\pi}|\mathfrak{x}_r\times\mathfrak{x}_\theta|\,d\theta\,dr$$

$$\le \frac{1}{2\pi^2}\int_0^{2\pi}\int_0^{2\pi}\cdot\left[\int_\sigma^\rho\int_0^{2\pi}\frac{(1-r^2)\,|\sin\frac{1}{2}(\phi-\psi)|\,d\theta\,dr}{\Omega(r,\theta;\phi)\Omega(r,\theta;\psi)}\right]\cdot|d\mathfrak{x}(\phi)|\cdot|d\mathfrak{x}(\psi)|$$

and we can evaluate the integral

$$I = \frac{1}{2\pi^2}\int_\sigma^\rho\int_0^{2\pi}\frac{(1-r^2)\,d\theta\,dr}{\Omega(r,\theta;\phi)\Omega(r,\theta;\psi)}.$$

The inner integral,

$$(3.47)\qquad \frac{1}{2\pi}\int_0^{2\pi}\frac{1-r^2}{\Omega(r,\theta;\phi)\Omega(r,\theta;\psi)}\,d\theta,$$

could be calculated explicitly by elementary integration,[25] but it is simpler to find its value by the following reasoning. The expression

[25] E.g. by expanding the integrand into the product of two Fourier series in θ.

(3.48) $$\frac{1}{\Omega(r, \theta; \phi)} = \frac{1}{1 - 2r \cos(\psi - \theta) + r^2},$$

as a function of θ, can be considered as representing the boundary values for $R = 1$ of a regular, harmonic function $f(R,\theta)$ in the unit circle $R \leq 1$. (Note that this harmonic function depends on the values of r and ψ.) Poisson's integral (3.47) then yields the value of this function $f(R,\theta)$ at the point $R = r$, $\theta = \phi$. In order to find the function $f(R,\theta)$ we remember that, for $r < 1$, the Poisson kernel

$$\frac{R^2 - r^2}{R^2 - 2Rr\cos(\psi - \theta) + r^2}$$

is a harmonic function of R and θ which is regular in the exterior of the unit circle. The harmonic function with the same values on the unit circumference obtained from it by replacing R by $1/R$ is therefore regular inside the unit circle. Hence the required harmonic function with boundary values (3.48) is

$$f(R, \theta) = \frac{1}{1 - r^2} \frac{\left(\frac{1}{R}\right)^2 - r^2}{\left(\frac{1}{R}\right)^2 - \frac{2r}{R}\cos(\psi - \theta) + r^2},$$

and its value for $R = r$, $\theta = \phi$ is

$$\frac{1 + r^2}{1 - 2r^2\cos(\psi - \phi) + r^4};$$

this must be the value of the integral (3.47).

By elementary integration we obtain[26]

[26] The calculation proceeds as follows:

$$I = \frac{1}{2\pi}\int_\sigma^\rho \left[\frac{1}{1 - 2r\cos\frac{1}{2}(\psi - \phi) + r^2} + \frac{1}{1 + 2r\cos\frac{1}{2}(\psi - \phi) + r^2}\right] dr$$
$$= \frac{1}{2\pi}\frac{1}{|\sin\frac{1}{2}(\psi - \phi)|}\left[\arctan\frac{r - \cos\frac{1}{2}(\psi - \phi)}{|\sin\frac{1}{2}(\psi - \phi)|} + \arctan\frac{r + \cos\frac{1}{2}(\psi - \phi)}{|\sin\frac{1}{2}(\psi - \phi)|}\right]_\sigma^\rho.$$

If we use the addition formula for the arc tan,

$$\arctan a + \arctan b = \arctan\frac{a + b}{1 - ab},$$

valid for all a and b, the expression for I reduces to (3.49).

$$
(3.49)\qquad
\begin{aligned}
I &= \frac{1}{\pi}\int_\sigma^\rho \frac{1+r^2}{1-2r^2\cos(\psi-\phi)+r^4}\,dr\\
&= \left[\frac{1}{2\pi}\frac{1}{|\sin\frac{1}{2}(\psi-\phi)|}\arctan\frac{2r\,|\sin\frac{1}{2}(\psi-\phi)|}{1-r^2}\right]_\sigma^\rho .
\end{aligned}
$$

Hence

$$
(3.50)\qquad
\begin{aligned}
&A(\sigma,\rho)\\
&\leq \frac{1}{2\pi}\int_0^{2\pi}\int_0^{2\pi}\left[\arctan\frac{2r\,|\sin\frac{1}{2}(\psi-\phi)|}{1-r^2}\right]_\sigma^\rho |\,d\mathfrak{x}(\phi)\,|\cdot|\,d\mathfrak{x}(\psi)\,|.
\end{aligned}
$$

From this, letting $\sigma \to 0$, $\rho \to 1$, we find

$$A \leq \frac{1}{4}L^2,$$

since $|\,d\mathfrak{x}(\phi)\,|$ is the differential of the arc length on $\mathfrak{x}(\phi)$.

Now we might proceed as in article 4, but it is just as easy to establish the continuity theorem directly. Consider all those arcs of the curves $\mathfrak{x}^{(n)}(\theta)$, $n = 1, 2, \cdots$, and $\mathfrak{x}(\theta)$ that are images of arcs of length α of the unit circumference. Let $\eta(\alpha)$ be the maximum length of all such arcs. Since the length of $\mathfrak{x}^{(n)}(\theta)$ tends to the length of $\mathfrak{x}(\theta)$, we know that $\eta(\alpha) \to 0$ as $\alpha \to 0$.[27] Let M be an upper bound of the lengths of $\mathfrak{x}^{(n)}(\theta)$. In estimating $A_n(\sigma,\rho)$, decompose the region of integration into the two parts $|\,\psi - \phi\,| \leq \alpha/2$, called D_1, and $|\,\psi - \phi\,| \geq \alpha/2$, called D_2, where $|\,\psi - \phi\,|$ is the length of the shorter arc joining ψ, ϕ. The integral corresponding to the one in (3.50) for $\mathfrak{x}^{(n)}$ instead of $\mathfrak{x}$, extended over the first region D_1, is then less than or equal to

$$\frac{1}{2\pi}\cdot\frac{\pi}{2}\iint_{D_1}|\,d\mathfrak{x}^{(n)}(\phi)\,|\cdot|\,d\mathfrak{x}^{(n)}(\psi)\,| \leq \tfrac{1}{4}\eta(\alpha)\cdot M.$$

The integral over the second region is seen to be less than or equal to

$$\frac{1}{2\pi}\left[\frac{\pi}{2} - \arctan\frac{2\sigma\,|\sin(\alpha/4)|}{1-\sigma^2}\right]M^2 \leq \frac{M^2}{2\pi}\operatorname{arc\,cot}\frac{\sigma\alpha}{4(1-\sigma^2)},$$

in virtue of the fact that $\sin\alpha/4 \geq \alpha/8$. Therefore, for each of the surfaces $\mathfrak{x}^{(n)}$ and for $\mathfrak{x}$, the area A satisfies

[27] See footnote 24.

$$A(\sigma, 1) \leq \frac{M}{4}\eta(\alpha) + \frac{M^2}{2\pi} \operatorname{arc\,cot} \frac{\sigma\alpha}{2(1-\sigma^2)}. \tag{3.51}$$

For an arbitrary $\epsilon > 0$ select α so that $\eta(\alpha) < \epsilon/2M$ and choose $\sigma < 1$ so that

$$\frac{M^2}{2\pi} \operatorname{arc\,cot} \frac{\sigma\alpha}{2(1-\sigma^2)} < \frac{\epsilon}{8}.$$

Then (3.51) gives

$$A(\sigma, 1) \leq \frac{\epsilon}{4},$$

for all $\mathfrak{x}^{(n)}$ and for $\mathfrak{x}$.

On the other hand we know that

$$A_n(0,\sigma) \to A(0,\sigma)$$

as $n \to \infty$, because the convergence of $\mathfrak{x}^{(n)}$ to $\mathfrak{x}$ is uniform in every closed subdomain of B. Therefore an N can be found such that

$$| A_n(0, \sigma) - A(0, \sigma) | < \frac{\epsilon}{2}$$

for all $n \geq N$. Hence

$$|A_n - A| < \frac{\epsilon}{2} + \frac{\epsilon}{4} + \frac{\epsilon}{4} = \epsilon,$$

for all $n \geq N$. This completes the proof of the continuity theorem.

CHAPTER IV

The General Problem of Douglas

1. *Introduction*

In contours γ consisting of several closed curves, soap film experiments produce multiply connected minimal surfaces, as for example in the classical case of surfaces of revolution of least area. Moreover, for suitable contours γ, soap film experiments indicate the existence of minimal surfaces of higher topological structure such as surfaces not of genus zero or non-orientable surfaces,[1] see Figures 4.1 and 4.2. Such more complicated surfaces are formed by the soap film whenever they have greater stability—i.e. smaller area—than simply connected surfaces spanned in the same contour.

We formulate the general

Problem of Douglas:[2] *Given, in the m-dimensional* $\mathfrak{x}$*-space, a system* γ *of* k *Jordan curves* $\gamma_1, \gamma_2, \cdots, \gamma_k$ *: to construct a minimal surface of prescribed Euler characteristic and prescribed character of orientability, bounded by* γ.

The possibility of solving this problem depends on certain conditions, as may be illustrated by the example of minimal surfaces of revolution bounded by two parallel circles. Certainly this boundary γ is spanned by a doubly connected minimal surface—a catenoid—if the two circles are so near each other that the admissible class includes doubly connected surfaces with area less than the sum of the areas of the two boundary circles. If the two circles are too far apart, no doubly connected stable minimal surface is bounded by them, and solutions exist only if the degenerated surface consisting of the two disks is considered to be a solution.

In more detail, consider the problem with the distance a between the planes of the boundary circles as a parameter. For small values

[1] A number of such experiments are described in Courant [14].

[2] This general problem was first clearly stated and attacked by J. Douglas, see [3], [4], [5], [7], and particularly [8].

of a the catenoid exists and furnishes the absolute minimum. As a increases to a certain value a_0 the area of the catenoid becomes equal to the sum of the areas of the disks spanned in the two boundary circles. For a slightly greater than a_0 the catenoid still furnishes a relative minimum of the area, while the absolute minimum (or rather greatest lower bound) is given by the sum of the areas of the two disks. If a increases to a certain value a_1 determining the "conjugate" position of the boundary circles, the catenoid ceases to yield even a relative minimum, though it still furnishes a stationary value for the area.[3]

A similar situation occurs in the case of Douglas' general problem: a sufficient, but not necessary, condition can be formulated for the

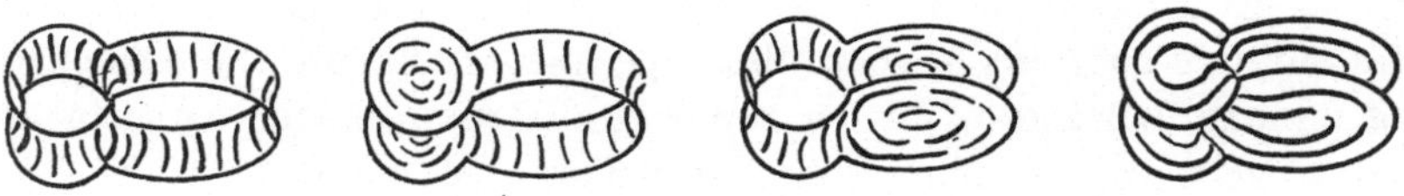

Figure 4.1. Minimal surface of different genus through same contour.

Figure 4.2. Minimal surface forming a Moebius strip and simply connected minimal surface through same contour.

existence of a solution. A surface bounded by the prescribed system γ of curves γ_ν is called of lower type if either or both of the following occur:

1) it has a smaller Euler characteristic than prescribed;
2) it is degenerate, consisting of two or more surfaces of total characteristic not greater than prescribed, bounded by complementary subsets of the set of curves γ_ν .

With this definition we state

Theorem 4.1: A sufficient condition for the existence of a solution of Douglas' problem is that the greatest lower bound d of Dirichlet's

[3] See the discussion in Bliss [1], Chapter IV.

integral for non-degenerate admissible vectors $\mathfrak{x}$ is actually less than the greatest lower bound d^* of Dirichlet's integral or of the combined Dirichlet integrals for surfaces of lower type bounded by γ.

Examples are indicated in Figures 4.1 and 4.2; they picture surfaces of genus 1 and of the type of the Moebius strip, respectively, furnishing a smaller minimum value d of the area than surfaces of genus zero. In such cases the system γ bounds not only the simply connected minimal surfaces shown to solve Plateau's problem in Chapter III, but also those of higher type, and the surfaces of higher topological structure have a greater degree of stability.

It should be emphasized that in the formulation of Douglas' problem no orientation was stipulated for the curves γ_ν. As the example of plane boundaries γ shows, such an additional stipulation may imply degeneration. We shall, however, not pursue this subject further.

The proof of theorem 4.1 proceeds by a variational method similar to that of Chapter III. Again we minimize Dirichlet's integral $D_B[\mathfrak{x}]$; now, however, the parameter domain B for $\mathfrak{x}(u, v)$ not only is required to be of prescribed topological structure, but cannot be a fixed domain chosen in advance. For instance, in the case of a doubly connected surface we cannot arbitrarily choose B as a fixed circular ring, since two such rings are not necessarily conformally equivalent, therefore need not be conformal images of the same minimal surface.

Accordingly we shall admit, as parameter domain B in our variational problem, any member of a fixed class $\mathfrak{N}$ of domains. For minimal surfaces not of genus zero the simple plane domains B used before must be replaced by Riemann domains. The proper choice of this class of domains B is decisive. It must be sufficiently restricted to permit, as in Chapter III, the construction of a convergent minimizing sequence, but wide enough to permit a variational proof that the limit is a minimal surface.

If we know beforehand that every Riemann domain of the prescribed structure is conformally equivalent to some member of our class $\mathfrak{N}$ of admissible domains B—a fact which is assured if we use slit domains as in Chapter II—the characteristic condition $\phi(w) = 0$ for a minimal surface can be proved exactly as in Chapter III, §3. If, however, mapping theorems are not presupposed, the identity $\phi(w) = 0$ must be deduced from the variational conditions expressing not only the freedom of choice in mapping the boundary β on γ, but also the permissible variations of the domain B itself. In this connection the case of a two-dimensional $\mathfrak{x}$-space is of particular interest. Here naturally only surfaces of genus zero are to be considered, so that Douglas' problem reduces to the question of mapping the parameter domain B conformally onto a plane k-fold connected domain G bounded by the curves γ_ν. Chapter V will deal with these applications to the theory of conformal mapping of k-fold connected domains on classes $\mathfrak{N}$ of "normal domains."

In the present chapter we will be primarily concerned with minimal surface theory. The previous argument (Chapter III, §5, 4) in favor of presupposing the theory of conformal mapping and the minimum value d are even more pertinent to the higher topological cases of Douglas' problem than to Plateau's original problem. It is conceivable that different types of parameter domains B might lead to different minimal surfaces as solutions. To show that the resulting surface is independent of the choice of domains, we must make use of theorems concerning conformal equivalence of the various classes of domains B. Moreover, only on the basis of such theorems can we establish equivalence of the variational problem for the area with that for Dirichlet's integral. In this chapter we shall, therefore, identify the solutions of our variational problem as minimal surfaces by reference to mapping theorems, while giving in the next chapter a proof of the decisive relation $\phi(w) = 0$, independent of such theorems.

2. *Solution of Variational Problem for k-fold Connected Domains*

1. *Formulation of Problem.* To construct minimal surfaces of genus zero bounded by k given oriented curves γ_ν, we formulate and solve our variational problem for any of three different classes $\mathfrak{N}$ of parameter domains B:

a) The domain B consists of the whole u, v-plane except the interiors of k circles whose radii and centers are not preassigned (Class $\mathfrak{N}_a$).
b) The domain B is a slit domain in the sense of Chapter II (Class $\mathfrak{N}_b$).
c) The domain B is a Riemann domain over the u, v-plane bounded by k unit circles and having branch points of the total multiplicity $2k - 2$ (Class $\mathfrak{N}_c$).

To describe the latter class intuitively, we take k unit circles over the u, v-plane and connect each, say, with the following one by two branch points whose positions are not prescribed, obtaining a domain with k separate boundary curves. For $k > 2$ we also permit some branch points to coalesce into branch points of higher multiplicity. Moreover we permit connection among the k sheets in any order, as long as the total multiplicity of the branch points is $2k - 2$.

For each of our classes $\mathfrak{N}$ the boundary β of B is a system of k curves $\beta_1, \beta_2, \cdots, \beta_k$. A vector $\mathfrak{x}(u, v)$ in B will be called admissible if it has the following properties:

1) $\mathfrak{x}$ is continuous in $B + \beta$ and has piecewise continuous first derivatives in B.

2) $\mathfrak{x}$ maps the curves β_v in a continuous and monotonic way on the prescribed curves γ_v.

After choosing one of the classes $\mathfrak{N}$, we formulate

Variational Problem IV: To find in the class $\mathfrak{N}$ a domain B, and in B an admissible vector $\mathfrak{x}$, for which $D[\mathfrak{x}]$ attains the least possible value d.

Once and for all we assume the existence of admissible vectors for which $D[\mathfrak{x}] < \infty$, which is certainly justified for rectifiable curves γ_ν.[4]

2. *Condition of Cohesion.* As mentioned in §1, we must ensure the possibility of solving problem IV by additional restrictions. We formulate below a "condition of cohesion" which postulates the existence of minimizing sequences not degenerating into separated surfaces. For higher topological structure the same condition also excludes degeneration of the minimizing sequence into surfaces of lower type.

A sequence $\mathfrak{x}_n$ of surfaces in $\mathfrak{x}$-space is said to be cohesive if there exists a fixed positive α such that each closed curve on $\mathfrak{x}_n$ of diameter less than α can be contracted continuously to a point on the surface. Otherwise the sequence is said to *tend to degeneration* or to separation, a terminology indicating the intuitive meaning of our condition. If problem IV admits a cohesive minimizing sequence we shall say that the *problem* satisfies the condition of cohesion. Our aim is to show that *problem IV can be solved if the condition of cohesion is satisfied.*

Since the converse is obvious, it will be seen that cohesion is necessary and sufficient for the existence of a solution. Still the condition of cohesion is not equivalent to the condition for existence

[4] See Chapter III, §8, 3. The proof given there for one contour can be immediately extended to the present case.

of a solution given by theorem 4.1. By way of example we recall the problem of the minimal surface of revolution and suppose that the minimizing catenoid and the two disks have the same area. Then there exist minimizing sequences tending to the catenoid and satisfying the condition of cohesion; on the other hand different minimizing sequences tending to the two disks plus a connecting line do *not* satisfy this condition. Hence theorem 4.1 can yield no more than a sufficient condition for a minimum.

After solving problem IV under the assumption of cohesion we shall prove that cohesion is a consequence of the condition of theorem 4.1, which is more restrictive, yet more easily verifiable in concrete cases.

3. *Solution of Variational Problem for k-fold Connected Domains G and Parameter Domains Bounded by Circles.* We first construct the solution for the class $\mathfrak{N}_a$ of parameter domains. Consider a cohesive minimizing sequence $\mathfrak{x}_n$; we have

$$D_{B_n}[\mathfrak{x}_n] \to d$$

where B_n is the parameter domain for $\mathfrak{x}_n$ and d is the greatest lower bound for Dirichlet's integral in problem IV. We shall select a subsequence of the vectors $\mathfrak{x}_n$ for which the corresponding domains B_n tend to a limiting domain B belonging to the class $\mathfrak{N}_a$ of domains, and for which the values $\mathfrak{x}_n$ on the boundaries $\beta^{(n)}$ of B_n form an equicontinuous set of functions. Subsequently the reasoning may proceed exactly as in Chapter III, §2.

By a linear transformation of the w-plane we map B_n onto a domain bounded by the unit circle $\beta_1^{(n)}$, a smaller concentric circle $\beta_2^{(n)}$, and $k - 2$ circles lying in the ring between $\beta_1^{(n)}$ and $\beta_2^{(n)}$. This normalization replaces the three point condition of Chapter III, §2, 2. The domains B_n define a limiting k-fold connected domain B if, for $n \to \infty$, B_n does not degenerate in one of the following ways:

1) Two circles, e.g. $\beta_1^{(n)}$ and $\beta_\mu^{(n)}$ come arbitrarily near each other at a point P while their radii remain above a positive bound η.
2) Two circles come arbitrarily near each other at a point P, but the radius of one of them, say $\beta_\mu^{(n)}$, shrinks to zero.
3) One or more circles, e.g. $\beta_3^{(n)}, \beta_4^{(n)}, \cdots$, shrink to the same point P while the point remains bounded away from the non-shrinking circles.

We begin by excluding the first type of degeneration. Denote by

M a common upper bound for $D[\mathfrak{x}_n]$; choosing a small constant δ, $0 < \delta < 1$, set $\epsilon(\delta) = 2M/(\log 1/\delta)$. According to lemma 3.1[5] there is for each value of n a circle κ_n about P of radius ρ_n, $\delta \leq \rho_n \leq \sqrt{\delta}$, such that the length L_n of the image by $\mathfrak{x}_n$ of any arc of κ_n lying in B_n satisfies the inequality $L_n \leq \sqrt{2\pi\epsilon(\delta)}$. The assumed degeneration implies, for sufficiently large n, the existence of an arc of κ_n joining a point on $\beta_1^{(n)}$ with one on $\beta_\mu^{(n)}$, and therefore the existence of an arc in $\mathfrak{x}$-space of length L_n joining a point on γ_1 with one on γ_μ. The distance between points on these curves is bounded away from zero; but choosing δ suitably, we can make ϵ—and hence L_n—so small as to yield a contradiction. The first type of degeneration cannot occur.

By a similar argument the third type of degeneration is shown to be impossible. In this case we could, for sufficiently large fixed n, include the circles shrinking to P in a circle κ_ρ of radius ρ about P,

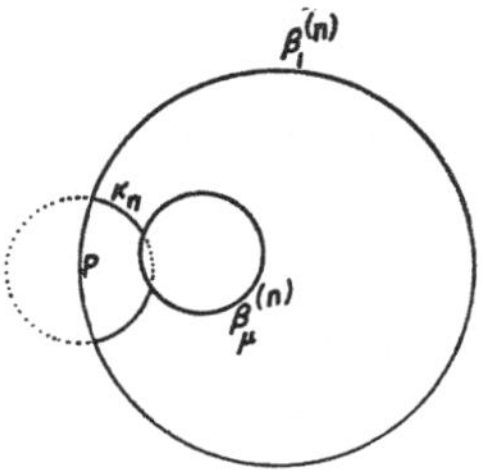

Figure 4.3

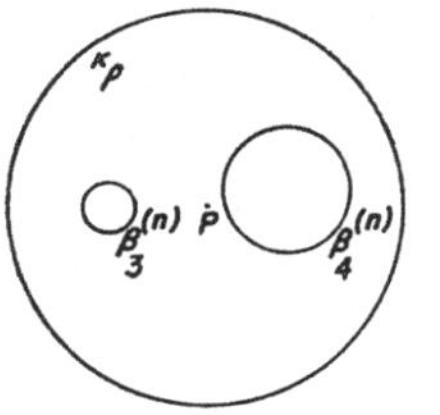

Figure 4.4

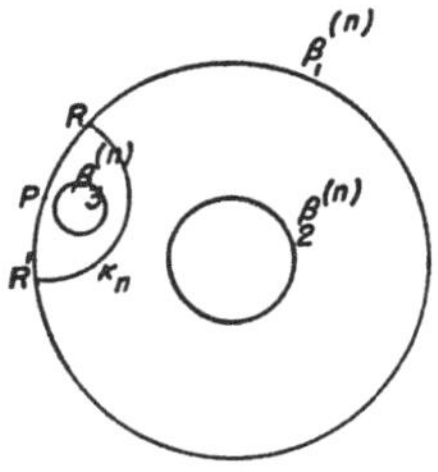

Figure 4.5

in such a way that the length L_ρ of the image of κ_ρ by $\mathfrak{x}_n$ is not greater than $\sqrt{2\pi\epsilon(\delta)}$. This inequality would imply that $\mathfrak{x}_n$ tends to degeneration, in contradiction to the condition of cohesion.

Finally, to exclude the second type of degeneration we consider the typical case where a circle $\beta_3^{(n)}$ shrinks to a point P on $\beta_1^{(n)}$ while $\beta_2^{(n)}$ is bounded away from $\beta_1^{(n)}$. Again by lemma 3.1 we can, for suitably small fixed δ and sufficiently large n, draw a circular arc κ_n about P which joins two points R and R' on $\beta_1^{(n)}$ and separates $\beta_3^{(n)}$ from $\beta_2^{(n)}$, in such a way that the length L_n of the image λ_n of κ_n by $\mathfrak{x}_n$ is less than $\sqrt{2\pi\epsilon(\delta)} = \eta(\delta)$. The two arcs on $\beta_1^{(n)}$ determined by R and R' are mapped onto two complementary arcs of γ_1 whose end-

[5] This lemma was proved under the assumption that the parameter domain does not depend on n. The proof for the more general case needed here is literally the same.

points Q and Q'—the images of R and R'—lie at a distance less than $\eta(\delta)$, so that the diameter of one of these two arcs is arbitrarily small for sufficiently small δ. Hence δ can be chosen so small that this arc, together with the arc λ_n, defines on the surface $\mathfrak{x}_n$ a closed curve ω_n of arbitrarily small total diameter. The curve ω_n cannot be contracted to a point on the surface $\mathfrak{x}_n$, since it separates γ_2 from γ_3; consequently $\mathfrak{x}_n$ tends to degeneration, contrary to our assumption.

Summarizing, we state: no degeneration of B_n can occur; it follows that the sequence B_n or a suitable subsequence tends to a domain B again of type $\mathfrak{N}_a$.

Next we prove the equicontinuity of the functions $\mathfrak{x}_n$ on each boundary circle, say on $\beta_1^{(n)}$; the proof follows the pattern of Chapter III, §2, 2. Accordingly we again define a quantity $\tau(\sigma)$ (inherent in the geometrical shape of γ) with the following properties: if Q, Q' are two points on one of the boundary curves γ_ν at a distance less than σ, the diameter of one of the two arcs on γ_ν between Q and Q' is not greater than τ; obviously τ is greater than σ, but tends to zero with σ, and for given τ a suitably small σ can always be found.

Given an arbitrarily small positive constant ϵ we choose σ such that $\tau(\sigma) < \epsilon$ and a suitably small positive constant

$$\delta < \exp\left\{-\frac{2\pi M}{\sigma^2}\right\},$$

M denoting a common upper bound for $D[\mathfrak{x}_n]$. By lemma 3.1 we can draw, about an arbitrary point P_n on $\beta_1^{(n)}$, a circular arc κ_n lying in B_n with radius ρ_n, $\delta \leq \rho_n \leq \sqrt{\delta}$, joining two points R and R' on $\beta_1^{(n)}$ in such a way that the oscillation of $\mathfrak{x}_n$ along κ_n is less than σ. The smaller arc β_1^* connecting R and R' on the circle $\beta_1^{(n)}$ is mapped on either the larger or the smaller arc of γ_1 connecting the respective image points Q and Q' of R and R'. The larger arc γ_1^{**} is excluded: for otherwise, the image γ_1^* of the larger arc β_1^{**} connecting Q and Q' on β_1 would have diameter not exceeding τ, and γ_1^* together with the image λ_n of κ_n would form a closed curve on $\mathfrak{x}_n$ of diameter not exceeding $\tau + \sigma$. This curve, while not contractable to a point on $\mathfrak{x}_n$, could be made to have arbitrarily small diameter for sufficiently small τ. We have arrived at a contradiction to the assumption of cohesion. It follows that the oscillation of $\mathfrak{x}_n$ on β_1^* is less than τ, hence less than ϵ, and $\mathfrak{x}_n$ is equicontinuous on β_1. The same proof holds for all the other boundary circles.

After replacing the vectors $\mathfrak{x}_n$ by harmonic vectors $\mathfrak{z}_n$ with the same boundary values on the circles $\beta_\nu^{(n)}$, we complete the existence proof exactly as in Chapter III: a suitable subsequence of the $\mathfrak{z}_n$ converges to an admissible harmonic vector $\mathfrak{x}$ in B for which $D[\mathfrak{x}] = d$.

4. *Solution of Variational Problem for Other Classes of Normal Domains.* The arguments of the preceding section remain almost literally valid for the class $\mathfrak{N}_b$ of parameter domains; for the class $\mathfrak{N}_c$ a slight modification is necessary. It suffices to explain the method for doubly connected domains B, consisting of two unit circles connected by two branch points A and A'. We fix A' at $w = 1/2$, leaving free the other branch point A. To prove that the sequence B_n has as limit a domain B of class $\mathfrak{N}_c$ we must show that the branch point A in B_n can tend neither to A' nor to the boundary. The first degeneration is excluded as above by lemma 3.1,[6] the second by the following

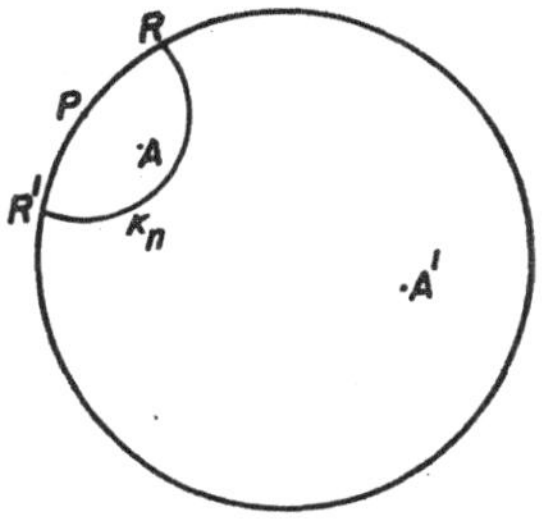

Figure 4.6

argument. If A tends to a boundary point P on the unit circle β_1, we draw about P a circular arc κ_n in B with radius ρ, $\delta \leq \rho \leq \sqrt{\delta}$, δ being chosen sufficiently small as in our lemma. For large n the point A is separated from A' by κ_n, so that κ_n joins a point R on β_1 with a point R' on β_2. The oscillation of $\mathfrak{x}_n$ on κ_n becomes arbitrarily small with δ, while the images of R and R' must remain at a distance no smaller than the minimum distance between γ_1 and γ_2; the second type of degeneration is thus likewise seen to be impossible.

3. Further Discussion of Solution

1. *Douglas' Sufficient Condition.* The condition of cohesion, necessary and sufficient for the existence of a solution of variational

[6] It is obvious that such a lemma is valid also for Riemann domains B.

problem IV, is not easily verifiable in particular cases. The restrictive, merely sufficient, but more intuitive condition of theorem 4.1 has been mentioned in the introduction to this chapter. We repeat and amplify the statements for k-fold connected domains.

A k-fold connected surface $\mathfrak{x}$ of genus zero bounded by γ is said to be degenerate if it consists of two separate surfaces (which may possibly intersect) whose boundaries together form γ. The greatest lower bound of $D[\mathfrak{x}]$ for such degenerate surfaces spanned in γ may be called d^*. Let $\mathfrak{x}$ be decomposed into two surfaces $\mathfrak{x}'$ and $\mathfrak{x}''$ bounded by γ' and γ'', where γ' consists of k', γ'' of k'' curves γ_ν and $k' + k'' = k$. The Dirichlet integrals over k'-fold and k''-fold connected parameter domains, for surfaces bounded respectively by γ' and γ'' in $\mathfrak{x}$-space, are denoted by $D'[\mathfrak{x}]$ and $D''[\mathfrak{x}]$, the corresponding greatest lower bounds by d' and d''. Noting

$$d^* = d' + d''$$

we shall prove

Theorem 4.2: For every possible type of degeneration of a surface $\mathfrak{x}$ we have

$$d \leq d^*. \tag{4.1}$$

Furthermore we restate theorem 4.1 (Douglas' condition): The condition that

$$d < d^* \tag{4.2}$$

for all possible degenerations is sufficient for the existence of a solution of variational problem IV (and hence for the existence of a non-degenerate k-fold connected minimal surface bounded by γ).

Simultaneously with the proof of theorem 4.1 it is convenient to establish

Theorem 4.3: The greatest lower bound d in variational problem IV depends semicontinuously on the boundary γ. More precisely, if a sequence $\gamma^{(n)}$ of boundary curves tends to γ in the strong (Fréchet) sense, we have

$$d \leq \lim\inf. d_n ,$$

where d_n is the greatest lower bound for $D[\mathfrak{x}]$ for surfaces $\mathfrak{x}$ spanned in $\gamma^{(n)}$. The theorem remains valid if $\gamma^{(n)}$ has multiple points. In Chapter III the statement was proved for $k = 1$; it will now be proved for every k by induction.

The proof of these theorems would be facilitated by accepting from the outset the fact that d, d^*, d', d'' can be interpreted as greatest lower bounds of area — an immediate consequence of theorems on conformal mapping. However, since we intend in the next chapter to develop a method for deriving mapping theorems from the solution of problem IV, we shall prove theorems 4.1–4.3 merely by discussing the Dirichlet integral, making no use of its connection with area. The proofs will depend on properties of the Dirichlet integral formulated in two lemmas.

2. *Lemma 4.1 and Proof of Theorem 4.2.* Our two lemmas express the fact that vectors $\mathfrak{x}$ can be subjected to certain locally large variations without an essential increase in $D[\mathfrak{x}]$. The first lemma means that a variation of replacing $\mathfrak{x}$ by any prescribed constant vector in the neighborhood of a point P in B and affecting only a slightly larger neighborhood of P can be so constructed as to change $D[\mathfrak{x}]$ by arbitrarily little. Precisely we state

Lemma 4.1:[7] Let $\mathfrak{z}(u, v)$ be an arbitrary vector satisfying $|\mathfrak{z}| < M$, continuous in $B + \beta$ with piecewise continuous first derivatives in B, and for which $D[\mathfrak{z}] < \infty$. Given an arbitrarily small positive constant α and any point P in B—say the origin—we can find a sufficiently small constant η, $0 < \eta < 1$, and a vector $\mathfrak{y}(u, v)$ identical with $\mathfrak{z}$ outside the neighborhood $u^2 + v^2 \leq \eta^2$ of P, having a prescribed constant value — say $\mathfrak{y} = 0$ — inside the smaller neighborhood $u^2 + v^2 \leq \eta^4$ of P, and satisfying the inequality

$$D[\mathfrak{y}] < D[\mathfrak{z}] + \alpha. \tag{4.3}$$

In other words: without essentially increasing Dirichlet's integral, we can pull out a spine from the surface $\mathfrak{z}$ reaching to a given point.

Corollary: If B contains the point at infinity, we can similarly find a sufficiently large constant η' and a vector $\mathfrak{y}(u, v)$ having the values of

[7] This lemma, in connection with the theorems of Chapter III, guarantees the existence of d for rectifiable boundary curves.

$\mathfrak{z}$ within the domain $u^2 + v^2 \leq \eta'^2$ and the value $\mathfrak{y} = 0$ outside the larger circle $u^2 + v^2 \leq \eta'^4$, such that inequality (4.3) holds.

Proof:[8] To prove the lemma we form the function $p(u, v)$ with $r^2 = u^2 + v^2$ and any constant $\eta < 1$,

$$p(u, v) = p(r) = \begin{cases} 1, & r > \eta, \\ 1 + \dfrac{\log \eta/r}{\log \eta}, & \eta^2 \leq r \leq \eta, \\ 0, & r < \eta^2, \end{cases}$$

and define the vector $\mathfrak{y}$ by

$$\mathfrak{y}(u, v) = p(u, v)\mathfrak{z}(u, v). \tag{4.4}$$

Set $\epsilon = D[p] = \dfrac{1}{2}\iint_{\eta^2 < r < \eta} (p_u^2 + p_v^2)\, du\, dv = -\pi/\log \eta.$

Using $|p| < 1$, $|\mathfrak{z}| < M$, and Schwarz' inequality, we have

$$\begin{aligned} D[\mathfrak{y}] &\leqq D[\mathfrak{z}] + M^2 D[p] + \iint_{\eta^2 < r < \eta} p(p_u \mathfrak{z}_u \mathfrak{z} + p_v \mathfrak{z}_v \mathfrak{z})\, du\, dv \\ &\leq D[\mathfrak{z}] + M^2\epsilon + 2M\sqrt{\epsilon D[\mathfrak{z}]} \end{aligned}$$

or

$$D[\mathfrak{y}] \leq (\sqrt{D[\mathfrak{z}]} + M\sqrt{\epsilon})^2.$$

Since we can satisfy (4.3) by making η and hence ϵ sufficiently small, the lemma is proved. The proof of the corollary is essentially the same.

Theorem 4.2 follows easily from lemma 4.1. To prove it, say for the class $\mathfrak{N}_a$ of domains B, we consider, in circular domains B' and B'', admissible vectors $\mathfrak{x}'$ and $\mathfrak{x}''$ referring to the systems γ' and γ'' of boundary curves, respectively. Without loss of generality we may assume that B' lies inside one of its boundary circles while B'' contains the point at infinity, and that both contain the origin O. We select an arbitrarily small positive constant ϵ and restrict ourselves to admissible vectors for which

$$D_{B'}[\mathfrak{x}'] < d' + \epsilon, \qquad D_{B''}[\mathfrak{x}''] < d'' + \epsilon.$$

[8] See Courant [3], p. 685 ff.

According to lemma 4.1 we replace the vector $\mathfrak{x}'$ in B' by another admissible vector $\mathfrak{y}'$ which is zero in a small disk K of radius η about O and for which $D_{B'}[\mathfrak{y}'] < d' + 2\epsilon$. By the corollary we construct, for sufficiently large η', an admissible vector $\mathfrak{y}''$ in B'' such that $\mathfrak{y}'' = 0$ for $r > \eta'$ and $D_{B''}[\mathfrak{y}''] < d'' + 2\epsilon$. Next we subject B'' to a similarity contraction from the origin, reducing r by the factor η/η' so that the circle of radius η' shrinks to one of radius η. Dirichlet's integral is not changed by this contraction of B''. We retain the notations B'' and $\mathfrak{y}''$ with reference to the contracted domain and denote by B^* the domain $B' - K$ and by B^{**} the part of B'' which is interior to K. Finally we define a new k-fold connected circular domain $B = B^* + B^{**}$, and an admissible vector $\mathfrak{y}$ in B by

$$\mathfrak{y} = \begin{cases} \mathfrak{y}' & \text{in } B^*, \\ \mathfrak{y}'' & \text{in } B^{**}, \\ 0 & \text{on } \kappa, \end{cases}$$

where κ denotes the periphery of K. Then $D_{B^*}[\mathfrak{y}'] = D_{B'}[\mathfrak{y}']$ and $D_{B^{**}}[\mathfrak{y}''] = D_{B''}[\mathfrak{y}'']$, so that

$$D[\mathfrak{y}] = D_{B^*}[\mathfrak{y}'] + D_{B^{**}}[\mathfrak{y}''] < d' + d'' + 4\epsilon.$$

Since $d \leq D[\mathfrak{y}]$ and since ϵ can be chosen arbitrarily small, theorem 4.2 is proved for parameter domains bounded by circles. An entirely similar proof holds for the other types $\mathfrak{N}_b$ and $\mathfrak{N}_c$ of domains defined in §2.

3. *Lemma 4.2 and Proof of Theorem 4.1.* Whereas lemma 4.1 refers to a variation of the vector $\mathfrak{z}(u,v)$ in the parameter domain B, the following lemma appraises the effect on vectors $\mathfrak{z}$ of particular transformations defined in the $\mathfrak{z}$-space. This transformation, which we shall call "pinching," contracts a whole spherical neighborhood of a point into the point while leaving fixed the points outside a larger neighborhood. Let A be a point represented by an arbitrarily chosen constant vector $\mathfrak{a}$ in $\mathfrak{z}$-space, ϵ a small positive constant. We transform the whole $\mathfrak{z}$-space into a $\mathfrak{y}$-space of vectors $\mathfrak{y}$ by contracting the sphere of radius η^2 about the point A into that point while leaving unchanged the space outside the concentric sphere with radius η; here $\eta = e^{-1/\epsilon}$. Let the surface $\mathfrak{z}$ be represented by a vector $\mathfrak{z}(u, v)$ defined in the parameter domain B; then the "pinching" transforms

$\mathfrak{z}(u, v)$ into another vector $\mathfrak{y}(u, v)$ which is identical with $\mathfrak{a}$ wherever $|\mathfrak{z} - \mathfrak{a}| < \eta^2$ and identical with $\mathfrak{z}$ wherever $|\mathfrak{z} - \mathfrak{a}| > \eta$. We prove

Lemma 4.2: By a suitable "pinching" process any vector $\mathfrak{z}$ with finite Dirichlet integral is transformed into a vector $\mathfrak{y}$ for which

$$D[\mathfrak{y}] \leq (1 + \epsilon)^2 D[\mathfrak{z}].$$

In other words, without essentially increasing the Dirichlet integral of $\mathfrak{z}$ and without essentially modifying the values of $\mathfrak{z}$ we can deform the surface $\mathfrak{z}$ so that all parts of the surface near a point $\mathfrak{a}$ are pinched into this point.

We shall apply the lemma to a sequence of surfaces $\mathfrak{z}$ tending to degeneration. Specifically we shall suppose that $\mathfrak{z}$ is "almost degenerate" in the neighborhood of a point A; that is, we assume the existence of a closed curve α on the surface, lying entirely within a sphere of radius η^2 about A, which separates the surface into two parts each bounded not only by α but also by a part of the boundary γ. Then our pinching transforms the surface $\mathfrak{z}$ into an actually degenerate surface.

Proof: We may suppose $\mathfrak{a} = 0$, and contract the whole m-dimensional $\mathfrak{z}$-space into a $\mathfrak{y}$-space by the transformation

$$y_\mu = p(r) z_\mu, \qquad \mu = 1, 2, \cdots, m$$

or

$$\mathfrak{y} = p(r)\mathfrak{z}, \tag{4.5}$$

where p as a function of the distance $r = (z_1^2 + z_2^2 + \cdots + z_m^2)^{1/2}$ and the parameter $\eta = e^{-1/\epsilon}$ is given[9] by

$$p(r) = \begin{cases} 1, & r > \eta, \\ 1 + \dfrac{\log \eta/r}{\log \eta}, & \eta^2 \leq r \leq \eta, \\ 0, & r < \eta^2. \end{cases}$$

In (4.5) we substitute for $\mathfrak{z}$ the surface vector $\mathfrak{z}(u, v)$; then the vector $\mathfrak{y}$ represents a surface parametrically in B.

[9] This function is essentially different from that used for lemma 4.1 because it refers to distance in the vector space instead of in the parameter domain B.

Consider the open point set B^*: $\eta^2 < |\mathfrak{z}| = r < \eta$ contained in B. The Dirichlet integral of $\mathfrak{y}$ is given by

$$D[\mathfrak{y}] = \frac{1}{2}\iint_B [(p\mathfrak{z}_u + p_u\mathfrak{z})^2 + (p\mathfrak{z}_v + p_v\mathfrak{z})^2]\, du\, dv = a + b + c.$$

Since $p \leq 1$ we have the inequality

$$a \equiv \frac{1}{2}\iint_B p^2(\mathfrak{z}_u^2 + \mathfrak{z}_v^2)\, du\, dv \leq D[\mathfrak{z}];$$

furthermore, from the definition of p we find

$$b \equiv \frac{1}{2}\iint_B \mathfrak{z}^2(p_u^2 + p_v^2)\, du\, dv = \frac{1}{2}\iint_{B^*} \mathfrak{z}^2(p_u^2 + p_v^2)\, du\, dv,$$

and

$$c \equiv \iint_B (pp_u\mathfrak{z}\mathfrak{z}_u + pp_v\mathfrak{z}\mathfrak{z}_v)\, du\, dv$$

$$= \iint_{B^*} (pp_u\mathfrak{z}\mathfrak{z}_u + pp_v\mathfrak{z}\mathfrak{z}_v)\, du\, dv.$$

The derivatives of p in B^* have the bounds

$$|p_u| \leq \frac{\epsilon}{r}|\mathfrak{z}_u|, \qquad |p_v| \leq \frac{\epsilon}{r}|\mathfrak{z}_v|.$$

Recalling that $|\mathfrak{z}| = r$ we derive

$$|p\, p_u\, \mathfrak{z}\mathfrak{z}_u| \leq \epsilon\mathfrak{z}_u^2, \qquad |p\, p_v\, \mathfrak{z}\mathfrak{z}_v| \leq \epsilon\mathfrak{z}_v^2,$$

$$\mathfrak{z}^2(p_u^2 + p_v^2) \leq \epsilon^2(\mathfrak{z}_u^2 + \mathfrak{z}_v^2),$$

and consequently

$$b \leq \epsilon^2 D_{B^*}[\mathfrak{z}] \leq \epsilon^2 D[\mathfrak{z}],$$

$$c \leq 2\epsilon D_{B^*}[\mathfrak{z}] \leq 2\epsilon D[\mathfrak{z}].$$

Collecting the results, we have

$$D[\mathfrak{y}] \leq D[\mathfrak{z}](1 + 2\epsilon + \epsilon^2) = (1 + \epsilon)^2 D[\mathfrak{z}],$$

as stated in the lemma.

The proof of theorem 4.1 is carried out in two steps: First we realize that a system γ of curves for which problem IV has no solution bounds minimizing sequence B_n, $\mathfrak{x}_n$ tending to degeneration; for convenience we suppose that the parameter domains B_n are of class $\mathfrak{N}_a$,[10] again normalized by the stipulation that the outer boundary circle is the unit circle and that another boundary circle has the same center. Secondly we show that the existence of a degenerating minimizing sequence implies an equality of the form $d = d' + d''$, contradicting Douglas' condition $d < d' + d''$. This second step is a proof by induction based on lemma 4.2 and on a semicontinuity theorem which generalizes, to k-fold connected domains, theorem 3.1 for simply connected domains. We shall prove this semicontinuity theorem simultaneously with theorem 4.1, again using induction.

Theorem 4.4: Let $\gamma^{(n)}$ be a sequence of systems of $\bar{k}$ continuous contours—possibly having multiple points—converging to γ in the sense of Fréchet as $n \to \infty$, and let $\mathfrak{x}_n$ be an admissible vector defined in a parameter domain B_n of the prescribed class, mapping the boundary of B_n onto $\gamma^{(n)}$; then the greatest lower bound d for the Dirichlet integrals of surfaces spanned in γ satisfies the inequality

$$d \leq \liminf. D_{B_n}[\mathfrak{x}_n]. \tag{4.6}$$

Relation (4.6) expresses the semicontinuous dependence of d on γ.

Using our induction assumption, we suppose that theorem 4.4 holds for $\bar{k} = 1, 2, \cdots, k - 1$ and turn to the second step mentioned above. Consider a sequence $\mathfrak{x}_n$, B_n of vectors and corresponding parameter domains (not necessarily a minimizing sequence) admissible in problem IV for the boundary $\gamma^{(n)}$, and suppose that the surfaces $\mathfrak{x}_n$ tend to degeneration. For a suitably chosen subsequence we may assume that separation of the surfaces $\mathfrak{x}_n$ occurs at the origin. In other words, a closed curve $\tau^{(n)}$ on the surface $\mathfrak{x}_n$ separates two parts of the boundary $\gamma^{(n)}$, and the curves $\tau^{(n)}$ shrink uniformly to the point O; that is, we can find a sequence η_n of positive numbers tending to zero such that

$$|\mathfrak{x}_n| < \eta_n \qquad \text{on } \tau^{(n)}.$$

The curve $\tau^{(n)}$ is defined as the image of a simple closed Jordan curve $t^{(n)}$ in B_n separating B_n into two domains B_n' and B_n'' bounded

[10] For other types of domains B the proof is entirely similar.

by $t^{(n)}$ and by two complementary systems $\beta^{(n)'}$ and $\beta^{(n)''}$ of k' and k'' boundary components respectively.

According to lemma 4.2 we replace the surface $\mathfrak{x}_n$ defined in B_n by a degenerate surface $\mathfrak{y}_n$ in such a manner that

$$D_{B_n}[\mathfrak{y}_n] < D_{B_n}[\mathfrak{x}_n] + \alpha_n$$

with $\alpha_n \to 0$; the vector $\mathfrak{y}_n$ maps the boundary $\beta^{(n)}$ of B_n onto a system $\gamma^{(n)*}$ of curves which tend to γ as do the boundaries $\gamma^{(n)}$.[11] We have evidently

$$\text{(4.7)} \qquad \liminf. D_{B_n}[\mathfrak{y}_n] \leq \liminf. D_{B_n}[\mathfrak{x}_n].$$

Assume that the curve $t^{(n)}$ in $B_n + \beta^{(n)}$, on which $\mathfrak{y}_n$ vanishes, has B_n' in its interior and B_n'' in its exterior. We denote by B_n^* the sum of B_n' and the whole exterior of $t^{(n)}$ and by B_n^{**} the sum of B_n'' and the whole interior of $t^{(n)}$, and define vectors $\mathfrak{y}_n^*$ and $\mathfrak{y}_n^{**}$ by

$$\mathfrak{y}_n^* = \begin{cases} \mathfrak{y}_n & \text{in } B_n', \\ 0 & \text{outside } t^{(n)}; \end{cases} \qquad \mathfrak{y}_n^{**} = \begin{cases} \mathfrak{y}_n & \text{in } B_n'', \\ 0 & \text{inside } t^{(n)}. \end{cases}$$

The vectors $\mathfrak{y}_n^*$ and $\mathfrak{y}_n^{**}$ are piecewise smooth in B_n^* and B_n^{**}, respectively, and their Dirichlet integrals satisfy the relation

$$\text{(4.8)} \qquad D_{B_n^*}[\mathfrak{y}_n^*] + D_{B_n^{**}}[\mathfrak{y}_n^{**}] = D_{B_n}[\mathfrak{y}_n].$$

Furthermore, $\mathfrak{y}_n^*$ and $\mathfrak{y}_n^{**}$ map the boundaries of B_n' and B_n'', respectively, on sets of curves $\gamma^{(n)*}$ and $\gamma^{(n)**}$ where $\gamma^{(n)*} \to \gamma'$, $\gamma^{(n)**} \to \gamma''$. Thus they are admissible in variational problems IV relating to lower numbers k' and k'' of boundary curves. For such lower numbers semicontinuity of d has been assumed; consequently we have

$$\liminf. D_{B_n^*}[\mathfrak{y}_n^*] \geq d', \qquad \liminf. D_{B_n^{**}}[\mathfrak{y}_n^{**}] \geq d'',$$

where d' and d'' refer to the partition of the boundary γ into γ' and γ'', respectively. We obtain by (4.7) and (4.8)

$$\text{(4.9)} \qquad d' + d'' \leq \liminf. D_{B_n}[\mathfrak{x}_n].$$

[11] The system $\gamma^{(n)*}$ is identical with $\gamma^{(n)}$ if the origin O is not on γ; for then η_n can be chosen so small that the transformation (4.5) does not affect the boundary curve. The slight complication of the proof of theorem 4.1 arising from the need of ascertaining semicontinuity is due to the fact that a position of the point O of separation on γ cannot *a priori* be excluded.

Applying inequality (4.9) to a minimizing sequence B_n, $\mathfrak{x}_n$ for which $D_{B_n}[\mathfrak{x}_n] \to d$, we find $d \geq d' + d''$, whereas theorem 4.2 states that $d \leq d' + d''$; hence $d = d' + d''$. Furthermore, the inequality $d \geq d' + d''$ contradicts condition (4.2)

$$d < d' + d'' = d^*;$$

therefore, this condition precludes the existence of a degenerating minimizing sequence, and is sufficient for the existence of a solution of problem IV.

To complete the proof by induction we have still to deduce theorem 4.4 (semicontinuity) for $\tilde{k} = k$ under the assumption that it is proved for $\tilde{k} < k$. We distinguish two cases. If the vectors $\mathfrak{x}_n$ referred to in the theorem tend to degeneration, (4.9) and (4.1) establish semicontinuity. If, on the other hand, the vectors $\mathfrak{x}_n$ satisfy the condition of cohesion, the reasoning of Chapter III, §2, 3, holds almost literally: the domains B_n form a compact set and define a limiting domain B, the vectors $\mathfrak{x}_n$ are equicontinuous on the boundaries $\beta^{(n)}$, and the harmonic vectors $\mathfrak{z}_n$ with the same boundary values tend for $n \to \infty$ to an admissible vector $\mathfrak{x}$, mapping the boundary β of B on γ, for which

$$D[\mathfrak{x}] \leq \text{lim inf. } D_{B_n}[\mathfrak{x}_n].$$

In particular we have

$$d \leq \text{lim inf. } d_n,$$

where d_n is the minimum value of Dirichlet's integral for $\gamma^{(n)}$, B_n, so that semicontinuity is established for k-fold connected domains in both cases.

Again it should be emphasized that the analysis of the preceding sections remains valid for the various types of parameter domains described in §2, and that no essential modifications are necessary for still other classes of normal domains.

4. *Remarks and Examples.* In the preceding proofs, use of theorems on conformal mapping was avoided so that we can utilize theorem 4.2 in Chapter V for the proof of a number of general mapping theorems. There the number of space dimensions will be $m = 2$.

On the other hand, in the general case—in particular for $m > 2$—we cannot remove certain unsatisfactory features of the solution of problem IV without making use of mapping theorems. For example, the preceding solution was given with specific reference to a prescribed class $\mathfrak{N}$ of parameter domains B. Without theorems concerning the conformal equivalence of an arbitrary k-fold connected Riemann domain with a member of $\mathfrak{N}$, the characterization of d as the minimum of $D[\mathfrak{x}]$ retains an artificial element, even though the class $\mathfrak{N}$ is wide enough to permit identification of the solution as a minimal surface by variational methods. The simplest choice of the class $\mathfrak{N}$ is that of half-plane slit domains, which according to Chapter II are conformal representatives of all Riemann domains of finite Euler characteristic. Our remark applies even more to the case of surfaces not of genus zero discussed in the next section.

As in Chapter III theorems on conformal mapping allow us to identify the solution of problem IV with the surfaces of least area under the prescribed boundary conditions. With such an interpretation of d, d', d'' as greatest lower bounds of areas of surfaces, the theorems of this chapter become intuitively clearer, and the conditions easier to verify in particular cases.

Examples:

1) Douglas' sufficient condition $d < d^*$ is certainly satisfied if the lower type surface of least area is self-intersecting. This is intuitively evident: by cutting the surface along the lines of self-intersection, establishing suitable new identifications of the edges of the cuts, and slightly deforming the new surfaces, one can remove self-intersections and obtain a surface of smaller area. This new surface will sometimes have a higher topological structure; when it does, it is plausible that the given system of curves bounds a non-degenerate minimal surface of higher topological structure more stable than the original self-intersecting solution of Douglas' problem.

2) Consider the case $k = 2$ for surfaces of genus zero. Assume that the two boundary curves γ_1 and γ_2 are interlocking. It is immediately seen that the degenerate surface consisting of the two simply connected minimal surfaces bounded by γ_1 and γ_2 has greater area than the doubly connected surfaces obtained from an intersecting pair of surfaces by a deformation that eliminates the self-intersection. Hence, Douglas' condition is satisfied and a non-degenerate solution exists. Similarly one can see that a knotted curve γ always bounds, besides the self-intersecting simply connected minimal surface, a surface of higher structure.[12] In general the condition is satisfied

[12] The result that two interlocking curves always define a doubly connected minimal surface was first obtained by Douglas.

for $k = 2$ if the two simply connected surfaces of least area through γ_1 and γ_2 intersect in a closed curve. It is an interesting problem to prove these facts without using the interpretation of $D[\mathfrak{x}]$ as area.

3) We consider an example that dispenses with the identification of d, d', d'' as areas. In Chapter V, we shall verify the condition $d < d^*$ directly by analyzing Dirichlet's integral in the case where the system γ of curves bounds a plane domain. Using an obvious generalization of theorem 3.3 to k-fold connected domains, we establish the sufficient condition by a slight deformation of the boundary system γ. Hence problem IV has a solution for contours γ lying sufficiently close to a plane.

4. Generalization to Higher Topological Structure

Finally we turn to the problem of minimal surfaces not of genus zero. As stated, soap film experiments indicate that for certain suitably twisted contours γ the solution of least area need not be orientable (if γ is a single curve) or simply connected (if γ is a system of curves); instead we obtain surfaces of the type of a Moebius strip or other non-orientable surfaces, or surfaces of higher genus, see Figures 4.2, 4.1.[13]

1. *Existence of Solution.* To formulate variational problem IV for surfaces not of genus zero we must choose parameter domains B of the same topological structure. The most convenient class of such domains B are the normalized half-plane slit domains with properly coordinated infinite interior slits, as studied in Chapter II, §7; it was shown that these domains form sets of normal domains.

As in the case of genus zero, Douglas' problem for prescribed higher topological structure need not have a solution. For example, a single plane boundary curve γ bounds neither a minimal surface of genus one nor one of the type of a Moebius strip.[14] The same condition of cohesion as for genus zero is necessary and sufficient, and it is formulated exactly as in §2, 2. Likewise the same condition $d < d^*$ is sufficient for the existence of a solution, where the quantity d^* refers to degenerate surfaces having γ as boundary. Such surfaces either consist of two separate surfaces the sum of whose Euler characteristics is not greater than that prescribed, or are degenerate merely

[13] See footnote 1.

[14] The only minimal surface bounded by a plane contour is a part of that plane.

by having a smaller Euler characteristic. If d and d^* are identified with minimum area, the condition $d < d^*$ assures existence of a minimal surface of given topological structure whenever the area for such surfaces can be made smaller than the greatest lower bound of the area for surfaces of lower structure.

2. *Proof for Topological Type of Moebius Strip.* The proof proceeds exactly along the lines of the preceding sections; we shall be content to give it in detail for the typical case of a minimal surface of the type of a *Moebius strip bounded by one* contour γ.[15]

Here the parameter domain is given by Figure 2.24 on page 85 and the limiting Figure 2.25 on page 85, where A is the fixed point $u = 0$, $v = 1$, while the ordinate of A' is a parameter. Let $\mathfrak{x}_n$, B_n denote a minimizing sequence of vectors and domains admissible in variational problem IV, normalized by the stipulation that the image of the point at infinity $u = \infty$ in the mapping of β on γ is fixed. We suppose $d < d^*$, where d^* is the lower bound of Dirichlet's integral for simply connected surfaces spanned by γ. Assuming that the domains B_n, or a subsequence, tend to a limiting domain B, we establish the equicontinuity of the vectors $\mathfrak{x}_n$ on β, supposing B to be of the type of Figure 2.24 (for a domain B such as in Figure 2.25 the reasoning is only slightly different).

We reason as before, *cf.* lemma 3.1. For every positive σ there is a quantity $\tau(\sigma)$ such that, if two points Q, Q' on γ are at a distance less than σ, one of the two arcs QQ' on γ has diameter less than τ. By choosing σ properly, τ can be made arbitrarily small. If P is a point on β, we can accordingly find half-squares $R_n T R_n'$ as in Figure 4.7 such that the oscillation of $\mathfrak{x}_n$ on $R_n T R_n'$ remains less than a quantity σ while

$$\delta < \overline{R_n P R_n'} < \sqrt{\delta},$$

δ depending only on σ and tending to zero with σ. Since $\mathfrak{x}_n$ maps β monotonically on the Jordan curve γ, the oscillation of $\mathfrak{x}_n$ is less than τ either on the segment $R_n P R_n'$ or on the complementary infinite part of β. The second alternative is easily ruled out: it would imply that the simply connected domain B_n^* bounded by the half-square is mapped by $\mathfrak{x}_n$ onto a simply connected surface, bounded by a curve $\gamma^{(n)*}$ which with increasing n tends to γ in the strong (Fréchet) sense. We would have

[15] For the reasoning in the general case, where the technical details are quite similar though more cumbersome, see Shiffman [3].

(4.10) $$\liminf D_{B_n^*}[\mathfrak{x}_n] \leq \lim D_{B_n}[\mathfrak{x}_n] = d$$

and by the semicontinuity of d^* for simply connected regions,

(4.11) $$d^* \leq \liminf D_{B_n^*}[\mathfrak{x}_n];$$

combining inequalities (4.10) and (4.11) we would have

$$d^* \leq d,$$

contrary to our assumption $d^* > d$.

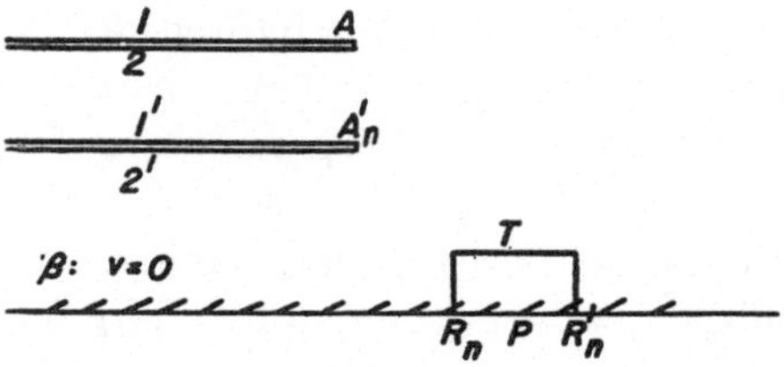

Figure 4.7

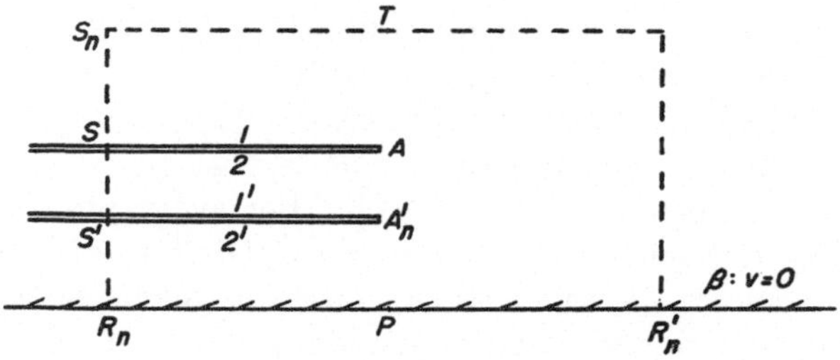

Figure 4.8

This statement of equicontinuity must be supplemented by a proof of equicontinuity at $u = \infty$; in the preceding notation, we must show that for $|u| > 1/\sqrt{\delta}$ the oscillation of $\mathfrak{x}_n$ on β remains less than $\tau(\sigma)$. For the proof we construct a large half-square with an edge larger than $1/\sqrt{\delta}$, as $R_n\, T\, R_n'$ in Figure 4.8, where the side R_nS_n consists of three properly coordinated segments, and where the total variation of $\mathfrak{x}_n$ is less than σ. We must rule out the possibility that the oscillation of $\mathfrak{x}_n$ on the segment $R_n\, P\, R_n'$ of β is less than τ. In this case the coordination of the edges would make the part B_n^* of B_n outside the half-square a simply connected domain—the orientability being safeguarded by the barrier SS' which excludes paths along which the orientation might change. The boundary of B_n^* is again mapped by

$\mathfrak{x}_n$ onto a curve γ^* which tends to γ in the strong sense and for which $D_{B_n^*}[\mathfrak{x}_n] < D_{B_n}[\mathfrak{x}_n]$; thus we obtain the same contradiction of $d < d^*$ as above.

Instead of using the simply connected domain B_n^* defined by boundary coordination, we could use the following alternative construction. First we replace $\mathfrak{x}_n$ according to lemma 4.2 by an admissible vector $\mathfrak{y}_n$ which is constant along the contour $R_n TR_n'$ and for which $D[\mathfrak{y}_n] < D[\mathfrak{x}_n] + \alpha$; for sufficiently small δ, the quantity α can be made arbitrarily small. In the part B_n^* of B_n outside the half-square we reflect the strip C between the edges 2 and 1′ in the middle line μ, so that 1′ is adjacent to 1 and 2 is adjacent to 2′ on the

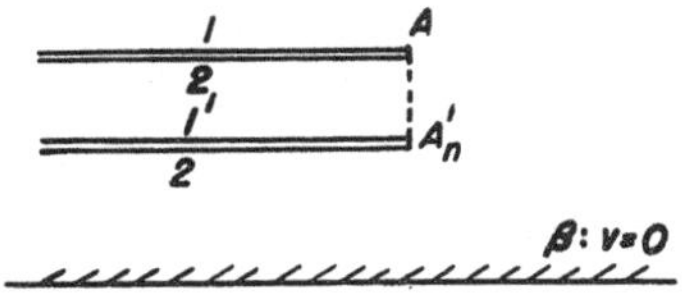

Figure 4.9

boundaries of C. Denote the reflected vector by $\mathfrak{y}_n^*$. We define a new vector $\mathfrak{y}_n^0$ by

$$\mathfrak{y}_n^0 = \begin{cases} \mathfrak{y}_n^* & \text{in } C, \\ \mathfrak{y}_n & \text{in } B_n^* - C, \end{cases} \tag{4.12}$$

so that it remains continuous on crossing the slits. Removing the slits we have in the exterior B_n^* of the half-square a plane simply connected domain B_n^0 and a vector $\mathfrak{y}_n^0$, continuous in B_n^0, for which $D[\mathfrak{y}_n^0] = D[\mathfrak{y}_n]$; the preceding reasoning may be applied to prove equicontinuity. The remaining details of the proof for the existence of a solution $\mathfrak{x}$ are the same as in the case of simply connected domains.

We have yet to show that our assumption $d < d^*$ ensures non-degeneration of the sequence B_n, i.e. that the point A_n' remains bounded away from the fixed point A. Suppose that $A_n' \to A$; then according to lemma 3.1 we can replace $\mathfrak{x}_n$ by an admissible vector $\mathfrak{y}_n$ in B_n, identical with $\mathfrak{x}_n$ outside a fixed circle about A, and identically zero within a smaller fixed concentric circle, for which $D[\mathfrak{y}_n] < D[\mathfrak{x}_n] + \alpha$, with arbitrarily small preassigned α. In particular $\mathfrak{y}_n$ is zero along the straight segment AA_n', for sufficiently large n. Again we replace $\mathfrak{y}_n$ in the strip C between the edges 2 and 1′ by a vector $\mathfrak{y}_n^*$, and define a vector $\mathfrak{y}_n^0$ as in (4.12). Obviously $\mathfrak{y}_n^0$ is continuous in the whole half-plane H: $v > 0$ with the cuts removed, and

$$D_H[\mathfrak{y}_n^0] = D[\mathfrak{y}_n] < D[\mathfrak{x}_n] + \alpha.$$

Since $\mathfrak{y}_n^0$ is an admissible vector in the variational problem for the simply connected surfaces spanned by γ, we have $d^* \leq D_H[\mathfrak{y}_n^0]$, and consequently letting n tend to ∞

$$d^* \leq d + \alpha;$$

since α can be chosen arbitrarily small we conclude

$$d^* \leq d,$$

contrary to our assumption. Degeneration has been shown impossible, and the proof is complete.

3. *Other Types of Parameter Domains.* Many other classes of parameter domains B could be used in the general problem of Douglas. For example, to obtain surfaces of genus p with k boundary curves γ_ν one might consider Riemann surfaces B consisting of the interior of k unit circles β_ν and one full plane, these $k + 1$ sheets being connected by $2k - 2 + 4p$ branch points;[16] the circle β_ν is to correspond to the boundary curve γ_ν.

By reflecting such Riemann surface domains simultaneously in all the boundary circles one obtains symmetric closed Riemann surfaces $\bar{B}$ which may again serve as parameter domains, carrying the same values for admissible vectors in points corresponding by symmetry. For the study of non-orientable minimal surfaces we may use as parameter domains closed symmetric surfaces which, unlike those just described, are not dissected by the symmetry lines corresponding to γ.[17] Another possible class of parameter domains B for surfaces of genus p is provided by fundamental domains of automorphic groups generated by linear transformations in the complex plane, k circular disks being removed from the fundamental domain. Apart from the intrinsic interest one may have in linking the theory of minimal surfaces with the concepts of classical function theory, the value of such parameter domains for our theory is questionable. Even more than in the case of genus zero we cannot

[16] See Courant [6].

[17] Douglas [4] has based his analysis on such symmetric Riemann surfaces.

expect a satisfactory result from the theory unless we know beforehand that the class of parameter domains used is a class of normal domains of the prescribed structure.

The difficulty encountered if one does not wish to presuppose mapping theorems may be illustrated by the case $k = 1, p = 1$. We choose for B a two-sheeted surface consisting of a unit circle and a full plane, the two sheets being connected by four branch points within the unit circle. If a minimizing sequence B_n degenerates in such a way that two branch points tend to the same point—cancelling each other so that only two branch points remain in the limit—B would degenerate to a simply connected domain with $p = 0$. This domain, however, consisting of a unit circle with a full plane affixed by two branch points, is of a different type from the domains B originally used to define the lower limit d for simply connected domains of genus zero. Conformal mapping must therefore be applied to show that domains obtained by such processes of degeneration are equivalent to domains of the originally admitted type.

The same situation arises for other types of domains B, e.g. for plane domains defined by fundamental domains of automorphic groups of linear substitutions with p generating transformations. Since the group and the boundary circles depend only on a finite number of parameters, the reasoning concerning the solution of the variational problem follows our previous pattern. In this case B_n may so degenerate that the limit domain B is of lower genus while still defined by a group with p generating transformations. A degeneration of this kind occurs, for example, if two corresponding boundaries of the fundamental domains touch in corresponding points or, as one says, if the fundamental domain of the limiting group has a "parabolic vertex." The genus of the limit domain is lowered, so that this domain no longer belongs to the type admitted for the lower genus; again equivalence must be established by mapping theorems. Corresponding considerations hold for slit domains.

4. *Identification of Solutions as Minimal Surfaces. Properties of Solution.* If we know beforehand that our class $\mathfrak{N}$ of parameter domains contains conformal representatives of all Riemann domains of the prescribed structure, the identification of the solution $\mathfrak{x}$ as a minimal surface is immediate. Exactly as in Chapter III, §3 for the case of genus zero, we may use the sewing theorem; alter-

natively we may base the relation $\phi(w) = 0$ on the vanishing of the first variation (3.17)

$$V = \iint_B [(\lambda_u - \mu_v)p - (\lambda_v + \mu_u)q] \, du \, dv$$

for arbitrary λ, μ as described in Chapter III, §4.

From the fact that polyhedral surfaces of the given structure are conformally equivalent to domains B it follows as before that our solutions are the surfaces of least area under the given conditions. A further rather obvious remark is that one-to-one correspondence of β and γ can again be proved for the solutions of Douglas' problem. Likewise the theorems on semicontinuous and continuous dependence of d on the boundary are easily extended to the solutions of Douglas' general problem.

CHAPTER V

Conformal Mapping of Multiply Connected Domains

1. Introduction

1. *Objective.* In Chapters I and II, the conformal mapping of general Riemann domains G on slit domains B was obtained by employing Dirichlet's Principle to construct functions in G mapping G onto B. A different approach[1] to the problem of conformal mapping of arbitrary domains G on individuals of any of three classes $\mathfrak{N}$ of normal domains B is provided by the methods of Chapters III and IV; Dirichlet's Principle is used there to construct functions in B giving the inverse mapping of B onto G. Restricting ourselves to k-fold connected plane domains G,[2] we shall pursue this latter approach to obtain a variety of mapping theorems stating that arbitrary[3] k-fold connected domains can be mapped conformally onto individuals of a great variety of specific classes $\mathfrak{N}$ of domains.

To establish such mapping theorems, two steps are necessary: First, with domains B of $\mathfrak{N}$ as parameter domains, we must prove the existence of a solution of variational problem IV, formulated in the preceding chapter. Secondly we must deduce the relation $\phi(w) = 0$ for the solution $\mathfrak{x}$ of this problem as a consequence of the vanishing of the first variation V. We postpone the first step to the end of this chapter; assuming that the variational problem is solved by the vector $\mathfrak{x}$ and the domain B, we concentrate on the proof of the relation $\phi(w) = 0$.

[1] See Douglas [2].

[2] According to Chapter II any k-fold connected Riemann domain is conformally equivalent to a domain in the plane; therefore the restriction to plane domains G in this chapter does not impair the generality of the results.

[3] For convenience we disregard degenerate domains G having isolated points as boundary components. Such boundary elements may simply be omitted from the reasoning, and the theorems are then equivalent to those for a smaller value of k.

2. *First Variation.* As stated in Chapter III, §4, 1, the expression (3.17)

$$V(\mathfrak{x}, \lambda, \mu) = \iint_B [p(\lambda_u - \mu_v) - q(\lambda_v + \mu_u)]\, du\, dv,$$

$$p + iq = (\mathfrak{x}_u^2 - \mathfrak{x}_v^2) - 2i\mathfrak{x}_u\mathfrak{x}_v = \phi(w)$$

for the first variation of $D[\mathfrak{x}]$ is valid not merely for simply connected domains B and for variations that transform B into itself, but more generally for any domain of the prescribed class $\mathfrak{N}$ and for any variation Λ, M satisfying conditions (3.14) and transforming B into another admissible domain B'.

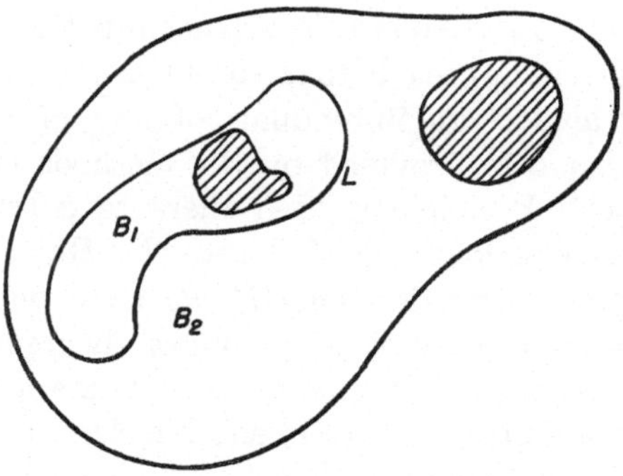

Figure 5.1

The expression (3.17) may be modified. Let B be divided into two parts, $B = B_1 + B_2$, by a piecewise smooth boundary line L in the interior of B. Suppose that $\Lambda + i\mathrm{M}$, and hence $\lambda + i\mu$, is an analytic function of w in B_1. Then the variation $w = w' + \epsilon(\Lambda + i\mathrm{M})$ maps the domain B_1 conformally onto a domain B_1'; therefore the corresponding part of Dirichlet's integral remains invariant, and we obtain V by considering B_2 alone.[4] Instead of (3.17) we have (3.23)

$$V(\mathfrak{x}, \lambda, \mu) = \iint_{B_2} [p(\lambda_u - \mu_v) - q(\lambda_v + \mu_u)]\, du\, dv.$$

Now we assume that in (3.17) $\mathfrak{x}$ is harmonic and consequently that $p + iq = \phi(w)$ is an analytic function. Under the further assumption

[4] For B_1 we may therefore drop the assumption that Λ, M have bounded first derivatives, but not for B_2.

that λ and μ vanish in a neighborhood of the boundary lines of B_2—except L—the relation (3.23) after integration by parts becomes

$$V(\mathfrak{x}, \lambda, \mu) = \int_L [\lambda(p\,dv + q\,du) + \mu(p\,du - q\,dv)]; \tag{5.1}$$

since p and q satisfy the Cauchy-Riemann equations the domain integral over B_2 vanishes and only the contour integral over L remains. The expression (5.1) can also be written in the form

$$V(\mathfrak{x}, \lambda, \mu) = \mathcal{I}m\left[\int_L (\lambda + i\mu)\phi(w)\,dw\right]. \tag{5.2}$$

If $\mathfrak{r}$ and B solve variational problem IV, V vanishes for all admissible variations of $\mathfrak{r}$ and B, as was seen in the preceding chapters. Our task is to deduce from the vanishing of the first variation the relation $\phi(w) = 0$ for various classes $\mathfrak{N}$. To this end we need not consider "arbitrary" variations but only a suitable subset of admissible variations depending on a finite number of parameters.[5]

2. *Conformal Mapping on Circular Domains*

1. *Statement of Theorem.*

Theorem 5.1: Any k-fold connected domain G can be mapped conformally on a "circular" domain B consisting of the whole plane with k circular disks removed[6] or on the half-plane $v > 0$ with $k - 1$ circular disks removed.

Corollary: One of the boundary circles β_ν of B, the antecedent on γ_ν of a fixed point on β_ν, and the center of a second boundary circle β_μ may be prescribed arbitrarily.

2. *Statement and Discussion of Variational Conditions.* In the following sections we assume variational problem IV solved by a circular domain B in the u,v-plane and a vector $\mathfrak{x}(u, v)$, postponing to §7 the proof that the solution exists for $k \geqq 2$. First we derive and exploit the variational conditions for the solution (independently of the number of dimensions of the $\mathfrak{x}$-space).

We denote by ω_ν the center of the boundary circle β_ν in the plane

[5] The relation $V = 0$ for all other admissible variations is then a consequence.

[6] As said before, the trivial exceptional case in which G has isolated boundary points will be disregarded in the following sections.

of $w = u + iv$. We assume that the point w^* on β_1 corresponds to a fixed point on γ_1; w^* will be called a fixed point. If B is a half-plane with circular disks removed, we choose the real axis $v = 0$ as the boundary β_1.

a) By variation of B into itself, i.e. of the representation of the boundary γ on β, we shall obtain the result: on the boundary circle β_ν in the w-plane the function $(w - \omega_\nu)^2\phi(w)$ is real and regular analytic except possibly for one pole of first order at the fixed point w^* on β_1.

If β_1 is a straight line, we shall find the variational condition: $\phi(w)$ is real on β_1 and regular analytic except possibly for a pole of first order at w^*. In this case, the function $\phi(w)$ vanishes to at least the fourth order at infinity. If, in particular, $w^* = \infty$, this zero merges with the pole at w^* to a zero of at least the third order.

b) By variation of the position of the center ω_ν of the circle β_ν we shall obtain the relation

$$\int_{\beta'_\nu} \phi(w)\, dw = 0$$

for every curve β'_ν homologous to β_ν in B.

c) By dilatation and rotation of β_ν we shall obtain the condition

$$\int_{\beta'_\nu} (w - \omega_\nu)\phi(w)\, dw = 0.$$

If β_ν is a straight line in the u,v-plane, say $v = 0$, conditions b) and c) assume the forms

$$\int_{\beta'_\nu} \phi(w)\, dw = 0, \qquad \int_{\beta'_\nu} w\, \phi(w)\, dw = 0.$$

The conditions a) for a straight line β_ν as boundary are immediately obtained from those for a circle. Consider, for example, the unit circle in the w-plane, which is transformed into the real ζ-axis by the function $w = (\zeta - i)/(\zeta + i)$. In the plane of the variable $\zeta = \xi + i\eta$ we must consider, instead of $\phi(w)$, the function

$$\chi(\zeta) = (\mathfrak{x}_\xi - i\mathfrak{x}_\eta)^2 = \phi(w)\left(\frac{dw}{d\zeta}\right)^2.$$

Since $dw/d\zeta = 2i/(\zeta + i)^2$, the function $(dw/d\zeta)^2$ has a zero of order 4 at $\zeta = \infty$; furthermore

$$w^2\phi(w) = \chi(\zeta)(\zeta^2 + 1)^2/4,$$

so that $\chi(\zeta)$ is real on the real axis, $w^2\phi(w)$ being real for $|w| = 1$.

Remarks: Condition a) for the exterior of a circle is immediately reduced to that for the interior by the transformation $\zeta = \alpha/(z - \omega_\nu)$.[7] Furthermore condition a) is invariant under conformal mapping of a circle onto itself. For example, if we map the unit circle onto itself by a transformation of the form $w = (\zeta - a)/(\bar{a}\zeta - 1)$, the fact that $w^2\phi(w)$ is real on $|w| = 1$ implies that $\zeta^2\chi(\zeta)$ is real on $|\zeta| = 1$ and vice versa. As a matter of fact a simple calculation gives, for $|\zeta| = |w| = 1$, the relation

$$w^2\phi(w) = \zeta^2\chi(\zeta)\frac{|a - \zeta|^4}{(|a|^2 - 1)^2},$$

from which condition a) obviously follows.

3. *Proof of Variational Conditions.* Conditions b) and c) follow immediately from $V = 0$ if we apply special variations λ, μ in the expression (5.2) for V. We assume B to be inside the circle β_1, say the unit circle, with polar coordinates r, θ. A translation of the circle in the direction of the u-axis is effected by an admissible variation Λ, M for which $\Lambda + i\mathrm{M} = \lambda + i\mu = 1$ in a neighborhood of β_1 bounded by a curve L homotopic to β_1, and $\Lambda + i\mathrm{M} = 0$ in a neighborhood of all the other boundary components. Since the expression $\lambda + i\mu$ is constant, therefore an analytic function of w, in the ring B_1 between L and β_1, we can apply (5.2) and obtain immediately

$$\mathscr{Im}\left[\int_L \phi(w)\,dw\right] = 0.$$

In the same way we obtain, for $\lambda + i\mu = i$ in B_1, the equation

$$\mathscr{Re}\left[\int_L \phi(w)\,dw\right] = 0;$$

[7] If β_ν is the unit circle, we have $\zeta = 1/w$ and $\zeta^2\chi(\zeta) = w^2\phi(w)$; the condition for the exterior of β_ν follows from the condition for the interior.

combining the two results, we have

$$\int_L \phi(w)\, dw = 0. \tag{5.3}$$

A dilatation of β_1 can be represented by the transformation $w' = (1 - \epsilon)w$, a dilatation combined with a rotation by the transformation $w' = (1 - i\epsilon)w$ in B_1, yielding

$$\lambda + i\mu = w \quad \text{or} \quad \lambda + i\mu = iw.$$

As above we deduce from (5.2) and $V = 0$ the variational condition

$$\int_L w\, \phi(w)\, dw = 0 \tag{5.4}$$

for a variation of the radius.

Since $\phi(w)$ is regular in B, (5.3) and (5.4) are equivalent to

$$\int_{\beta_1'} \phi(w)\, dw = 0, \tag{5.5}$$

and

$$\int_{\beta_1'} w\, \phi(w)\, dw = 0, \tag{5.6}$$

for any curve β_1' in B homologous to β_1. Condition (5.5) holds equally for the boundary circles β_ν of B, with centers ω_ν, $\nu > 1$. Instead of (5.6) we find

$$\int_{\beta_\nu'} (w - \omega_\nu)\phi(w)\, dw = 0. \tag{5.6a}$$

Combining (5.6a) with (5.5) we have again (5.6).[8]

To prove a) we must refine somewhat the corresponding argument of Chapter III, §4, for the case $k = 1$, where $\phi(w)$ is regular everywhere within the circle β. Without loss of generality we may assume that β_1 is the unit circle and contains all the other boundary circles β_ν. As in Chapter III, we write the relation $V = 0$ in the form

$$\iint_{B_t} [p(\lambda_u - \mu_v) - q(\lambda_v + \mu_u)]\, du\, dv \to 0 \tag{5.7}$$

[8] Conditions (5.5) and (5.6) are equivalent to the statement: The function $\phi(w)$ is the second derivative $\psi''(w)$ of an analytic single-valued function $\psi(w)$ in B.

as $t \to 1$, where B_t is the subdomain of B inside the circle of radius $r = t < 1$ about the origin. The limit is uniform for all variations λ, μ for which $|\lambda_u|, |\lambda_v|, |\mu_u|, |\mu_v|$ are uniformly bounded.

We choose a fixed value $r_0 < 1$ such that all boundary circles $\beta_2, \beta_3, \cdots, \beta_k$ are within the circle of radius $r = r_0$ and assume $t > r_0$. As in Chapter III we choose a function $\alpha(r, \theta)$ which vanishes for $r < r_0$ and for which $w' = w \exp\{i\epsilon\alpha(r, \theta)\}$ is an admissible variation of B into itself. Then (5.7) implies

$$\int_0^{2\pi} \alpha(t, \theta) H(t, \theta)\, d\theta \to 0 \tag{5.7a}$$

as $t \to 1$, where

$$H(r, \theta) = \mathcal{I}m\,[w^2 \phi(w)] = -2r\mathfrak{x}_r\mathfrak{x}_\theta \tag{5.8}$$

is harmonic in B.

We choose $h > 0$ in such a way that $1 - h > r_0$ and consider in B a point Q with polar coordinates ρ, η, where $\rho < 1 - h$. We define again

$$\alpha(r, \theta) = K(r, \theta; Q)P(r) = \frac{1}{2\pi} \frac{r^2 - \rho^2}{r^2 - 2r\rho\cos(\theta - \eta) + \rho^2} P(r),$$

where

$$P(r) = \begin{cases} 0, & r < 1 - h, \\ \dfrac{2}{h}(r - 1 + h), & 1 - h \leq r \leq 1 - \dfrac{h}{2}, \\ 1, & r > 1 - \dfrac{h}{2}. \end{cases}$$

Condition (5.7a) expresses the following fact: Let

$$H(Q; t) = H(\rho, \eta; t)$$

be the harmonic function, defined and regular in the disk $0 \leq \rho < t$, having for $\rho = t$ the same values as $H(Q)$ defined by (5.8). The function $H(Q; t)$ tends to zero for $t \to 1$, and this convergence is uniform for all points with $\rho < 1 - h$. (The uniformity of the convergence is a consequence of the fact that our choice of α implies the uniform boundedness of $|\lambda_u|, |\lambda_v|, |\mu_u|, |\mu_v|$.) Writing r, θ instead of ρ, η we consider the function $C(r, \theta; t) = H(r, \theta) - H(r, \theta; t)$

which is harmonic for $r_0 < r < t$, and vanishes for $r = t$, hence can by reflection be extended as a harmonic function to the region $r_0 < r < t^2/r_0$.

We proceed to show that $H(r, \theta)$ approaches zero for $r \to 1$. Denote by $A/2$ an upper bound for $|H(r_0, \theta)|$ and set $2d = 1 - r_0$. With an arbitrarily prescribed ϵ, choose h such that

$$h < \frac{d}{4}, \qquad h\frac{A}{d} < \frac{\epsilon}{2}$$

and restrict t to values in the range $1 - h/2 < t < 1$. Using the relation $H(\rho, \eta; t) \to 0$ for $t \to 1$, we keep t large enough to ensure $|H(r_0, \theta; t)| < A/2$ uniformly in θ and conclude that $|C(r_0, \theta; t)| = |C(t^2/r_0, \theta; t)| < A$. As a consequence, we have, in the whole ring $t^2/r_0 > r > r_0$, the inequality $|C(r, \theta; t)| < A$. For the derivative C_r we have, in the smaller ring $t \geq r \geq 1 - h$, the inequality $|C_r| <$ A/d since all points of the ring are centers of disks of radius d in

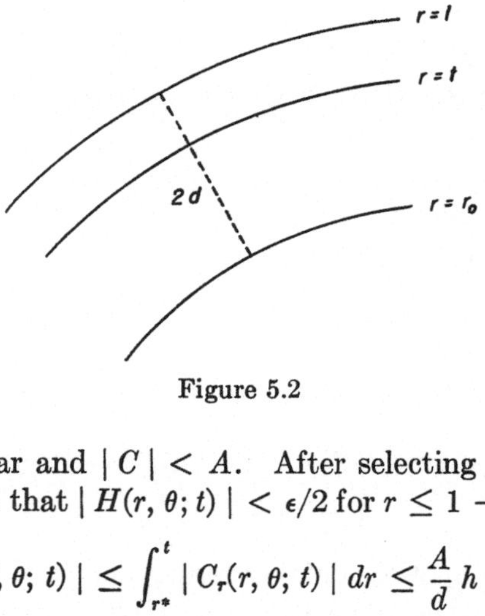

Figure 5.2

which C is regular and $|C| < A$. After selecting h we are free to choose t so near 1 that $|H(r, \theta; t)| < \epsilon/2$ for $r \leq 1 - h = r^*$. Then

$$|C(r^*, \theta; t)| \leq \int_{r^*}^{t} |C_r(r, \theta; t)| \, dr \leq \frac{A}{d} h < \frac{\epsilon}{2};$$

hence $|H(r^*, \theta)| \leq |C(r^*, \theta; t)| + |H(r^*, \theta; t)| < 2(\epsilon/2) = \epsilon$. Therefore for given ϵ we have $|H(r, \theta)| < \epsilon$ if $1 - r = h$ is chosen sufficiently small; condition a) is established, provided that no restrictions at the boundary are imposed on the variation. A consequence of the vanishing of $H(r, \theta)$ for $r \to 1$ is that H, and therefore $w^2\phi(w)$ and $\phi(w)$, can be analytically extended beyond the boundary β.

If a point F: $r = 1$, $\theta = \theta_0$, on β_1 is fixed in the sense that F has a prescribed image on γ, we are no longer permitted to choose $\alpha(r, \theta)$ as above since now $\alpha(1, \theta_0)$ must be zero. An admissible variation in this case is provided by

$$\alpha(r, \theta) = [K(r, \theta; Q) - K(r, \theta_0; Q)]P(r).$$

Instead of $H(Q; t) \to 0$ we obtain the relation

$$H(Q; t) - c\, K(t, \theta_0; Q) \to 0$$

where $c/2\pi$ is the mean value of $H(r, \theta)$ on $r = t$, or on any concentric circle. Substituting in the preceding reasoning

$$C(r, \theta; t) = H(r, \theta) - H(r, \theta; t) - c\, K(t, \theta_0; r, \theta),$$

we find by the same argument that

$$H(r, \theta) - c\, K(1, \theta_0; r, \theta)$$

is a regular harmonic function for $r_0 < r < 1$ which tends uniformly to zero for $r \to 1$. Since $K(1, \theta_0; r, \theta)$ is the imaginary part of $(i/2\pi)(e^{i\theta_0} + w)/(e^{i\theta_0} - w)$, an analytic function of w which is regular except for a pole at $w = e^{i\theta_0}$ and real for $|w| = 1$, statement a) concerning the function $w^2\, \phi(w)$ is proved.

4. *Proof that* $\phi(w) = 0$. To complete the proof of the mapping theorem we have to deduce the relation $\phi(w) = 0$ from the preceding variational conditions. The function $\phi(w)$, being regular in $B + \beta$, is either identically zero or has only a finite number of zeros in $B + \beta$. By counting these zeros in B and on β we shall exclude the second alternative.[9]

First we prove that on each freely variable boundary circle[10] β_ν the function $(w - \omega_\nu)^2\phi(w)$ has at least four zeros. Assuming again that β_ν is the unit circle we have $w = e^{i\theta}$, $dw = iw\, d\theta$, and

$$w^2\, \phi(w) = f(\theta)$$

on β_ν, where $f(\theta)$ is a real, periodic function of the angle θ for which, by (5.5) and (5.6),

[9] This method was suggested to the author by Hans Lewy.

[10] According to footnote 3 p. 167 we assume that none of these circles degenerates to a point.

$$\int_0^{2\pi} f(\theta)\, d\theta = 0 \tag{5.9}$$

and

$$\int_0^{2\pi} f(\theta) \cos\theta\, d\theta = 0, \qquad \int_0^{2\pi} f(\theta) \sin\theta\, d\theta = 0. \tag{5.10}$$

Equation (5.9) shows that there are at least two arcs on β_ν, separated by zeros of $f(\theta)$ where $f(\theta) > 0$ and $f(\theta) < 0$, respectively. If there were only two changes of sign of $f(\theta)$ on β_ν, we could suppose them to occur at points $\theta = \theta_0$ and $\theta = -\theta_0$, respectively. As a consequence the function $(\cos\theta - \cos\theta_0) f(\theta)$ would have the same sign for all values of θ, an impossibility since formulas (5.9) and (5.10) imply

$$\int_0^{2\pi} (\cos\theta - \cos\theta_0) f(\theta)\, d\theta = 0.$$

The number of changes of sign, i.e. of zeros of $f(\theta)$, on β_ν is therefore at least four, since it must be even. This result immediately extends to the other boundary circles β_ν of B, if the function $(w - \omega_\nu)^2 \phi(w)$ is considered instead of $w^2 \phi(w)$.

For a boundary circle β_ν whose center is fixed, only variational condition c) or (5.9) is established. Consequently we can prove only that $f(\theta)$ or $\phi(w)$ has two zeros. If a boundary circle β_ν is fixed, neither (5.9) nor (5.10) is proved, since these conditions follow from the variation of the boundaries. Moreover, if a point F: $w = w^*$ on β_ν is fixed, the function $(w - \omega_\nu)^2 \phi(w)$ may have a pole of first order at that point. Since the function is real on β_ν it must vary from $+\infty$ to $-\infty$ as w describes the circle starting from w^*, so that $f(\theta)$ has at least one zero on the circle. Consequently the number of zeros minus the number of poles on the circle is not negative.

For symmetry we assume that B is the domain outside all the k circles β_ν. Let β_1 be fixed, F a fixed point on β_1 (i.e. having a fixed image on γ), and let the center of β_2 be fixed. The function $\phi(w)$, regular in the infinite domain B, has a zero of at least the fourth order at $w = \infty$.[11] For the total number N of zeros of $\phi(w)$ in B

[11] The function $\phi(w)$ can be written in the form

$$\phi(w) = \sum_{\mu=1}^{m} f'_\mu(w)^2 = \sum_{\mu=1}^{m} \left(\frac{\partial x_\mu}{\partial u} - i\,\frac{\partial x_\mu}{\partial v}\right)^2$$

(see Chapter III, §1), where $f_\mu(w)$ is regular at infinity and $f'_\mu(w)$ has therefore a zero of at least the second order at infinity.

we have therefore

$$4 \leq N = \frac{1}{2\pi i} \sum \int_{\beta'_\nu} d \log \phi(w)$$

where β'_ν denotes a path of integration coincident with β_ν except for arbitrarily small semicircular detours about the zeros and possibly the pole of $\phi(w)$ on β_1 (see Figure 5.3). The contribution to the integral from a small semicircle about a zero or pole of order r is equal to $-\pi ir$ or πir, respectively, with arbitrarily close approximation for sufficiently small radii of the semicircles. Let N_ν and P_ν denote the

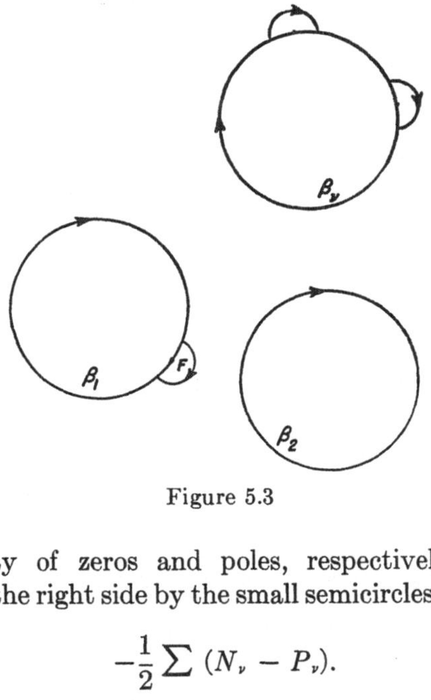

Figure 5.3

total multiplicity of zeros and poles, respectively, on β_ν. The contribution on the right side by the small semicircles is approximately

$$-\frac{1}{2} \sum (N_\nu - P_\nu).$$

The contribution to $(1/2\pi i) \int_{\beta'_\nu} d \log \phi(w)$ by the path of integration along β_ν is given by the expression

$$\frac{1}{2\pi i} \left\{ \int d \log [(w - \omega_\nu)^2 \phi(w)] - \int d \log (w - \omega_\nu)^2 \right\}.$$

Since $(w - \omega_\nu)^2 \phi(w)$ is real on β_ν, the contribution of the first integral is zero, while the contribution of the second integral on each circle is (approximately) two. Combining all these facts and realizing that only integers N, N_ν, P_ν are involved, we find

$$4 \leq N = -\frac{1}{2}\sum (N_\nu - P_\nu) + 2k.$$

Since $N_1 - P_1 \geq 0$, $P_2 = P_3 = \cdots = P_k = 0$, $N_2 \geq 2$, and $N_\nu \geq 4$ for $\nu > 2$, we obtain

$$4 \leq N \leq 3.$$

This result is absurd and the only alternative left is that $\phi(w)$ vanishes identically; theorem 5.1 is proved.

In the case of a finite domain B inside a circle β_1 or of an infinite half-plane domain where β_1 is a straight line, the reasoning remains practically unchanged and need not be repeated explicitly.

3. *Mapping Theorems for a General Class of Normal Domains*

1. *Formulation of Theorem.* The methods and results of the preceding section can be generalized to other classes $\mathfrak{N}$ of normal domains; a very general mapping theorem results, involving $3k - 6$ parameters ($k > 2$).

We consider plane domains B with boundaries $\beta_1, \beta_2, \cdots, \beta_k$. For β_1 we may choose a straight line or an arbitrary fixed closed Jordan curve; one fixed point F on β_1 corresponds to a given point P on γ_1. The boundary β_2 is an arbitrary Jordan curve, starshaped from a fixed center σ and permitted to vary by expansion or contraction from this center. The curves $\beta_3, \beta_4, \cdots, \beta_k$ are convex, each of arbitrarily prescribed shape, and each permitted to vary by expansion or contraction from an arbitrary center, including the limiting case of parallel displacement. Leaving aside other possible variants, we deal with the following two classes of parameter domains B:
a) Domains $\mathfrak{N}_a$; β_1 is closed, B is inside β_1 and outside $\beta_2, \beta_3, \cdots, \beta_k$, and these curves have no points in common.
b) Domains $\mathfrak{N}_b$; β_1 is a straight line and B is outside $\beta_2, \beta_3, \cdots, \beta_k$.
We state:

Theorem 5.2: Any k-fold connected domain G can be mapped conformally onto a domain B of class $\mathfrak{N}$, if $\mathfrak{N}$ is either $\mathfrak{N}_a$ or $\mathfrak{N}_b$, in such a way that F and P are corresponding points.

First we shall assume that our curves β_ν are analytic; later in §6, 2, however, we employ a simple passage to limit to extend the result to non-analytic boundaries, e.g. to boundary curves with vertices, or to straight slits each of which has a prescribed direction.

The proof of the theorem again proceeds from the assumption (to be verified in §7) that variational problem IV for our classes $\mathfrak{N}$ of parameter domains is solved by a domain B and a vector $\mathfrak{x}$. Our task is merely to prove the identity $\phi(w) = 0$ as a consequence of the condition $V = 0$.

2. *Variational Conditions.* The evaluation of $V = 0$ is easily reduced to the special case of §2, 3. As a preliminary step we consider a single boundary curve β_ν. By a function $w = w(\zeta)$ we map the whole w-plane outside β_ν onto the exterior of the unit circle in the plane of an auxiliary complex variable $\zeta = re^{i\theta}$, in such a way that $\zeta = \infty$ and $w = \infty$ correspond to each other. Because of the supposed analytic character of β_ν, the function $w(\zeta)$ is analytic in the domain $|\zeta| \geq 1$ and its derivative $dw/d\zeta$ is bounded there; likewise the inverse of w is analytic on and beyond β_ν and $dw/d\zeta$ cannot vanish there. By this device the variational condition corresponding to the variations of β_ν into itself is reduced to the condition corresponding to the variation of the unit circle $|\zeta| = 1$ into itself.

According to §2, 2 this condition implies that $\zeta^2\phi(w)(dw/d\zeta)^2 = \zeta^2\chi(\zeta)$ is real, and hence regular, on $|\zeta| = 1$. The function $\chi(\zeta)$ is regular on $|\zeta| = 1$; therefore $\phi(w) = p + iq$ is regular on β_ν. Since a fixed point F on β_1 corresponds to a fixed point P on γ_1, $\phi(w)$ may again have a pole of the first order at F; by the same argument as in §2 it must have at least one zero on β_1.

Knowing that p and q are regular on β_ν, except possibly at F on β_1, we may directly exploit the condition

$$V = \iint_B [p(\lambda_u - \mu_v) - q(\lambda_v + \mu_u)]\, du\, dv = 0$$

by assuming $\lambda = \mu = 0$ on β_κ for $\kappa \neq \nu$. Observing that $p_u - q_v = p_v + q_u = 0$ we obtain after integration by parts

$$\int_{\beta_\nu} [(p\lambda - q\mu)\, dv + (p\mu + q\lambda)\, du] = 0.$$

Equivalently, considering θ as a parameter on β_ν and denoting differentiation with respect to θ by a prime, we have

$$\int_0^{2\pi} [\lambda(pv' + qu') + \mu(pu' - qv')]\, d\theta = 0.$$

On β_ν the variations λ, μ are the components of a displacement vector $\mathfrak{v}$, and the variations of the boundary curve β_ν into itself correspond to tangential vectors (λ, μ), i.e. to $\lambda = Au'$, $\mu = Av'$ where A is an arbitrary (piecewise smooth) function of θ on β_ν. A slight modification of the argument in §2 establishes the following conditions:

a) The variation of β_ν into itself leads to the condition that the vector $\mathfrak{w}$ with the components $pv' + qu'$ and $pu' - qv'$ is everywhere orthogonal to the curve β_ν. An equivalent statement is that $\zeta^2\chi(\zeta)$ is real on β_ν.

b) Variation of β_ν by a similarity transformation from the origin corresponds to $\lambda = u$, $\mu = v$ and leads to the condition

$$\int_0^{2\pi} \mathfrak{r}\mathfrak{w}\, d\theta = 0,$$

where $\mathfrak{r}$ is the vector with the components u, v.

c) Variation by parallel displacement of β_ν corresponds to arbitrary constant values of λ and μ on β_ν and leads to the condition

$$\int_0^{2\pi} \mathfrak{w}\mathfrak{a}\, d\theta = 0,$$

where $\mathfrak{a}$ is an arbitrary constant vector with components a, b. Conditions b) and c) together yield the relation

$$\int_0^{2\pi} \mathfrak{w}(\mathfrak{r} - \mathfrak{a})d\theta = 0, \tag{5.11}$$

which can also be obtained directly by similarity transformation of β_ν from the center at $u = a$, $v = b$.

At any point, $\mathfrak{w} = 0$ implies $\phi(w) = p + iq = 0$ since $u'^2 + v'^2 \neq 0$. As in §2, either the function $\phi(w)$ is identically zero in B or it has only a finite number N of zeros in B and a finite number N_ν of zeros on β_ν.

3. *Proof that* $\phi(w) = 0$. To show that $\phi(w)$ cannot have a finite number N of zeros, we proceed as in §2. Consider the continuous

vector $\mathfrak{w}$, normal to β_2 by a). Since β_2 is assumed starshaped from the origin, the inner product of the vector $\mathfrak{r}$ with $\mathfrak{w}$ does not change sign along β_2 unless $\mathfrak{w}$ passes through the value zero and changes direction. The vanishing of the integral of $\mathfrak{r}\mathfrak{w}$, assured by b), thus implies at least two zeros of $\mathfrak{w}$, and hence of $\phi(w)$, on β_2.

The same reasoning holds for the remaining curves β_ν. However, condition c) ensures the existence of two additional zeros of $\phi(w)$ on β_ν. Let A_1 and A_2 be the two zeros on β_ν required by condition b), Z the point of intersection of the tangents to β_ν drawn from A_1 and A_2, and denote by $\mathfrak{a}$ the vector from O to Z. If the two tangents are parallel, denote by $\mathfrak{a}$ an arbitrary parallel vector. From (5.11) we infer that the vector $\mathfrak{w}$ vanishes in at least two points on β_ν other

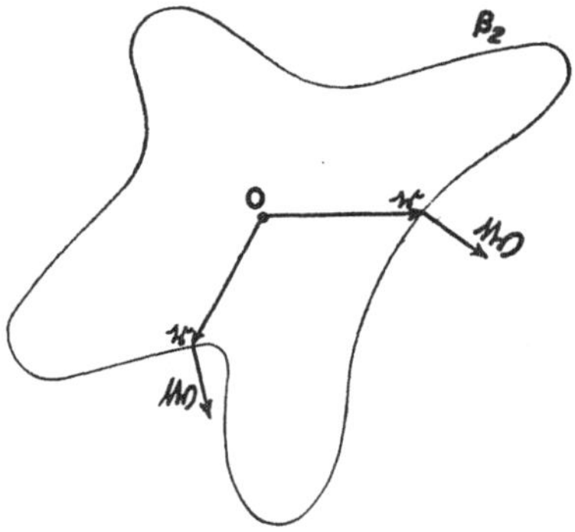

Figure 5.4

than A_1 and A_2. Otherwise $\mathfrak{w}$ would change from the inward normal direction to the outward normal direction along β_ν in A_1 and A_2 alone. The vector $\mathfrak{r} - \mathfrak{a}$ connects Z with a point traversing β_ν. Since β_ν is assumed convex for $\nu > 2$ the product $\mathfrak{w}(\mathfrak{r} - \mathfrak{a})$ would have no change of sign along the whole curve. This contradiction to (5.11) shows that $\phi(w)$ has at least four zeros on β_ν.

From this point on the proof that $\phi(w) = 0$ proceeds almost exactly as in §2. As before we assume that $\phi(w)$ has only a finite number $N \geq 0$ of zeros in B and denote by N_ν the number of zeros on β_ν. Again

$$N = \frac{1}{2\pi i} \sum_\nu \int_{\beta'_\nu} d \log \phi(w)$$

where β_ν' is identical with β_ν except for small semicircular detours in B circumventing the zeros on β_ν and the pole at F on β_1; the paths of integration leave B on the left side. Let us assume that β_1 is the real axis, F a finite point on β_1. Since $\phi(w)$ is real on β_1 with a pole of order $P_1 \leq 1$ at F and a zero of not less than fourth order at $w = \infty$, there must be at least one more zero of $\phi(w)$ on β_1; otherwise $\phi(w)$ could not approach the value $+\infty$ on the one side, the value $-\infty$

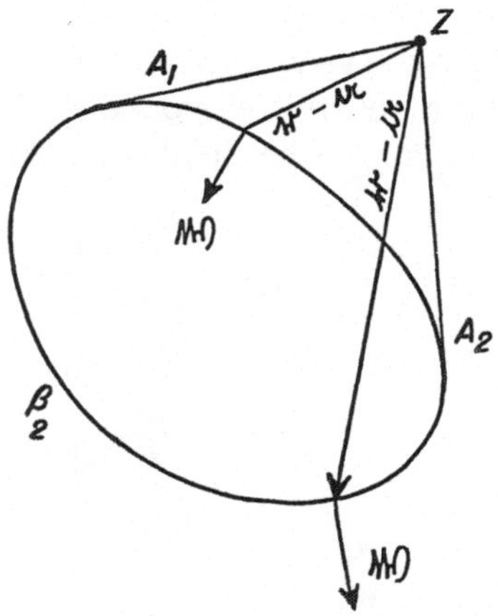

Figure 5.5

on the other side of F. We conclude for β_1 that $N_1 - P_1 \geq 4$ and therefore

$$\frac{1}{2\pi i}\int_{\beta_1'} d \log \phi(w) \leq -2.$$

Employing the function $w(\zeta)$ of article 2 and the relation

$$\chi(\zeta) = \phi(w)\left(\frac{dw}{d\zeta}\right)^2,$$

we write for β_2

$$\frac{1}{2\pi i}\int_{\beta_2'} d \log \phi(w) = \frac{1}{2\pi i}\int_{\beta_2'} d \log \chi(\zeta) - \frac{2}{2\pi i}\int_{\beta_2'} d \log w'(\zeta),$$

where the unit circle in the ζ-plane is taken to correspond to β_2. Since $w'(\zeta)$ is regular for $|\zeta| \geq 1$ and has no zeros, the second integral vanishes; the first, written in the form

$$\frac{1}{2\pi i}\int_{\beta_2'} d\log \zeta^2\chi(\zeta) - \frac{1}{2\pi i}\int_{\beta_2'} d\log \zeta^2,$$

yields, as before in §2, the value $2 - \frac{1}{2}N_2$. Since $N_2 \geq 2$, we have

$$\frac{1}{2\pi i}\int_{\beta_2'} d\log \phi(w) \leq 1.$$

In the same way we obtain from the condition $N_\nu \geq 4$ for $\nu > 2$,

$$\frac{1}{2\pi i}\int_{\beta_\nu'} d\log \phi(w) \leq 0.$$

Adding the results, we find

$$N \leq -1,$$

an absurdity which leaves only the alternative that $\phi(w)$ is identically zero. The proof of theorem 5.2 is complete.

For the other variants of the theorem the same reasoning holds with obvious modifications which need not be described in detail. That the boundary curves β_ν are analytic was essential for the proof. Using analytic boundaries as approximations, we can easily extend the result to non-analytic boundaries (see §6, 2).

It may be stated that the theorems of the preceding section can be considerably extended. Instead of considering variations of the boundary curve β_ν by similarity transformations we may permit β_ν to vary within a more general prescribed set of curves.

4. *Conformal Mapping on Riemann Surfaces Bounded by Unit Circles*

1. *Formulation of Theorem.*

Theorem 5.3: Every plane[12] k-fold connected domain G having no isolated boundary points can be mapped conformally onto a Riemann surface B consisting of k identical disks, e.g. interiors of unit circles, connected by branch points[13] of total multiplicity $2k - 2$. Moreover,

[12] As said before, in view of the general result of Chapter II the assumption that G is a plane domain is not an essential restriction.

[13] The conformality of the mapping is of course interrupted at the branch points.

an arbitrarily fixed point F_ν on each boundary circle β_ν can be made to correspond to a fixed boundary point P_ν on the boundary continuum γ_ν of G, and the position of one simple branch point in B may be prescribed. The class $\mathfrak{N}$ of these domains depends on $3k - 6$ real parameters: the $2k - 3$ freely variable branch points represent $4k - 6$ coordinates, while fixing the points F_ν reduces the number of parameters by k.

In §7 we shall prove existence of a solution of the corresponding variational problem IV by a vector $\mathfrak{x}(u,v)$ in a domain B of the class $\mathfrak{N}$. Anticipating this fact we need only deduce $\phi(w) = 0$ from $V = 0$, by a procedure similar to that employed in §2 and §3.

2. *Variational Conditions. Variation of Branchpoints.* Repeating the proof of §2, 4 we obtain the same conclusion. The analytic function

$$\psi(w) = w^2 \phi(w)$$

is regular in the neighborhood of each unit circle β_ν, except possibly for a pole of the first order at F_ν, and ψ is real on β_ν.

Further variational conditions are obtained by variations of branch points, the only possible type of variation of the domain. Let P be a branch point which is not fixed for the class $\mathfrak{N}$, say the point $w = 0$, and let D be a neighborhood of P in B containing no other branch point. Consider a subdomain B_1 of D containing P and bounded by a curve L, closed in B, e.g. by a multiple circle enclosing P. We apply the general expression (3.23) for V, choosing λ and μ as zero outside D and $\lambda + i\mu$ as an analytic function of w in B_1; then formula (5.2) holds. In particular, we choose $\lambda + i\mu = 1$ or $\lambda + i\mu = i$ in B_1. As before, we can deduce the variational condition (5.3)

$$\int_L \phi(w)\, dw = 0,$$

for any closed curve L enclosing no other branch point than P.

In the case of a simple branch point, (5.3) is the only variational condition. If P is a branch point of higher order—say of order j—we must supplement condition (5.3) by others corresponding to a resolution of P into branch points of lower order. In a small simply connected neighborhood B_1 of P, bounded by L, such a resolution is effected by an analytic transformation of the form

(5.12) $$w' = w + \epsilon w^{\nu/(j+1)}$$

or

(5.13) $$w' = w + i\epsilon w^{\nu/(j+1)},$$

where ν may range from 0 to $j - 1$ and ϵ is a parameter. To prove this fact we first use the transformation $w = \sigma^{j+1}$ to map the neighborhood B_1 of P on a simple neighborhood of $\sigma = 0$ in a σ-plane; transformation (5.12) takes the form $w' = \sigma^{j+1} + \epsilon\sigma^{\nu}$. The branch points of the transformed neighborhood B_1' over the w'-plane correspond to the zeros of

$$\frac{dw'}{d\sigma} = \sigma^{\nu-1}[(j + 1)\sigma^{j+1-\nu} + \nu\epsilon].$$

By way of example we see that (5.12) resolves the branch point $w = 0$ into $j + 1 - \nu$ simple branch points and one $(\nu - 1)$-fold branch point at $w' = 0$.

The variation (5.12) corresponds to $\lambda + i\mu = w^{\nu/(j+1)}$ in B_1, and (5.13) similarly corresponds to $\lambda + i\mu = iw^{\nu/(j+1)}$ in B_1. In either case we define $\lambda + i\mu = 0$ outside a wider simply connected neighborhood D about P. In the strip $D - B_1$ the expression $\lambda + i\mu$ can be defined arbitrarily, with the sole restriction that $\lambda + i\mu$ is piecewise smooth in B. Then formula (5.2) and the condition $V = 0$ immediately yield

$$\mathcal{R}e\left[\int_L w^{\nu/(j+1)}\phi(w)\,dw\right] = 0, \qquad \mathcal{I}m\left[\int_L w^{\nu/(j+1)}\phi(w)\,dw\right] = 0,$$

or

(5.14) $$\int_L w^{\nu/(j+1)}\phi(w)\,dw = 0, \qquad \nu = 0, 1, \cdots, j - 1.$$

Conditions (5.14) yield

Lemma 5.1: At a branch point of order j, the function $\phi(w)$ has a pole at most of order j.

Proof: The function $\phi(w)$ is of the form $\phi(w) = \sum_{\mu=1}^{m} f_\mu'^2(w)$, where

$\mathcal{R}e\,[f_\mu(w)] = x_\mu$. The transformation $\sigma = w^{1/(j+1)}$ maps the neighborhood of the branch point $w = 0$ onto a simple neighborhood of $\sigma = 0$. Since the functions $f_\mu(w)$ have bounded real parts in the neighborhood of $w = 0$, they are regular functions of σ, and so is $A(\sigma) = \sum_{\mu=1}^{m} (df_\mu/d\sigma)^2$. Consequently,

$$\chi(\sigma) = \phi(w) = \left(\frac{d\sigma}{dw}\right)^2 \sum_{\mu=1}^{m} \left(\frac{df_\mu}{d\sigma}\right)^2 = \frac{1}{(j+1)^2\sigma^{2j}}\,A(\sigma),$$

where $A(\sigma)$ is regular for $\sigma = 0$. Thus $\phi(w)$ appears to have a pole of order $2j$ at $\sigma = 0$. Consider, however, conditions (5.14); expressed with reference to the σ-plane, they become

$$(5.15) \qquad \int_{L'} \sigma^{\nu+j}\phi\,d\sigma = \int_{L'} \sigma^{\nu+j}\chi(\sigma)\,d\sigma = 0, \quad \nu = 0, 1, \cdots, j-1,$$

where L' is a simple closed curve about the pole of $\chi(\sigma)$ at $\sigma = 0$. Expanding $\chi(\sigma)$ in powers of σ,

$$\phi(w) = \chi(\sigma) = \cdots + \frac{a_1}{\sigma} + \cdots + \frac{a_j}{\sigma^j} + \frac{b_1}{\sigma^{j+1}} + \cdots + \frac{b_j}{\sigma^{2j}};$$

we see by (5.15) that all the coefficients b_ν vanish, and $\phi(w)$ has a pole at most of order j at $\sigma = 0$, as stated.

3. *Proof that* $\phi(w) = 0$. The relation $\phi(w) = 0$ is proved by the same method employed before. Unless $\phi(w)$ and hence $\psi(w) = w^2\phi(w)$ vanish identically, the function $\psi(w)$ has a finite number N of zeros in B, and, since it is regular in the neighborhood of the boundary circles, a finite number N_ν of zeros on β_ν. At the fixed point F_ν there is a pole of multiplicity $P_\nu = 1$ or 0. As before, $P_\nu = 1$ implies $N_\nu \geq 1$, so that $N_\nu - P_\nu \geq 0$ for every boundary circle β_ν.

Because of the factor w^2 the total multiplicity N of zeros of $\psi(w)$ in B is at least $2k$. According to the preceding article, the total multiplicity of the poles of $\psi(w)$ at the variable branch points is not greater than $2k - 3$. The pole at the fixed branch point may be of second order, so that the total multiplicity P of poles of $\psi(w)$ is at most $2k - 1$. Combining these results we have

$$N - P \geq 1,$$

a formula which obviously remains true if zeros and poles merge.

On the other hand

$$1 \leq N - P = \frac{1}{2\pi i} \sum_{\nu} \int_{\beta'_\nu} d \log \psi(w),$$

where again β'_ν denotes the circle β_ν described in the negative sense and modified by small semicircular detours in B about the poles and zeros of $\psi(w)$ on β_ν. As before

$$\frac{1}{2\pi i} \int_{\beta'_\nu} d \log \psi \leq 0$$

since $N_\nu - P_\nu \geq 0$ and since $\psi(w)$ is real on β_ν. We are led to the absurdity $1 \leq N - P \leq 0$; this contradiction proves that $\psi(w)$ and consequently $\phi(w)$ vanish identically. The proof of theorem 5.3 is complete.[14]

The preceding method can be used to establish a similar theorem for the more general domains obtained by adjoining to the k-fold connected plane domain G a given number of full planes, each plane being attached by two branch points.

5. *Uniqueness Theorems*

Theorem 5.4: For a given domain G the conformal mapping described in theorems 5.1–5.3 is uniquely determined.

1. *Method of Uniqueness Proof.*[15] Suppose a domain G is mapped on two domains B and B^* of the class $\mathfrak{N}$ under consideration; this assumption would imply a conformal mapping of B onto B^* by an analytic function $w^* = f(w)$. The uniqueness theorem is tantamount to the statement that $w^* = w$, or

$$g(w) = f(w) - w \equiv 0.$$

The proof of this identity will be like that used for the relation $\phi(w) = 0$. We assume that $g(w)$ is not identically zero, consequently has a finite number N of zeros and P of poles in B. An investigation of the number $N - P$ will lead to a contradiction. Since the respec-

[14] The theorem is due to Riemann. For different proofs see Bieberbach [1] and Grunsky [1].

[15] The method was first proposed by Carleman and later found independently by Shiffman, see Carleman [2] and Shiffman [6].

tive boundaries β and β^* of B and B^* may be assumed to be analytic, the function $g(w)$ is analytic on β and we have

$$N - P = \frac{1}{2\pi i} \sum_{\nu=1}^{k} \int_{\beta'_\nu} d \log g(w) = \sum_{\nu=1}^{k} I_\nu \tag{5.16}$$

where β'_ν coincides with β_ν except for small semicircular detours in B about the zeros and poles of $g(w)$ on β_ν; the sense of integration is such that B remains to the left.

2. *Uniqueness for Riemann Surfaces with Branch Points.* First we consider the k-fold unit circles of §4. We suppose that we have normalized the domains of this class by fixing one branch point and prescribing, for one point F_ν on each boundary circle β_ν, the corresponding point P_ν on γ_ν. At the fixed branch point we have $f(w) = w$, i.e. $g(w) = 0$, so that $N \geq 1$; it is immediately obvious that $P = 0$.

To evaluate the integral in (5.16) we observe that the complex number $g(w) = f(w) - w$ is represented geometrically as a vector with initial point at w and terminal point at $w^* = f(w)$. The integral

$$I_\nu = \frac{1}{2\pi i} \int_{\beta'_\nu} d \log g(w)$$

gives the number of full rotations of this vector as the point w describes β'_ν. If $g(w)$ had no zeros on β_ν, the path of integration would be $\beta'_\nu = \beta_\nu$, and the end point of the vector would always be on β_ν, so that $I_\nu = +1$; but we know that $g(w) = 0$ at the point F_ν of β_ν. The circumvention of a zero of n-th order contributes $-n/2$ to I_ν; consequently $I_\nu \leq 1 - \frac{1}{2}$, and since I_ν must be an integer we have $I_\nu \leq 0$. From (5.16) we obtain the inequality $N \leq 0$, contrary to the assumption $N \geq 1$. It follows that $f(w) \equiv w$; the uniqueness proof is complete.

3. *Uniqueness for Classes $\mathfrak{N}$ of Plane Domains.* We next prove the uniqueness theorem for the class $\mathfrak{N}_a$ described in §3. For the number of poles we have again

$$P = 0. \tag{5.17}$$

The contribution I_1 of β_1 to the right member of (5.16) can be discussed exactly as in article 2; accordingly

$$I_1 \leq 0. \tag{5.18}$$

Since β_2^* is obtained from β_2 by a radial transformation relative to an interior point, and since β_2 is star-shaped relative to this point, β_2 and β_2^* have no common points unless they coincide. If they coincide, we again reason as in the preceding article (except that the sense of integration is now different) and conclude that $I_2 = -1 - n/2$, where n is the total multiplicity of the zeros of $g(w)$ on β_2. If β_2^* surrounds β_2, we obviously have $I_2 = -1$, and the same is true if β_2^* is interior to β_2 : in all cases

$$(5.19) \qquad I_2 \leq -1.$$

Consider I_ν for $\nu > 2$. Each of the three situations enumerated for I_2 may arise, yielding $I_\nu \leq -1$; but three additional situations are possible for β_ν^* with respect to β_ν :

1) β_ν^* and its interior lie completely outside β_ν; in this case $I_\nu = 0$.

2) β_ν^* is tangent to β_ν. The function $g(w)$ either has no zero at all on β_ν, or it has a zero at the point of contact of β_ν and β_ν^*. If $g(w)$ has no zero we displace β_ν^* slightly so as not to intersect β_ν. In this displacement the total angle of rotation of the vector attached to the points of β_ν is changed continuously. Since it must always be a multiple of 2π it cannot change at all: in this case $I_\nu = 0$ or -1, according as the tangency is external or internal.

If $g(w)$ vanishes at the point of contact, the zero contributes at most $-\frac{1}{2}$ to I_ν. Together with the contribution of the remaining arc of β_ν, we have $I_\nu \leq -\frac{1}{2}$ or $I_\nu \leq -\frac{3}{2}$, according as the tangency is external or internal. These inequalities can be replaced by $I_\nu \leq -1$ or $I_\nu \leq -2$, respectively, since I_ν is an integer.

3) β_ν intersects β_ν^*. Then, as will be proved in the next paragraph, there are only two points of intersection by virtue of the convex and homothetic character of β_ν and β_ν^*. In this case we again have $I_\nu \leq 0$, as can be seen geometrically. In the self-explanatory diagram we distinguish three possibilities; the letters A and B refer to the points on β_ν in which β_ν intersects β_ν^*, while A^*, B^* are the corresponding points on β_ν^*. (The cases where $A = A^*$ or $B = B^*$ can be reduced to those of the diagram by a slight circumvention as previously.)

To prove that β_ν and β_ν^* have only two points in common, assume that they intersect in the points A, B, C, considered as points of β_ν. Three points A^*, B^*, C^* on β_ν^* correspond to these points, by the homothetic relation between β_ν and β_ν^*. The lines AA^*, BB^*, CC^* are concurrent at a point P. One of the lines PA, PB, PC, say PA, passes through the interior of the triangle $A^*B^*C^*$ and intersects B^*C^* at a point Q^* between B^* and C^*; therefore, PA intersects BC

at a point Q between B and C. From the homothetic relation between the triangles ABC and $A^*B^*C^*$ it follows that one of the two points Q and Q^*, say Q^*, does not lie between A and A^*. Being an interior point of β_ν^*, Q lies between two boundary points of β_ν^* on the line PA. Hence there exists a third point A^{**} of β_ν^* on PA in addition to A and A^*. Since β_ν^* is convex, there can be no such point; β_ν and β_ν^* have only two points of intersection.

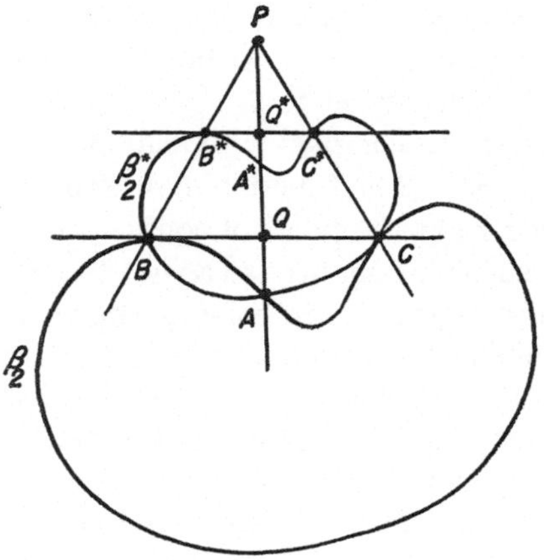

Figure 5.6

Collecting our results for $\nu > 2$, we have

$$I_\nu \leq 0, \qquad \nu = 3, 4, \cdots, k. \tag{5.20}$$

Substituting the relations (5.17–5.20) in equation (5.16), we obtain

$$N \leq -1,$$

in contradiction to $N \geq 0$.

4. *Uniqueness for Other Classes of Domains.* The same method yields a uniqueness proof for the other classes of domains considered in the previous sections. For domains B containing the point at infinity only a slight modification is necessary: we now have $P \leq 2$ instead of $P = 0$. The function $f(w)$ may take a finite point w_0 into $w^* = \infty$; then $g(w) = f(w) - w$ has a pole of first order at $w = w_0$ and another at $w = \infty$, so that $P = 2$. Otherwise $f(\infty) = \infty$; in this case $P = 0$ or 1 according as $f(w)$ has the residue 0 or 1 at infinity.

Hence $N - P \geq -2$, while the counting of the indices I_ν, since all curves β_ν are described in the same sense, yields $N - P < -2$. Again we must have $g(w) \equiv 0$.

If β_1 is the real axis, no new modification is necessary.

6. Supplementary Remarks

1. *First Continuity Theorem in Conformal Mapping.* Our results can be considerably extended on the basis of

Theorem 5.5: Let $w = g_n(z)$ be a function mapping a fixed k-fold connected domain G in the z-plane bounded by k Jordan curves $\gamma_1, \gamma_2, \cdots, \gamma_k$ conformally onto a domain B_n in the w-plane. Suppose that for $n \to \infty$ the domain B_n tends to a domain B in such a way that the k boundaries of B_n—assumed as Jordan curves—tend to k boundary curves of B in the strong sense. Then either $g_n(z)$ tends to a constant, or a subsequence of the g_n tends to a limit function $g(z)$ mapping G on B; the convergence is uniform in the closed domain G and the mapping by $g(z)$ as well as by $g_n(z)$ remains continuous and biunique at the boundary.

We may assume B as a bounded domain. Then the equicontinuity of a sequence $g_n(z)$ not tending to a constant follows exactly as in Chapter III, §2. Theorem 5.5 follows from this remark by the same arguments employed before.

2. *Second Continuity Theorem. Extension of Previous Mapping Theorems.* Suppose $\mathfrak{N}^{(n)}$ is a class of domains $B^{(n)}$ with analytic boundaries as considered in §3–§5, while $\mathfrak{N}$ is a class of domains whose boundary curves are limits in the strong sense, for $n \to \infty$, of the boundary curves of domains $B^{(n)}$. If G is a fixed domain, a mapping $w = g^{(n)}(z)$ of G onto a domain of $\mathfrak{N}^{(n)}$ is assured by our previous results. The same reasoning as that employed in Chapter IV, §2, 3 to exclude degenerations of the parameter domains, serves to prove that the domains $B^{(n)}$ corresponding to G cannot tend to degeneration. Hence we may assume that a suitable subsequence tends to a domain B of the class $\mathfrak{N}$. Theorem 5.5 then implies that G can be mapped on B. As a result we formulate

Theorem 5.6: The boundary curves of B need not be analytic functions; they may be merely Jordan curves.

No essential change is necessary in the proof. In the same way we find

Theorem 5.7: Instead of convex boundary curves, some or all the boundaries of B may be *straight slits whose directions are prescribed.*

3. *Further Observations on Conformal Mapping.* More mapping theorems can be proved by the methods in this chapter. For example, domains not of genus zero could be treated, but are beyond the scope of the present book. Attention should be given to the role of convexity in our proof: it is not difficult to show by examples that omission of this assumption might invalidate the uniqueness theorem. On the other hand, mapping theorems do not necessarily depend on convexity,[16] as may be seen from the following example, a proof of which is very simple on the basis of our methods. We consider a class $\mathfrak{N}$ of doubly connected domains B depending on one parameter. The domains B are ring-shaped. The outer boundary curve γ_1 may be an arbitrary Jordan curve. The inner curve γ_2 is any member of an arbitrarily chosen family of curves $\gamma(t)$ having the following properties: $\gamma(t_1)$ contains $\gamma(t_2)$ for $t_1 > t_2$, $\gamma(0)$ is a point, and $\gamma(1)$ has at least one point in common with γ_1. The mapping is uniquely determined if a fixed point on γ_1 is to have a fixed image.

Incidentally, our approach can also be used to study the behavior of mapping functions under continuous change of the domain. For example, consider the ring G between a unit circle and a concentric square in its interior, and map G onto a circular ring B. What happens in the limit if the square touches the unit circle? The answer is that the function mapping G onto B converges non-uniformly to constant values on the unit circle.

7. *Existence of Solution for Variational Problem in Two Dimensions*

To prove that the variational problems on which our mapping theorems are based possess a solution, we refer to the sufficient condition of theorem 4.1 and show that it is always satisfied for a two-dimensional $\mathfrak{x}$-space.

1. *Proof using Conformal Mapping of Doubly Connected Domains.* The mapping theorem for k-fold connected domains G will be proved

[16] See e.g. Manel [1], where non-convex slits are the boundary curves.

by induction;[17] for simply and doubly connected domains it holds by Riemann's mapping theorem. An important element in the proof is the fact that the minimum value d of Dirichlet's integral is equal to the area A of G, if a domain B can be mapped conformally onto G.

We shall prove in detail the existence of a solution of problem IV for circular domains B. The boundary β_1 may be chosen as the unit circle, with all the other boundary circles $\beta_2, \beta_3, \cdots, \beta_k$ in its interior. We note first that G may be assumed to be a finite plane domain bounded by k Jordan curves $\gamma_1, \gamma_2, \cdots, \gamma_k$, with γ_1 as the outer boundary. For if G were not plane, we could first map G conformally onto a plane domain G', as is always possible by theorem 2.2. Suppose that the boundary component γ_1' of G' is not a Jordan curve, but e.g. a boundary slit: a second conformal mapping will take the simply connected plane domain bounded by γ_1' and containing G' into a plane domain bounded by a circle γ_1''. The image of G' in this mapping is a plane domain G'' with the Jordan curve γ_1'' as one boundary. Treating the other boundaries successively in the same way, we obtain after k steps a domain of the desired type conformally equivalent to the original domain G.

Again let γ' and γ'' be any subdivision of the system γ of boundary curves into complementary subsets. Denote by d' and d'' respectively the lower bounds of Dirichlet's integral for the variational problems referring to γ' and γ'' as boundary systems, by d the corresponding bound for γ. As was shown in Chapter IV, to prove the existence of a solution of problem IV we need only verify the sufficient condition $d < d' + d''$. Let γ' consist of the k' curves $\gamma_1, \gamma_2, \cdots, \gamma_{k'}$, $k' < k$. By the induction assumption we have

$$d' = A_1 - A_2 - \cdots - A_{k'} = A' > A_1 - A_2 - \cdots - A_k = A,$$

where A_i denotes the area contained in γ_i and A, A' denote the areas respectively bounded by the systems γ, γ'. Consider a circle sufficiently small to be contained in the interior of any of the curves $\gamma_{k'+1}, \gamma_{k'+2}, \cdots, \gamma_k$, and let $2p$ be its area: this shows that

$$d' > A + 2p;$$

the sufficient condition $d < d' + d''$ is proved if we can establish the inequality

[17] Degenerate cases in which one or more of the boundary curves are isolated points may be disregarded by the induction assumption.

$$d \leq A + p. \tag{5.21}$$

We obtain (5.21) by constructing a domain B, and in B a vector $\mathfrak{y}(u,v)$, mapping B (non-conformally) onto G, for which $D[\mathfrak{y}] \leq A + p$. By Dirichlet's Principle we have $d \leq D[\mathfrak{x}] \leq A + p$ for the vector $\mathfrak{x}$ harmonic in B and coinciding with $\mathfrak{y}$ on β.

The presentation of the proof is simplified if we choose $\gamma_1, \gamma_2, \cdots, \gamma_{k-1}$ as circles. For since we assumed the mapping theorem to hold for domains bounded by $k - 1$ contours, we may first map the domain bounded by $\gamma_1, \gamma_2, \cdots, \gamma_{k-1}$ conformally onto a domain bounded by $k - 1$ circles. In other words we may from the outset assume $\gamma_1, \gamma_2, \cdots, \gamma_{k-1}$ to be circles, while the last contour γ_k, or

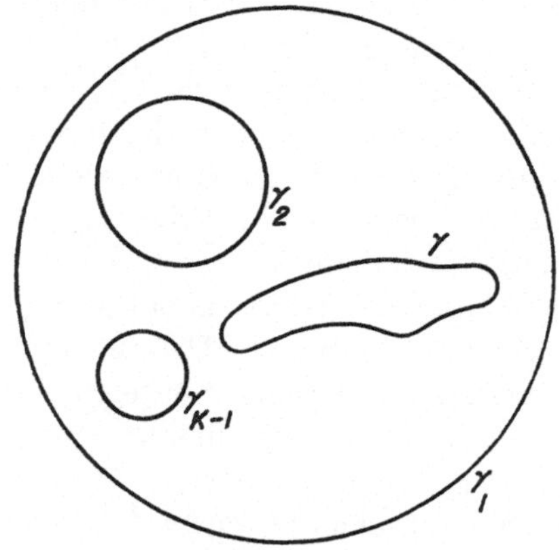

Figure 5.7

briefly γ, is any closed Jordan curve. For brevity we shall refer to β_k simply as β.

We proceed by a continuity method. Consider a closed set $\gamma(t)$ of curves, depending continuously on a parameter t, $0 \leq t \leq 1$, with the following properties:[18]

1) $\gamma(0)$ is a circle of area $2p$ inside γ.
2) The curves $\gamma(t)$ bound a monotone increasing sequence of domains.
3) $\gamma(1) = \gamma$.

[18] Such a set $\gamma(t)$ is for example furnished by the images of the circles that lie in a circular ring conformally equivalent to the doubly connected domain bounded by γ and $\gamma(0)$, and are concentric with the boundary circles of the ring.

The system $\gamma_1, \gamma_2, \cdots, \gamma_{k-1}, \gamma(t)$ bounds a continuous sequence $G(t)$ of domains, with area $A(t)$ and lower bound $d(t)$ of the variational integral. $G(1)$ is identical with G.

Since $G(0)$ is a circular domain, it is a domain for which both the mapping theorem and the existence of a solution of problem IV are already established. We have obviously $d(0) = A(0)$.

Inequality (5.21) will follow from

Lemma 5.2: Let τ be a value of the parameter t for which $G(t)$ can be mapped conformally on a domain B, so that problem IV for $G(\tau)$

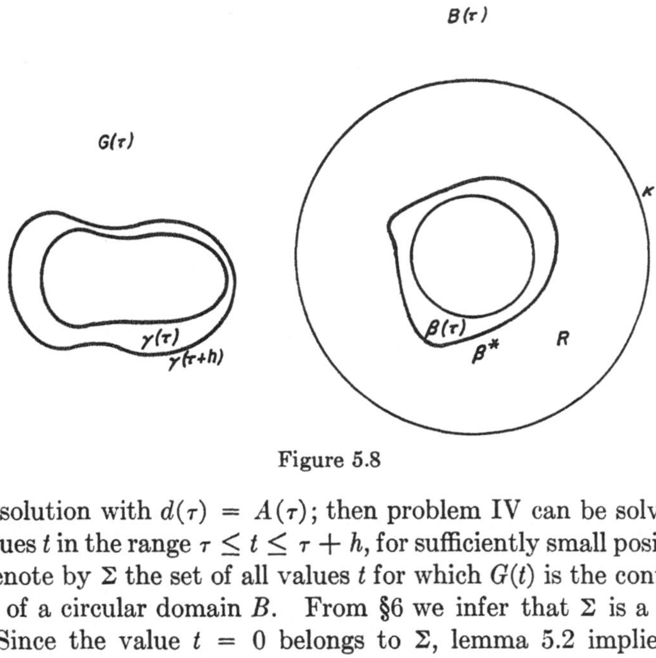

Figure 5.8

has a solution with $d(\tau) = A(\tau)$; then problem IV can be solved for all values t in the range $\tau \leq t \leq \tau + h$, for sufficiently small positive h.

Denote by Σ the set of all values t for which $G(t)$ is the conformal image of a circular domain B. From §6 we infer that Σ is a closed set. Since the value $t = 0$ belongs to Σ, lemma 5.2 implies that every value t, including $t = 1$, belongs to the set; hence G is conformally equivalent to a domain B.

Proof of Lemma 5.2: We assume that $G(\tau)$ is the conformal image of a circular domain $B(\tau)$, with $\gamma(\tau)$ corresponding to a circle $\beta(\tau)$. In Figure 5.8 the circles $\beta_1, \beta_2, \cdots, \beta_{k-1}$ and $\gamma_1, \gamma_2, \cdots, \gamma_{k-1}$ are not shown; they may be supposed to lie in the part of the diagram beyond the edge of the page.

Let $\mathfrak{t}(u, v)$ be the vector mapping $B(\tau)$ conformally on $G(\tau)$. We surround $\beta(\tau)$ by a fixed concentric circle κ of radius a, so chosen that there are no boundary points of $B(\tau)$ between $\beta(\tau)$ and its reflection in κ. To $\gamma(\tau + h)$ in $G(\tau)$ there corresponds, by our conformal mapping, a curve $\beta^*(\tau + h)$; for sufficiently small h, β^* is arbitrarily near $\beta(\tau)$ in $B(\tau)$ by the continuity of the mapping function $\mathfrak{t}$ at the boundary (proved in theorem 2.4). We denote by $B^*(\tau)$ the domain consisting of $B(\tau)$ with the interior of κ removed; R is the ring shaped domain between β^* and κ and $B'(\tau + h)$ denotes the noncircular domain $B^*(\tau) + R$ which is mapped conformally onto $G(\tau + h)$ by $\mathfrak{t}(u, v)$.

Our aim is to transform the domain $B'(\tau + h)$ into a circular domain $\bar{B}$ taking the point (u, v) of B' into $(\bar{u}, \bar{v})$ in $\bar{B}$, and to construct the transformation in such a way that the vector

$$\mathfrak{y}(\bar{u}, \bar{v}) = \mathfrak{t}(u, v)$$

in the circular domain $\bar{B}$ is admissible in problem IV for $G(\tau + h)$ and that $D_{\bar{B}}(\mathfrak{y}) < A(\tau + h) + p$. Then our previous remark shows that problem IV has a solution for $G(\tau + h)$, and the proof of the lemma is complete.

The construction leaves the region $B^*(\tau)$ outside κ untouched ($\bar{u} = u$, $\bar{v} = v$ in B^*) and affects only the noncircular ring R. We transform R into a circular ring R^* with the same outer radius a such that the points on κ remain fixed.

First we map R conformally onto a ring R^*, by a transformation that takes (u, v) in R into (u^*, v^*) in R^*.[19] In general this transformation carries points of κ into other points of κ. Let r, θ be polar coordinates in R, r^*, θ^* polar coordinates in R^*, and set $w = r\, e^{i\theta}$, $w^* = r^*\, e^{i\theta^*}$; the point $a\, e^{i\theta}$ on κ may correspond to the point

$$a \exp \{i\theta^*\} = a \exp \{i[\theta + \sigma(\theta)]\}.$$

Secondly we restore the points of κ to their original locations by the transformation

$$\bar{w} = w^*\, e^{-i\sigma(\theta)}.$$

[19] Here we are making use of the mapping theorem for two-fold connected domains. If we want to admit knowledge of the theorem for simply connected domains alone, we must adopt the modification described in article 2.

This new transformation is not conformal. However, the vector $\mathfrak{y}$ defined by $\mathfrak{y}(\bar{u}, \bar{v}) = \mathfrak{t}(u, v)$ is seen to be continuous and admissible in $\bar{B} = B^* + R^*$ if we realize that the function $\sigma(\theta)$ and its derivatives are continuous. For the conformal transformation taking R into R^* maps the circle κ into itself and can therefore be extended beyond κ by reflection. Moreover, h can be chosen so small that the transformation of R into R^* differs, on κ, arbitrarily little from the identity. Consequently $\sigma(\theta)$ and its first derivative can be made as small in absolute value as desired. Substituting $\mathfrak{y}(\bar{w}) = \mathfrak{t}(w^* e^{-i\sigma})$, we find[20] that $D_{R^*}[\mathfrak{y}]$ differs from $D_R[\mathfrak{t}]$ by less than $p/2$ if h is taken sufficiently small. We can further choose h so small that $A(\tau + h) > A(\tau) - p/2$. Recalling the assumption that $d(\tau) = A(\tau)$, we find

$$\begin{aligned} D_{\bar{B}}[\mathfrak{y}] &= D_{B^*}[\mathfrak{t}] \quad + D_{R^*}[\mathfrak{y}] \\ &< D_{B^*}[\mathfrak{t}] + D_R[\mathfrak{t}] + p/2 \\ &= A(\tau) + p/2 \\ &< A(\tau + h) + p. \end{aligned}$$

Lemma 5.2 is established: by our previous remark, we havé proved the existence of a solution of problem IV for the domain G.

2. *Alternative Proof. Supplementary Remarks.* We present a variant of the proof that makes no use of the mapping theorem for doubly connected regions. Instead of mapping R onto a circular ring, we map the whole simply connected exterior of β^* onto the exterior of a circle by a function $w^* = g(w)$ differing only slightly from the identity on κ. Thereby R is mapped onto a domain R^* bounded by the image κ^* of κ and by a circle. We can transform R^* (non-conformally) into another ring, in such a way that the points of κ^* are again taken into their original positions on κ. As above we obtain an admissible domain $\bar{B}$, and an admissible vector $\mathfrak{y}$ mapping $\bar{B}$ onto $G(\tau + h)$, for which $D_{\bar{B}}[\mathfrak{y}] < A(\tau + h) + p$ if h is chosen sufficiently small.

For the other classes $\mathfrak{N}$ of domains considered in this chapter the proof is quite similar to that for circular domains. Only one point requires attention: the choice of the domain $G(0)$ for the class of Riemann domains considered in §4. We must choose $G(0)$ as a domain in the x, y-plane which can be mapped onto a Riemann

[20] See similar estimates in Chapter IV, §3.

domain B by an explicit function. For this purpose we use the domain $B = B_0$ having $(k - 1)$-fold branch points at $w = 1/2$ and $w = -1/2$. The domain B_0 is mapped by the function

$$x + iy = \left(\frac{2w - 1}{2w + 1}\right)^{1/k}$$

on a k-fold connected plane domain, which by a linear transformation can be changed into a ring domain $G(0)$.

By showing that variational problem IV always possesses a solution, we have completed the proof of the mapping theorems of the present chapter.

CHAPTER VI

Minimal Surfaces with Free Boundaries and Unstable Minimal Surfaces

1. Introduction

In this chapter we shall discuss two extensions of the theory of minimal surfaces. First we shall solve the problem of finding minimal surfaces of least area when the whole boundary or part of it is not prescribed but left *free* on given manifolds.[1] Secondly we shall study minimal surfaces whose areas are not relative minima. Minimal surfaces of this type correspond to unstable equilibria of a soap film; they will therefore be referred to as unstable minimal surfaces.

1. *Free Boundary Problems.* Problems with free boundaries were discussed in the early nineteenth century by Gergonne, who proposed the problem: to find the surface of least area bounded by two opposite faces of a cube and by two skew diagonals of a second pair of opposite faces (see Figure 6.1). Riemann and Schwarz generalized the problem by asking for a simply connected surface of relative minimum area whose boundary consists alternately of straight line segments and of arcs on given planes. This problem can obviously be further generalized to that of finding a simply connected surface of least area whose boundary consists alternately of given Jordan arcs and of point sets on given manifolds; we call this type of boundary a "*chain.*" Furthermore we shall consider problems of the following kind: to find a doubly connected surface of minimal area one of whose boundaries is free on a given surface, say a sphere, while the other boundary is a prescribed Jordan curve. Finally, we may ask for surfaces of least area whose entire boundary is free on a closed manifold, e.g. a torus. In this case we must introduce an additional topological specification in order to exclude trivial solutions degenerated into points.

[1] The word manifold, in this connection, means a connected closed point set.

For problems involving free boundaries we shall be able to prove the existence of solutions of least area; concerning the "transversality properties" of the minimal surfaces on the free boundary, however, only partial results will be obtained.

2. *Unstable Minimal Surfaces.* The phenomenon of unstable minimal surfaces within prescribed contours has long been observed. It occurs, for example, in the problem of minimal surfaces of revolution if the two boundary circles are chosen at such a distance that the generating catenary contains the points conjugate to the end points; the corresponding catenoid furnishes a stationary value, but neither a maximum nor a minimum of the area. Other examples of unstable

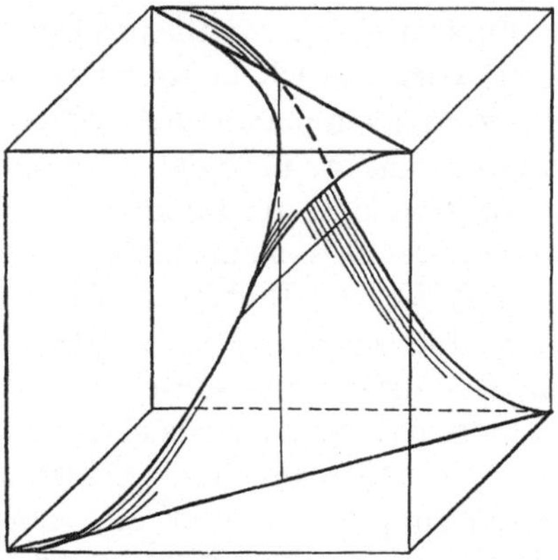

Figure 6.1 Gergonne's surface.

solutions may be found in Schwarz' investigations on minimal surfaces.[2]

In general, one might expect that the existence of two relative minima in Plateau's problem would guarantee the existence of another minimal surface of unstable character, just as the existence of two distinct relative minima of a differentiable function of a finite number of variables implies the existence of a stationary "mini-maximum." Theorems about the existence of such unstable minimal surfaces were first proved by M. Shiffman and at the same time by M. Morse and C. Tompkins. In addition, these papers propose the classification of unstable minimal surfaces by considering the individual features of the problem from within the framework of an abstract theory of

[2] See Schwarz [1].

critical points in function spaces. The discussions are based on Douglas' explicit expressions for the Dirichlet integral of harmonic vectors in terms of their boundary values. A different approach due to the author, using the method of Dirichlet's Principle, proceeds by reduction of the problem to that of stationary values of a differentiable function of a finite number of independent variables, provided that the boundary contour is polygonal. This theory was essentially extended by Shiffman. He observed that a passage to the limit from polygons to more general (rectifiable) closed curves is possible on the basis of the continuity theorem of Morse and Tompkins (theorem 3.6).

Attention may be called to unsolved problems of unstable minimal surfaces with free boundaries. Consider, for example, the following analogy: We join two given points A and B on a closed smooth curve by the shortest connection (i.e. a straight segment) and then seek a position of the points A and B for which this minimum length is a maximum. This maxi-minimum is furnished by the diameter, a stationary chord perpendicular to the curve at its two end points. In a similar way we consider a closed surface M and on it a closed curve μ in which we span a simply connected minimal surface of least area; then we seek a position of μ for which this minimal surface is a stationary "diameter surface" (not a maximum).

2. *Free Boundaries. Preparations*

1. *General Remarks.* In the case of fixed boundaries the solution of Plateau's problem is based on the fact of compactness. The reasoning used in Chapters III and IV shows that a minimizing sequence of harmonic vectors for Plateau's problem always contains a uniformly convergent subsequence, the limit vector being either a constant or an admissible vector. Equicontinuity of the boundary values of the admissible vectors $\mathfrak{x}(u, v)$ was the basis for the construction of the minimizing vector.

A new difficulty arises for free boundaries, since in this case it is no longer possible to require that admissible vectors have continuous boundary values on the given manifold M. In Plateau's problem we proved the convergence in the interior of B by first investigating the properties of the boundary values of the admissible vectors. However, it is possible to construct examples of surfaces of least area with free boundaries, for which these free boundaries are not continuous curves (see §5, 2). Hence the methods of Chapter III must be replaced by a reasoning which refers to the interior of the domain and starts from a suitable definition which does not use continuous boundary values.

Definition 6.1: Denote by $g_M[\mathfrak{x}]$ the shortest distance from the point $\mathfrak{x}$ to a closed manifold M. If $\mathfrak{x} = \mathfrak{x}(u, v)$ is a continuous surface defined in a parameter domain B with boundary β, then $g_M[\mathfrak{x}]$ is a function $g_M[\mathfrak{x}(u, v)]$ of u, v in B. If $g_M[\mathfrak{x}(u, v)]$ tends to zero as the point (u, v) in B tends to the boundary β of B, we say that the boundary of $\mathfrak{x}(u, v)$ is on M.

Remarks:
a) If the boundary of $\mathfrak{x}(u, v)$ is on M, the distance $g_M[\mathfrak{x}(u, v)]$ tends uniformly to zero as (u, v) tends to β; e.g. if β is the line $v = 0$, the distance $g_M[\mathfrak{x}(u, v)]$ tends to zero *uniformly* in u as v approaches zero. For otherwise there would exist an infinite sequence of points (u_i, v_i) in B with $v_i \to 0$ and a number $\alpha > 0$ such that $g_M[\mathfrak{x}(u_i, v_i)] > \alpha$. The points (u_i, v_i) contain a subsequence converging to a point of β (possibly the point at infinity) and for this subsequence we have by our definition $g_M[\mathfrak{x}(u_i, v_i)] \to 0$, in contradiction to the assumption.
b) In some of the subsequently treated cases, only a part of the boundary of the surface $\mathfrak{x}$ will be on M. By this we mean that the statement of the definition above is satisfied for a part β_1 of β alone. (For instance, if β is the line $v = 0$, β_1 will consist of one or more segments of the line.)

The manifold M may be a closed surface such as a sphere or a torus, or it may be a part of such a surface, for example the ring-shaped part cut out of a spherical surface by two parallel planes; also M may simply be a continuous closed curve. In more than three dimensions we have a correspondingly greater variety of possibilities.

2. *A Theorem on Boundary Values.* The treatment of free boundaries is based on a theorem of compactness which enables us to prove that the limit vector of a convergent minimizing sequence is admissible. For convenience we state this theorem under the assumption that B is the upper half-plane $v > 0$.

Theorem 6.1: Let $\mathfrak{x}_n(u,v)$ be a sequence of harmonic vectors in B whose boundaries are on closed manifolds M_n, and assume

$$D[\mathfrak{x}_n] \leq A^2,$$

A being a constant independent of n. Suppose that the vectors $\mathfrak{x}_n$ converge uniformly in each closed subdomain of B to a harmonic vector $\mathfrak{x}$ (for which automatically $D[\mathfrak{x}] \leq A^2$), while the manifolds M_n tend to a continuous manifold M in the sense that the greatest distance from points of M_n to M tends to zero. Then the boundary of the limiting vector $\mathfrak{x}$ is on M. Note that no assumptions are made concerning the dimensions of M_n and M. In our applications M_n will be a curve, M a surface.

Proof: We denote by $g_{M_n}[\mathfrak{x}_n(u, v)]$ the distance of the point $\mathfrak{x}_n(u, v)$ from M_n. We must prove, given the assumptions of the theorem, that the relation

$$g_{M_n}[\mathfrak{x}_n(u, v)] \to 0,$$

as $v \to 0$, implies

$$g_M[\mathfrak{x}(u, v)] \to 0,$$

as $v \to 0$. Let $g(M_n, M)$ be the largest distance from a point of M_n to M. By the triangle inequality we have

$$g_M[\mathfrak{x}(u, v)] \leq g_{M_n}[\mathfrak{x}(u, v)] + g(M_n, M),$$

hence

$$g_M[\mathfrak{x}(u, v)] \leq \liminf_{n \to \infty} . \, g_{M_n}[\mathfrak{x}(u, v)].$$

It will be sufficient, therefore, to investigate $g_{M_n}[\mathfrak{x}(u, v)]$.

Nothing is known about the continuity of the values of $\mathfrak{x}$ or $\mathfrak{x}_n$ on the boundary $v = 0$. Consider

$$\mathfrak{y}_n(u, v) = \mathfrak{x}_n(u, v + \delta_n),$$

δ_n being a given positive constant. This harmonic vector is continuous for $v = 0$, and its values $\mathfrak{y}_n(u, 0)$ lie on a curve M_n' whose distance from M_n can be made arbitrarily small if δ_n is chosen sufficiently small. Since $\mathfrak{y}_n(u, v) \to \mathfrak{x}(u, v)$ as $\delta_n \to 0$ we may prove our theorem by writing M_n instead of M_n' and $\mathfrak{x}_n$ instead of $\mathfrak{y}_n$; note that $\mathfrak{x}_n(u, v)$ now has continuous (even analytic) boundary values $\mathfrak{x}_n(u, 0)$ on β.

We need only prove that the distance $g_{M_n}[\mathfrak{x}(u, v)]$ from the limit vector to M_n is arbitrarily small, uniformly in n, for points with sufficiently small v-coordinate. To that end we choose $v = h$, where h is so small that, for the strip $B_h : 0 < v \leq 2h$, we have

$$D_{B_h}[\mathfrak{x}] < \epsilon^4,$$

where $\epsilon > 0$ is a preassigned small constant, as is possible since $D[\mathfrak{x}] < \infty$. Subsequently we choose a number $N(h)$ so large that for $n > N$ we have the inequality

$$(6.1) \qquad |\mathfrak{x}_n(u, h) - \mathfrak{x}(u, h)| < \epsilon.$$

To estimate $g_{M_n}[\mathfrak{x}(u, h)]$ without assuming that for fixed u and (sufficiently small) h, the distance $g_{M_n}[\mathfrak{x}_n(u, h)]$ is small uniformly in n, we consider a fixed point—say $(0, h)$. First we estimate the oscillation of $\mathfrak{x}(u, h)$ along the line $v = h$. By the mean value theorem for harmonic functions and by Schwarz' inequality we have on $v = h$

$$|\mathfrak{x}_u| < \frac{\sqrt{2}\epsilon^2}{\sqrt{\pi}h},$$

so that

$$|\mathfrak{x}(u, h) - \mathfrak{x}(0, h)| = \left| \int_0^u \mathfrak{x}_u(u, h)\, du \right| \le |u| \frac{\sqrt{2}\epsilon^2}{\sqrt{\pi}h}.$$

Hence for $|u| \le (\sqrt{\pi}\, h)/(\sqrt{2}\, \epsilon)$ we find

$$(6.2) \qquad |\mathfrak{x}(u, h) - \mathfrak{x}(0, h)| \le \epsilon;$$

the inequality shows that $\mathfrak{x}(u, h)$ varies little in an interval that is large relative to h. Consequently we have for $n > N$ and

$$|u| \le \frac{\sqrt{\pi}\, h}{\sqrt{2}\, \epsilon} = \tfrac{1}{2}s$$

the inequality

$$(6.3) \qquad |\mathfrak{x}(0, h) - \mathfrak{x}_n(u, h)| < 2\epsilon.$$

In at least some points of the interval (6.3) we can find a suitable estimate for the distance $g_{M_n}[\mathfrak{x}_n(u, h)]$. For according to the mean value theorem there exists a value u_0—possibly varying with n—in the u-interval of length s, for which

$$\frac{1}{2} \int_0^h \left[\frac{\partial \mathfrak{x}_n}{\partial v} (u_0, v) \right]^2 dv \le \frac{A^2}{s} = \sqrt{\frac{2}{\pi}} \frac{A^2 \epsilon}{h}.$$

Hence by Schwarz' inequality we have in the usual way

$$|\mathfrak{x}_n(u_0, h) - \mathfrak{x}_n(u_0, 0)| \le \sqrt{h} \sqrt{\frac{2A^2}{s}},$$

or, equivalently,

$$| \mathfrak{x}_n(u_0, h) - \mathfrak{x}_n(u_0, 0) | \leq \sqrt[4]{\frac{8}{\pi}} A\sqrt{\epsilon}.$$

Since $\mathfrak{x}_n(u_0, 0)$ is on M_n, we find

$$g_{M_n}[\mathfrak{x}_n(u_0, h)] \leq \sqrt[4]{\frac{8}{\pi}} A\sqrt{\epsilon}.$$

From (6.3) we have

$$| \mathfrak{x}_n(u_0, h) - \mathfrak{x}(0, h) | < 2\epsilon;$$

hence the triangle inequality yields

$$g_{M_n}[\mathfrak{x}(0, h)] < 2\epsilon + \sqrt[4]{\frac{8}{\pi}} A\sqrt{\epsilon},$$

where the right side is independent of n. The bound is uniformly small, and letting n go to infinity we obtain

$$g_M[\mathfrak{x}(0, h)] \leq 2\epsilon + \sqrt[4]{\frac{8}{\pi}} A\sqrt{\epsilon}.$$

The theorem is proved. At the same time it follows that $g_{M_n}[\mathfrak{x}_n(u, h)]$ tends to zero with h, uniformly in n.

Corollaries:

1. The theorem can be immediately extended, say by conformal mapping or by direct application of the same reasoning, to the interior B of a circle.

2. The theorem also holds when different parts of an admissible vector $\mathfrak{x}_n$ lie on different adjacent manifolds $M_n^{(i)}$, $i=1,2,\cdots, k$, which converge to manifolds $M^{(i)}$. More precisely: Let the boundary β: $v = 0$ of B be divided into segments $\beta_1, \beta_2, \cdots, \beta_k$, and assume that the distance from a point on the admissible surface $\mathfrak{x}_n(u, v)$ to the manifold $M_n^{(i)}$ tends to zero as the parameter point (u, v) tends to β_i. If the other assumptions of theorem 6.1 are satisfied, the part of the boundary of $\mathfrak{x}$ which corresponds to the segment β_i lies on $M^{(i)}$.

3. It is obvious that the theorem remains valid if one or more domains with boundaries $\beta_1, \beta_2, \cdots, \beta_k$ are omitted from the half-plane and the admissible vectors are required to map $\beta_1, \beta_2, \cdots, \beta_k$ continuously on prescribed Jordan curves $\gamma_1, \gamma_2, \cdots, \gamma_k$.

3. *Minimal Surfaces with Partly Free Boundaries*

1. *Only One Arc Fixed*

Theorem 6.2: Let T be a closed manifold of $m - 1$ or less dimensions, γ a Jordan arc joining two points P_1, P_2 of T. Then there exists a simply connected minimal surface of least area having its boundary on γ and on T.

Proof: We establish the theorem by minimizing Dirichlet's integral. Surfaces $\mathfrak{x}(u, v)$ defined in the interior B of the unit circle β are admissible if they satisfy the following conditions:
1) $\mathfrak{x}$ is continuous and piecewise smooth in B.
2) The boundary of $\mathfrak{x}$ lies on $M = T + \gamma$, and the Jordan arc γ corresponds in a continuous and monotonic way to an arc c of the unit circle β. (Naturally we assume that there exist admissible vectors with finite Dirichlet integral.)
a) Existence Proof: We shall construct an admissible vector $\mathfrak{x}$ for which Dirichlet's integral assumes its absolute minimum d. The proof that this vector represents a minimal surface is then the same as for Plateau's problem in Chapter III.

Starting with a minimizing sequence $\mathfrak{y}_n$, we replace it by a sequence of harmonic vectors with the same boundary values, again denoted by $\mathfrak{y}_n$: The end points of c are joined by an analytic arc c_n' in B which is so close to the complementary circular arc of c that the image γ_n' of c_n' on the surface $\mathfrak{y}_n$ is at a distance not more than $\epsilon = 1/n$ from T. Let $\mathfrak{y}_n'$ be the harmonic vector in B' enclosed by $c + c_n'$ whose values on $c + c_n'$ coincide with those of $\mathfrak{y}_n$. By Dirichlet's Principle we have $D_{B'}[\mathfrak{y}_n'] \leq D[\mathfrak{y}_n]$. A conformal mapping of B' onto the unit disk B transforms the vector $\mathfrak{y}_n'$ into a harmonic vector $\mathfrak{x}_n$, defined in B and spanning the closed curve $\gamma + \gamma_n'$. By a linear transformation of the unit circle into itself we may adjust $\mathfrak{x}_n$ so that a three-point condition is satisfied with reference to the end points and one interior point of c. We have the inequality

$$\liminf_{n\to\infty} D[\mathfrak{x}_n] \leq \lim_{n\to\infty} D[\mathfrak{y}_n]. \tag{6.4}$$

As in Chapter III, we conclude that the boundary values of $\mathfrak{x}_n$ are equicontinuous on c. Since $D[\mathfrak{x}_n]$ is bounded, at least a subsequence of the $\mathfrak{x}_n$ converges to a non-constant harmonic vector $\mathfrak{x}$, the convergence being uniform in every closed subdomain of $B + c$, in particular on c (see lemma 1.3b). By the semicontinuity of Dirichlet's integral for harmonic functions and inequality (6.4), we find

$$D[\mathfrak{x}] \leq \liminf_{n\to\infty} D[\mathfrak{x}_n] \leq \lim_{n\to\infty} D[\mathfrak{y}_n] = d.$$

The vector $\mathfrak{x}$ maps the arc c monotonically and continuously on γ. Theorem 6.1 shows that all other boundary points of $\mathfrak{x}$ are on T; hence $\mathfrak{x}$ is admissible, so that $D[\mathfrak{x}] \geq d$. Consequently

$$D[\mathfrak{x}] = d,$$

and $\mathfrak{x}$ solves our problem.

b) Transversality Relation: We establish a relation for the free boundary on T which expresses orthogonality in a weak sense between T and the minimal surface. For the proof we suppose T to have a continuous tangent plane. We further assume that we can transform the $\mathfrak{x}$-space in the neighborhood of T by transformations

$$x'_\mu = x_\mu + \epsilon\, \xi_\mu(x_1, x_2, \cdots, x_m; \epsilon) \tag{6.5}$$

depending on a parameter ϵ and satisfying the following conditions: The functions ξ_μ have piecewise continuous, bounded derivatives with respect to x_ν and ϵ, and are zero in a neighborhood of the arc γ; equations (6.5) transform T into itself. We write $\xi_\mu(x_1, x_2, \cdots, x_m; 0) = \eta_\mu$ and combine the x'_μ as a vector $\mathfrak{x}'$, the η_μ as a vector η. On substitution of the components $x_\nu = x_\nu(u, v)$ of the minimizing vector $\mathfrak{x}$, the vector $\mathfrak{x}'$ becomes admissible in B. Since

$$D[\mathfrak{x}'] \geq D[\mathfrak{x}],$$

we obtain in the usual way

$$D[\mathfrak{x}, \eta] = \frac{1}{2} \iint_B (\mathfrak{x}_u \eta_u + \mathfrak{x}_v \eta_v)\, du\, dv = 0;$$

consequently, if B_δ is a closed subdomain of B, bounded by a piecewise smooth curve β_δ which tends to β as $\delta \to 0$, we have

$$\iint_{B_\delta} (\mathfrak{x}_u \eta_u + \mathfrak{x}_v \eta_v)\, du\, dv \to 0,$$

as $\delta \to 0$. Since $\Delta \mathfrak{x} = 0$, we have by Green's formula

$$\int_{\beta_\delta} \eta \frac{\partial \mathfrak{x}}{\partial n} \, ds \to 0,$$

where $\partial/\partial n$ denotes differentiation along the normal to β_δ, s arc length on β_δ. If G is our minimal surface, γ_δ the image of β_δ on G, this formula may be interpreted on γ_δ to mean

$$\int_{\gamma_\delta} \eta \frac{\partial \mathfrak{x}}{\partial n} \, ds \to 0,$$

where $\partial/\partial n$ now denotes differentiation in G normal to γ_δ, s arc length on γ_δ; for the integral is invariant under conformal mapping and $\mathfrak{x}$ maps B conformally on G. Since η vanishes in a neighborhood of γ, the last formula can be replaced by

$$\int_{\gamma_\delta'} \eta \frac{\partial \mathfrak{x}}{\partial n} \, ds \to 0,$$

where γ_δ' tends to the free boundary as $\delta \to 0$.

Since $\partial \mathfrak{x}/\partial n$ on G is a vector tangent to G and η an arbitrary tangential vector field near T, this relation expresses what may be called a weak condition of orthogonality between T and G. The arcs γ_δ' may be chosen as any sequence of piecewise smooth curves on G whose end points tend to those of γ and whose greatest distance from T tends to zero.

2. *Remarks on Schwarz' Chains.*[3] We consider briefly the more general problem of simply connected minimal surfaces of least area in three dimensions whose boundaries consist of k parts which are alternately on Jordan arcs and on two-dimensional surfaces successively connected by these arcs. Boundaries of this type are frequently called "*Schwarz' chains*"; the problem was investigated by Riemann as well as by Schwarz. We consider in addition the special case where some of the surfaces coincide.

We subdivide the unit circle β into k segments β_i by points $A_1, A_2, \cdots, A_k$ corresponding to the angles $\theta_1, \theta_2, \cdots, \theta_k$, and consider as admissible all vectors $\mathfrak{x}$ which map these intervals consecutively onto the different boundary manifolds. The solution proceeds in two steps: First we solve the variational problem for fixed points $A_1, A_2, \cdots, A_k$. The minimum of Dirichlet's integral is a function

[3] See also I. Ritter [1].

$d(\theta_1, \theta_2, \cdots, \theta_k)$ of the angles θ_1, θ_2, $\cdots$, θ_k, and is attained by an admissible vector; but this vector does not necessarily represent a minimal surface. Secondly, to obtain a minimal surface which solves the problem we have to vary the angles θ_i, $i = 1, 2, \cdots, k$, and seek a set of angles for which $d(\theta_i)$ is a minimum. By a linear transformation of the unit circle into itself we can always fix the values of three of the θ_i. Again we have two alternatives. If $d(\theta_i)$ attains its minimum value for a set of distinct values θ_i, the corresponding vector $\mathfrak{x}$ represents a minimal surface and solves the problem of Schwarz' chains. Otherwise we obtain a sequence $\theta_i^{(n)}$ of sets for which

$$\lim_{n \to \infty} d(\theta_i^{(n)}) = d$$

(the lower limit of Dirichlet's integral for admissible vectors) and which tends to degeneration in that at least two distinct sequences $\theta_i^{(n)}$, $\theta_j^{(n)}$ approach the same limit θ. In this case the corresponding harmonic surfaces $\mathfrak{x}_n$ tend to degeneration as in Chapter IV, § 2. Exactly as before, it follows that the problem has a degenerate solution, consisting of several minimal surfaces each of which belongs to a chain formed by only a part of the given Jordan curves and surfaces. This degeneration is excluded if we know that the lower bound d of Dirichlet's integral for non-degenerate surfaces is less than that for degenerate surfaces.

A simple illustration is given by two vertical semicircular arcs standing on a horizontal plane M, the planes of the semicircles being perpendicular to the line joining their centers. The problem of Schwarz' chain is solved by a minimal surface consisting of a half-catenoid if the distance between the circles is sufficiently small; otherwise, we obtain the degenerate solution consisting of two half-disks.

3. *Doubly Connected Minimal Surfaces with One Free Boundary.* As an example of higher connectivity we consider a doubly connected minimal surface one of whose boundaries is free on a closed manifold M while the other is a monotonically described closed Jordan curve γ. In the variational problem we admit surfaces represented by continuous vectors $\mathfrak{x}(u, v)$ with piecewise continuous first derivatives in the circular ring B of the u, v-plane bounded by the unit circle β_1 and a smaller concentric circle β_2. We further assume that $\mathfrak{x}$ has continuous boundary values on β_1 which map the unit circle mono-

tonically on γ, and that the boundary of $\mathfrak{x}$ corresponding to β_2 is on M in the sense of definition 6.1.

We seek a doubly connected minimal surface of least area (or least Dirichlet integral) d, bounded by the curve γ and the manifold M. We shall prove

Theorem 6.3: Such a minimal surface exists, provided that the greatest lower bound d is smaller than the greatest lower bound d^* belonging to Plateau's problem for γ as the sole boundary.

Proof: We again obtain the minimal surface G by solving the variational problem of finding a domain B and an admissible vector $\mathfrak{x}$ for which $D[\mathfrak{x}] = d$ is a minimum. Accordingly we suppose

$$d < d^*$$

and consider a minimizing sequence consisting of admissible vectors $\mathfrak{x}_n$ defined in ring domains B_n for which $D_{B_n}[\mathfrak{x}_n] \to d$. First we show, similarly as in Chapter IV, § 2, that B_n cannot tend to degeneration: a) If the radius a_n of the inner circle $\beta_2^{(n)}$ of B_n were to approach 1, we would have exactly the same contradiction as in Chapter IV, § 2. b) The radius a_n cannot tend to zero for a subsequence B_n. For suppose $a_n \to 0$; according to lemma 4.1, we could replace the vector $\mathfrak{x}_n$ by another vector $\mathfrak{y}_n = p_n(u,v)\,\mathfrak{x}_n$ which is zero for $r^2 = u^2 + v^2 \leq a_n^2$ and equal to $\mathfrak{x}_n$ for $r = 1$, and for which

$$D[\mathfrak{y}_n] < D[\mathfrak{x}_n] + \alpha_n$$

with $\alpha_n \to 0$. We may consider $\mathfrak{y}_n$ as a vector in the whole unit circle, admissible in Plateau's problem for the single contour γ for which the minimum is d^*. Letting n tend to infinity we immediately obtain, since $d^* \leq \liminf_{n\to\infty} D[\mathfrak{y}_n]$,

$$d^* \leq d,$$

in contradiction to our assumption $d < d^*$. It follows that we can choose a subsequence of the domains B_n tending to a ring B with radii 1 and a, $0 < a < 1$.

In the same way as before we see that the boundary values of $\mathfrak{x}_n$ on β_1 are equicontinuous, so that, at least for a suitable subsequence, they converge uniformly to a monotonic, continuous representation of γ. The vectors $\mathfrak{x}_n$ are next replaced by harmonic vectors $\mathfrak{y}_n$ in B which have the boundary values $\mathfrak{x}_n(1, \theta)$ on the outer circle and the boundary values $\mathfrak{x}_n(r_n, \theta)$ on the inner circle, where $r_n > a_n$ is a sequence tending to a. The vector $\mathfrak{x}_n(r_n, \theta)$ as a function of θ defines a continuous curve M_n tending to M as $n \to \infty$. By Dirichlet's Principle we have

$$\lim \inf. D_{B_n}[\mathfrak{y}_n] \leq \lim \inf. D_{B_n}[\mathfrak{x}_n]$$

for the harmonic vectors $\mathfrak{y}_n$.

Since the $\mathfrak{y}_n$ have equally bounded Dirichlet integrals and converge on β_1 we conclude by lemma 1.3b that they converge to a limiting harmonic vector $\mathfrak{x}$ uniformly in every closed subdomain of $B + \beta_1$; by the convergence of $\mathfrak{y}_n$ alone, we see that $\mathfrak{x}$ maps β_1 monotonically on γ.

Theorem 6.1 and its corollary 3 imply that the boundary of the vector $\mathfrak{x}$ corresponding to β_2 is on M. Hence $\mathfrak{x}$ is admissible and $D[\mathfrak{x}] \geq d$. On the other hand, we have as in Chapter I, §2 the relation

$$D[\mathfrak{x}] \leq \lim \inf. D_{B_n}[\mathfrak{x}_n] = d,$$

and therefore

$$D[\mathfrak{x}] = d;$$

hence $\mathfrak{x}$ solves the variational problem. That $\mathfrak{x}$ is a minimal surface is again seen as in Chapter IV or V.

Along the free boundary of $\mathfrak{x}$ we have the same transversality condition as in article 1, i.e.

$$\int_{L_\epsilon} \eta \frac{\partial \mathfrak{x}}{\partial n} \, ds \to 0,$$

where L_ϵ is any set of piecewise smooth contours homologous to γ on the surface $\mathfrak{x}$ and tending to M as $\epsilon \to 0$, η is an arbitrary tangential vector field near M, and $\partial/\partial n$ denotes differentiation normal to L_ϵ in the surface $\mathfrak{x}$, s arc length on L_ϵ. The proof is almost literally the same as in article 1.

4. *Multiply Connected Minimal Surfaces with Free Boundaries.* Extension of the theorems of this section and their proofs to multiply connected minimal surfaces, some of whose boundaries are free on prescribed manifolds, offers no new difficulties. The decisive sufficient condition for the existence of minimal surfaces of given

topological structure with free boundaries on a closed manifold M is again that the greatest lower bound d for surfaces of the prescribed topological structure is smaller than that for surfaces with the same fixed boundaries but of lower topological type.

Significant geometrical facts are connected with conditions of this sort: Suppose for instance that M is a closed surface in space while γ, the fixed part of the boundary, lies entirely outside M. Suppose further that, among all possible different choices of topological structure, G is the minimal surface of least area bounded by γ and

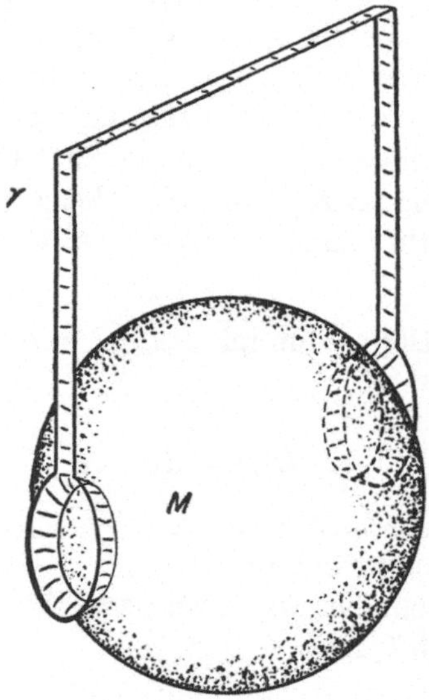

Figure 6.2 Illustrating the fact that demand for absolute minimum may imply higher connectivity.

by M; then G is *entirely outside* the surface M. For, otherwise, omitting from G all parts inside M, we would obtain a minimal surface G' bounded by γ and by M but having smaller area.

That we need restrictions to ensure that the minimal surface lies outside M is made plausible by the following example (see Figure 6.2). Consider the problem of article 3. Let M be a sphere: cut it by two planes perpendicular to a diameter and very close to its opposite ends. In each plane choose a circle concentric with the circle of intersection of the plane and the sphere, but of slightly greater radius. Finally, remove a small segment from each of these circles and connect the breaks by a narrow bridge consisting of two arcs that lead far away from M. The curve γ is to consist of the incomplete circles and their joining arcs. Obviously there is a triply connected surface of arbitrarily small area bounded by γ and by M, consisting of the narrow bridge plus the two plane rings interior to the circles and exterior to M. If, however, we stipulate *double* connectivity, it is plausible that we can bring the circles so close to M and the bridge so far from M that the greatest lower bound of the area is greater than for the corresponding triply connected domain. Moreover it is readily seen that to approximate the lower bound we must admit surfaces penetrating M.

The same is true for surfaces of simple connectivity as discussed in article 1. It is easy to construct examples for which doubly connected surfaces yield a smaller minimum; then the simply connected surface, whose existence was proved in article 1, may penetrate M.

4. Minimal Surfaces Spanning Closed Manifolds

1. *Introduction.* While in the preceding problems parts of the boundary of the minimal surface were given as Jordan curves, the minimal surfaces investigated in the present section have their entire boundaries free on prescribed surfaces.[4] As pointed out in § 1, a new element enters the problem and the existence proof: it is necessary to specify the topological position of the required solution relative to the boundary manifold M. For example, we may prescribe linking numbers between the boundaries of the surfaces under consideration and preassigned curves not on the manifold M.[5]

We limit ourselves to a typical problem. Let M be a torus. We shall seek a minimal surface of least area whose boundary lies on the torus and which covers the "hole" in the torus. For our topological specification we consider a fixed simple closed polygon H (see Figure 6.3), having no points in common with M, which is "linked" with M, in the sense that it is linked with every member of a class of homol-

[4] See also Courant and Davids [8] and the more detailed discussion in Davids [1].

[5] It should be stated that other topological specifications may be equally pertinent, cf. remark d) at end of section.

ogous non-bounding cycles on M.[6] Plainly such a polygon will lie either completely exterior to the torus, passing through the hole, or completely interior, winding around the hole on the inside.

To formulate our variational problem we suppose that the admissible surfaces are represented parametrically by continuous vectors $\mathfrak{x}(u,v)$ in the unit circle B of the u,v-plane. The vectors $\mathfrak{x}$ are assumed to have piecewise continuous first derivatives and bounded Dirichlet integral. Since the boundary of $\mathfrak{x}$ need not be a continuous curve on M we impose the following linking condition: the images by $\mathfrak{x}$ of simple closed curves in a subregion of B sufficiently close to the boundary β—precisely, the images of concentric circles near

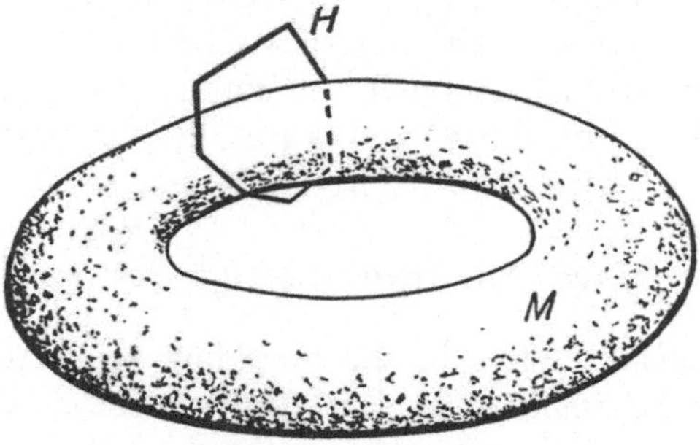

Figure 6.3 Polygon interlocked with torus.

β—are required to be curves linked with H. The variational problem is again to find an admissible vector $\mathfrak{x}$ for which

$$D[\mathfrak{x}] = d$$

is the smallest possible value. That a solution of this problem yields a minimal surface follows exactly as in the case of Plateau's problem.

2. *Existence Proof.* Denoting by d the greatest lower bound of $D[\mathfrak{x}]$ for admissible vectors, we define: a sequence $\mathfrak{x}_n$ of surfaces is termed an *admissible sequence* if its members satisfy all admissibility conditions except that the boundary of the surface $\mathfrak{x}_n$ is not necessarily on M, but on a manifold M_n, which tends to M as $n \to \infty$ in

[6] Then any orientable surface through a member of the class of non-bounding cycles intersects H, and the algebraic sum of the intersection numbers is different from zero.

the sense that the greatest distance from points of M_n to M goes to zero. Let δ denote the greatest lower bound of the lower limits of $D[\mathfrak{x}_n]$ for all such admissible sequences; an admissible sequence $\mathfrak{x}_n$ for which

$$D[\mathfrak{x}_n] \to \delta$$

is called a *generalized minimizing sequence.* Obviously $\delta \leq d$; we shall see that $\delta = d$.

Starting from an arbitrary generalized minimizing sequence $\mathfrak{y}_n$ we construct another consisting of harmonic vectors $\mathfrak{x}_n$ bounded by closed continuous curves M_n. Let r, θ be polar coordinates in B, and choose a radius r_n so close to 1 that the piecewise smooth curve M_n represented by the vector $\mathfrak{y}_n(r_n, \theta)$ lies at a distance from M going to 0 with n and is linked with H. Denote by $\mathfrak{x}_n^*$ the harmonic surface whose values on the boundary β of B are given by $\mathfrak{x}_n^*(1, \theta) = \mathfrak{y}_n(r_n, \theta)$. This surface has M_n as boundary. Since $D[\mathfrak{x}_n^*] \leq D[\mathfrak{y}_n]$, the vectors $\mathfrak{x}_n^*$ again form a generalized minimizing sequence. Since M_n and H are linked, there must exist at least one point (u_0, v_0) in B for which $\mathfrak{x}_n^*(u_0, v_0)$ is on H. By a linear transformation of B into itself we can map (u_0, v_0) into the origin; in this way we obtain a generalized minimizing sequence of harmonic vectors $\mathfrak{x}_n$ with $D[\mathfrak{x}_n] = D[\mathfrak{x}_n^*]$ and with the boundaries M_n.

With this normalized sequence $\mathfrak{x}_n$ we can reason as before (see Chapter III, § 2, 3). The boundary values of $\mathfrak{x}_n$ are bounded, hence also the values of $\mathfrak{x}_n$ in B. Since $D[\mathfrak{x}_n]$ is bounded, we can, by lemma 1.3b, choose a subsequence of the vectors $\mathfrak{x}_n$ tending uniformly to a harmonic vector $\mathfrak{x}$ in every closed subdomain of B; by the usual reasoning we have

$$D[\mathfrak{x}] \leq \delta.$$

Theorem 6.1 shows that the boundary of $\mathfrak{x}$ is on M. The point $\mathfrak{x}(0,0)$ is on H and therefore at a positive distance from M; thus $\mathfrak{x}$ is not constant.

To recognize $\mathfrak{x}$ as an admissible vector we must still prove that the boundary of $\mathfrak{x}$ is linked with H. This is the crucial point of the reasoning. Once the admissibility of $\mathfrak{x}$ is established, we have (since $D[\mathfrak{x}] \leq \delta$ and $\delta \leq d$) the relation

$$D[\mathfrak{x}] = \delta = d;$$

consequently $\mathfrak{x}$ solves our problem.

First mark off the points or arcs in B for which $\mathfrak{x}(u,v)$ is on the polygon H. Because $\mathfrak{x}$ is regular and harmonic in every closed subdomain of B, there can be only a finite number of "intersections"; for the boundary of $\mathfrak{x}$ is on M and therefore bounded away from H. (Note that the total number of intersections of H and $\mathfrak{x}_n$ has not been shown to be bounded.) Given a small positive constant ϵ we can choose a circle $r = r_\epsilon$, enclosing all the intersections and such that the vector $\mathfrak{x}(r_\epsilon, \theta)$ represents a curve M_ϵ whose distance from M is everywhere less than $\epsilon/2$. Then we choose n so large that

$$|\mathfrak{x}_n(r_\epsilon, \theta) - \mathfrak{x}(r_\epsilon, \theta)| < \frac{\epsilon}{2}$$

for all values of θ; consequently the curve $\mathfrak{x}_n(r_\epsilon, \theta)$ is at a distance not greater than ϵ from M. Suppose that the curve $\mathfrak{x}(r_\epsilon, \theta)$ were not linked with H; then $\mathfrak{x}_n(r_\epsilon, \theta)$ also would not be linked with H.[7] Since the curve $\mathfrak{x}_n(1, \theta)$ *is* linked with H, the algebraic sum of the intersection numbers of $\mathfrak{x}_n$ corresponding to the ring $R_\epsilon : r_\epsilon < r < 1$ would be different from zero.

For any subdomain of B, e.g., for the disk B': $r \leq 1/2$, we have

$$D_{B'}[\mathfrak{x}] = \lim D_{B'}[\mathfrak{x}_n];$$

furthermore, since the harmonic function $\mathfrak{x}$ cannot be constant in B', $D_{B'}[\mathfrak{x}]$ is not zero. Therefore

$$D_{B'}[\mathfrak{x}_n] > \alpha > 0, \tag{6.6}$$

where α is a fixed constant.

Secondly we appraise the radial oscillation of $\mathfrak{x}_n(r, \theta)$ (cf. Chapter III, § 8, 4). Let A be a uniform bound for all $D[\mathfrak{x}_n]$; then there exists, for each n, a value $\theta = \theta_n$ such that

$$\frac{1}{2}\int_{1/2}^{1}\left[\frac{\partial}{\partial r}\mathfrak{x}_n(r, \theta_n)\right]^2 dr \leq \int_{1/2}^{1}\left[\frac{\partial}{\partial r}\mathfrak{x}_n(r, \theta_n)\right]^2 r\, dr \leq \frac{1}{\pi}A.$$

Hence, by Schwarz' inequality, we have

$$|\mathfrak{x}_n(r, \theta_n) - \mathfrak{x}_n(1, \theta_n)|^2 \leq 2(1 - r)\frac{A}{\pi}, \qquad \tfrac{1}{2} \leq r < 1.$$

[7] For these curves can, for sufficiently small ϵ, be deformed into each other without meeting H.

Therefore the oscillation of $\mathfrak{x}_n(r, \theta_n)$ on the radial segment L_n: $\theta = \theta_n$, $r_\epsilon \leq r \leq 1$, is less than $\epsilon/2$, if r_ϵ is chosen sufficiently close to 1. The point $\mathfrak{x}_n(1, \theta_n)$ is on M_n, and the greatest distance from M_n to M goes to 0 with n; therefore the values of $\mathfrak{x}_n$ on the segment L_n are at a distance less than ϵ from M, provided n is large enough. Since r_ϵ is independent of n, we may rotate to obtain $L_n = L$. We cut the ring R_ϵ along L to obtain a simply connected domain $R^* = R_\epsilon^*$, whose boundary is mapped by $\mathfrak{x}_n$ on a continuous curve, at a distance less than ϵ from M, linked with H since the algebraic sum of the intersection numbers of $\mathfrak{x}_n$ in the ring R_ϵ differs from zero. (No intersection points of $\mathfrak{x}_n$ with H can correspond to points on L since the distance of such points from M is greater than ϵ.) We have

$$D_{R^*}[\mathfrak{x}_n] = D[\mathfrak{x}_n] - D_{B-R^*}[\mathfrak{x}_n]$$

and by (6.6)

$$D_{B-R^*}[\mathfrak{x}_n] > D_{B'}[\mathfrak{x}_n] > \alpha,$$

hence

$$D_{R^*}[\mathfrak{x}_n] < D[\mathfrak{x}_n] - \alpha.$$

Letting ϵ tend to zero and, accordingly, n to infinity and r_ϵ to 1, we obtain

$$\liminf . D_{R^*}[\mathfrak{x}_n] \leq \delta - \alpha. \tag{6.7}$$

By a conformal mapping we can transform R^* into the unit circle B and $\mathfrak{x}_n$ in R^* into a vector $\mathfrak{z}_n$ in B with $D[\mathfrak{z}_n] = D_{R^*}[\mathfrak{x}_n]$. The sequence $\mathfrak{z}_n$ is certainly an admissible sequence. Hence

$$\liminf . D[\mathfrak{z}_n] = \liminf . D_{R^*}[\mathfrak{x}_n] \leq \delta - \alpha.$$

in contradiction to (6.7); the vector $\mathfrak{x}$ is admissible. We have completed the existence proof with the additional result that $\delta = d$.

Remarks:

a) The solution of the variational problem satisfies a transversality condition analogous to that proved in § 3, 1, b.

b) In the special case where M degenerates to a Jordan curve γ, we obtain the solution of a problem similar to Plateau's problem but different in that a wider class of surfaces is admitted to competition: we no longer require that $\mathfrak{x}$ map the unit circle monotonically and continuously on the curve γ. It is sufficient to require that the boundary of $\mathfrak{x}$ be "on γ" in the sense of definition 6.1, and that $\mathfrak{x}$ be linked with the polygon H, as explained in article 1. The linking condition may be regarded as substitute for the monotonic mapping of β on γ.

The widening of the range of competition in this new problem might conceivably lead to a lower value for the minimum area. It has been proved, however, that the generalized problem has the same solutions as Plateau's problem.[8]

c) The problem and methods developed here for simply connected surfaces can be generalized to the case of minimal surfaces of higher topological structure. We shall dispense with a discussion of these generalizations.[9]

d) It should be stated that the assumption of a closed boundary surface M is not necessary, though convenient for the formulation of the proof. Obviously the character of a minimal surface $\mathfrak{x}$ as a relative minimum with a free boundary on M is not affected if M is replaced by an open or closed surface M' which is identical with M in a neighborhood of the set of boundary points of $\mathfrak{x}$ on M. Instead of considering linking numbers, we may restrict the free boundary set in other ways to obtain a compact closed class of admissible vectors.

5. *Properties of the Free Boundary. Transversality*

Beyond ascertaining the existence of minimal surfaces of least area with free boundaries, the preceding sections give little information about the nature of the "*trace*" of the minimal surface G, i.e. the set ρ of accumulation points of $\mathfrak{x}(u, v)$ as (u, v) approaches the part β_1 of β corresponding to the free boundary. To discuss properties of the trace under proper assumptions concerning M is a challenging task.

1. *Plane Boundary Surface. Reflection.* In the case where M is a plane the situation is quite simple. Transversality of a minimal surface G with respect to a plane M implies that G can be analytically extended by reflection in the plane (c.f. theorem 3.2); consequently we have

Theorem 6.4: The trace is an analytic curve and the tangent plane of G along the trace is orthogonal to the plane M.

[8] See Courant [12].

[9] See Davids [1].

Proof: Let $\mathfrak{x}(u,v)$ represent a minimal surface G having a free boundary on the plane $z = 0$ and furnishing a relative minimum for the Dirichlet integral. The free boundary may correspond to an interval $\langle u_1, u_2 \rangle$ on the boundary β: $v = 0$ of the parameter domain B: $v > 0$ in the u, v-plane. For the harmonic function $z(u,v)$ we have $z \to 0$ as $v \to 0$ for $u_1 < u < u_2$. Hence, by the principle of reflection for harmonic functions, z can be analytically extended across the interval $\langle u_1, u_2 \rangle$ so that $z(u, v) = -z(u, -v)$. Consider in B a semicircle H of radius a whose diameter lies in $\langle u_1, u_2 \rangle$. Since $\mathfrak{x}$ solves the minimum problem for the Dirichlet integral over the whole parameter domain, $D_H[\mathfrak{x}]$ is likewise a minimum among all vectors in H whose values on the semicircular boundary of H coincide with those of $\mathfrak{x}$ while the values corresponding to points on the diameter lie on $z = 0$. This condition does not affect the components $x(u,v)$ and $y(u,v)$; their boundary values for $v = 0$ are completely free. Writing

$$D_H[\mathfrak{x}] = D_H[x] + D_H[y] + D_H[z]$$

we see therefore that $x(u,v)$ is the solution of the problem of minimizing the Dirichlet integral in a semicircle when the boundary values are prescribed on the circular boundary, but free on the diameter. The same is true for $y(u,v)$ (with different prescribed boundary values). From theorem 1.3 it follows that $x(u,v)$, $y(u,v)$ are regular on the part of β bounding H and that $x_v = 0$, $y_v = 0$ there. By the principle of reflection we may extend $x(u,v)$ and $y(u,v)$ into the lower half-plane in such a way that $x(u,-v) = x(u,v)$, $y(u,-v) = y(u,v)$. Our statement is proved.

By way of example: the minimal surface bounded by a horizontal circle and having a free boundary on a horizontal plane is a catenoid, which meets the plane at right angles.

It should be noted that the principle of reflection and the orthogonality on a plane depend only on the *stationary character* of $D[\mathfrak{x}]$ (for a precise definition of the notion of stationary character see article 3 of this section); the result of this article is therefore generally valid for *stationary minimal surfaces*.

As a consequence of the principle of reflection we formulate

Lemma 6.1: Let G be a surface of least area bounded by the plane M and a Jordan arc γ joining two points of M. A plane E orthogonal to M and having no point in common with γ has also no point in common with G. In fact, G is half of a minimal surface G' whose boundary γ' consists of γ and the image of γ by reflection in M. If E is represented by $x = 0$, we may assume $x > 0$ on γ'. By the extreme value theorem for harmonic functions we know that $x > 0$ throughout G', hence throughout G.

Note that a minimal surface G of least area, bounded by a plane M: $z = 0$ and a given boundary curve γ in the half-space $z \geq 0$, lies entirely in this half-space. For the harmonic function $z(u,v)$, which is not a constant, has non-negative boundary values; hence $z > 0$ in the interior of G. Furthermore the trace of G on the plane M is a monotonic image of the corresponding segment of the u-axis in the parameter plane. For otherwise there would be a point P on this segment where $\mathfrak{x}_u^2 = \mathfrak{x}_v^2 = 0$. Let us assume that G is the image of the upper half of the disk $u^2 + v^2 < 1$. Then the harmonic function $z(u,v)$ must have a crossing point at P. Consequently a branch of the curve $z(u,v) = 0$ must issue from P into the interior of the half-disk, separating a region $z > 0$ from a region $z < 0$; this contradiction of the inequality $z > 0$ in H proves that the trace is a monotonic image of a segment of β.

2. *Surface of Least Area Whose Free Boundary Is Not a Continuous Curve.* The general question of characterizing the trace of G on M is quite complex. That there are inherent difficulties may be seen from the example sketched in the following paragraphs, which shows that the trace need not be a continuous curve.

To define the boundary surface M we take a cube one of whose faces is given by $z = 0$, $|x| \leq 2$, $|y| \leq 2$. We carve out of the interior of this cube an infinite number of straight ditches D_n, $n = 1,2,\cdots$, whose tops are the strips

$$z = 0, \qquad |x| \leq 1, \qquad \left| y - \frac{1}{n} \right| \leq \epsilon_n^3,$$

and whose bottoms are given by

$$z = -\epsilon_n, \qquad |x| \leq 1, \qquad \left| y - \frac{1}{n} \right| \leq \epsilon_n/2;$$

the side walls are formed by the connecting planes, the end walls by $|x| = 1$. We choose the positive quantities ϵ_n so that their sum is less than $1/4$. The top of each ditch is a plane rectangle of area $2\epsilon_n^3$; the longitudinal cross section area of the ditch is $2\epsilon_n$. We consider first the surface G_1 bounding the cube with a single ditch—say the ditch D_1—removed. As arc γ we choose the straight segment $|y| \leq \epsilon_1^3$, $x = 0$, $z = 0$, which bridges the ditch at the central line. It is easily seen that there are two symmetric solutions for the variational problem, each furnishing an absolute minimum: the two plane rectangular strips of area ϵ_1^3 covering half of the ditch. The free boundary is the corresponding rectangular half of the top edge of the ditch. We denote by G_1^+ the strip on which $x > 0$, by G_1^- the strip on which $x < 0$. A third solution is given by the central vertical cross section of the ditch; this solution, having an area larger than $\epsilon_1^2/2$, furnishes only a relative minimum.

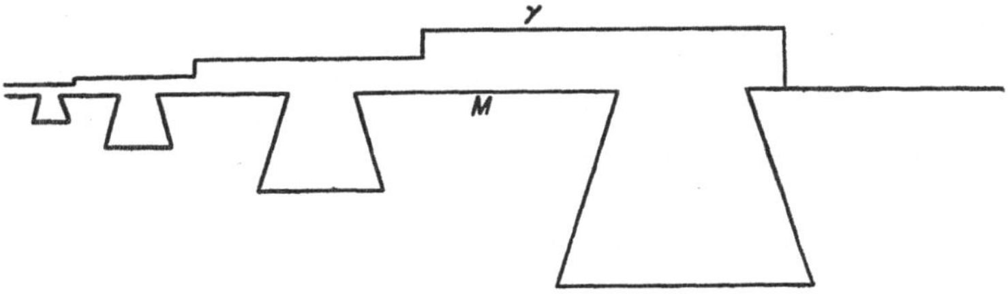

Figure 6.4 Problem with infinitely many solutions with discontinuous trace.

Next consider the continuous surface M bounding the cube with all the ditches D_n removed; the tops of the ditches have the segment $z = 0$, $y = 0$, $|x| \leq 1$ as limit line. As the curve γ we choose a polygon connecting the points $x = 0$, $y = 0$, $z = 0$, and $x = 0$, $y = 1 + \epsilon_1$, $z = 0$. The polygon γ is chosen to run closely above the center line of each ditch, and is composed alternately of vertical and horizontal segments. Above the ditch D_n there is a horizontal segment at a height η_n above the surface of the cube, with $\eta_n < \epsilon_n^4$; the vertical segments connect successive horizontal segments. As γ tends to a straight segment in the plane $z = 0$ (as $\eta_n \to 0$), the solutions G of the minimum problem converge to degenerate limit solutions. We denote by G_n^+ the surface for which $x > 0$, by G_n^- the surface for which $x < 0$, covering the ditch D_n. Each combination of surfaces G_n^+ and G_n^-, $n = 1, 2, \cdots$, is an absolute minimum for the degenerate case. Here the trace on M contains a sequence of rectangular edges of the respective halves of the ditches. Again we assume a continuity lemma similar to lemma 3.3 (a proof could be given by methods developed later in this chapter). We infer that for γ

sufficiently near the plane $z = 0$ there exist infinitely many minimal surfaces solving our minimum problem and that there is such a surface arbitrarily near any degenerate surface of the type just described. The traces of the latter accumulate on the line $|x| \leq 1$, $y = 0$, $z = 0$, or on half of it. The same is true of our non-degenerate minimal surfaces if γ is sufficiently close to M, i.e. if the η_n are chosen sufficiently small; hence the trace is not a continuous curve. The various solutions to the problem, moreover, form a non-denumerable set, a remarkable fact that supplements a corresponding observation in Chapter III, § 6, 2, with respect to non-uniqueness in the problem for fixed boundaries.

3. *Transversality*. From the preceding example we infer that, to ensure a trace with "reasonable" properties, we shall have to impose conditions on the boundary manifold M; such conditions are likewise needed to replace the weak transversality relation of § 3, 1, *b*, by the stronger statement that the free boundary is orthogonal in the ordinary sense.

From a systematic viewpoint it seems appropriate to introduce the general concept of transversality by a definition applicable to all free boundary problems in the calculus of variations. A minimal surface G with boundary ρ lying on a surface M is called *transversal* to M in the neighborhood of a point P of ρ where G remains on one side of M, if G has the following variational property: there exists a piece G' of G bounded by a Jordan arc γ and a portion ρ' of ρ containing P, and there exists an open portion M' of M containing ρ', such that G' furnishes the absolute minimum for the problem with boundary fixed on γ and free on M'. Employing transversality as a property in the small we define, in the large: a minimal surface G with boundary ρ on M is *stationary* with respect to M if it is transversal at every point P of ρ. A surface that is stationary in the large, however, need not furnish a minimum. The diameter surface G of a sphere M may serve as an example.

According to article 1, a minimal surface orthogonal to a plane M is transversal to M and automatically has an analytic trace ρ. Conversely, given a plane or, more generally, any analytic surface M, one can use the representation of minimal surfaces given in Chapter III, § 1, to construct a minimal surface orthogonal to M along a prescribed analytic trace ρ, as is known from differential geometry.[10]

[10] See e.g. Blaschke [1], Chapter VIII, § 111.

This surface can easily be seen to be transversal to M along ρ. If one were able to show that there are no other surfaces transversal to an analytic surface M, the result would be a proof that the trace ρ of surfaces solving the problem of the preceding articles is an analytic curve.

However, little progress has so far been made toward such a complete result. All that has been proved is that convexity of the boundary surface M is essentially[11] sufficient to ensure that the trace is a continuous and rectifiable curve, and that the minimal surface possesses almost everywhere on the trace a tangent plane orthogonal to a supporting plane of M.[12]

6. *Unstable Minimal Surfaces with Prescribed Polygonal Boundaries*

1. *Unstable Stationary Points for Functions of N Variables.* We turn to the subject of unstable minimal surfaces. As pointed out in §1, *a continuously differentiable function* $d(u_1, u_2, \cdots, u_N)$ *of* N *independent variables in a domain* C *possesses "unstable" stationary points (mini-maxima) if it possesses two separate relative minima at points* P_1 *and* P_2, *and if at the boundary of* C *the values of* d *become infinite.*[13] It is assumed that the two minima P_1 and P_2 are strict, i.e. that if $P \neq P_i$ is a point in the neighborhood of P_i, $d(P) > d(P_i)$, $i = 1, 2$. We shall not prove, however, that the unstable stationary point P_3 is a strict saddle. Generally it is true only that in every neighborhood of P_3 there exists a point P such that $d(P) < d(P_3)$.

We proceed to construct such a stationary point. Let P_1 and P_2 be two points in C where the relative minima d_1 and d_2 are attained. Join P_1 and P_2 in C by a continuous curve or, more generally, by a connected closed point set Σ. Denote by d_Σ the maximum value of $d(u_1, u_2, \cdots, u_N)$ in Σ, by P_Σ a point in Σ where this maximum is attained; the value d_Σ is greater than both d_1 and d_2. Finally we consider a sequence $\Sigma_1, \Sigma_2, \cdots$ of such connected closed point sets with maximum values $d_{\Sigma_1}, d_{\Sigma_2}, \cdots$ assumed at the points $P_{\Sigma_1}, P_{\Sigma_2}, \cdots$ so that

$$d_{\Sigma_n} \rightarrow d,$$

[11] The word "essentially" points to some slight further assumptions.

[12] See Courant [12].

[13] If C is unbounded, it is assumed that d becomes infinite also at infinity.

where d is the greatest lower bound of all possible maximum values d_Σ. Obviously $d > d_1$ and $d > d_2$.

Let $A_1, A_2, \cdots$ be a sequence of points in $\Sigma_1, \Sigma_2, \cdots$, respectively. The totality of all points of accumulation for all such sequences A_i defines a compact point set Σ which is connected and which contains P_1 and P_2. A suitable subsequence of the points P_{Σ_n} has, furthermore, a limit point

$$P_\Sigma = \lim_{n\to\infty} P_{\Sigma_n}.$$

The point P_Σ lies in Σ. By continuity $d(P_\Sigma) = d$, while for all other points of Σ the value $d(u_1, u_2, \cdots, u_N)$ does not exceed d; therefore $d_\Sigma = d$. We call Σ a "*minimizing connecting set*" joining P_1 and P_2.

In Σ there exists a closed set $\mathfrak{M}$ of points P in which the maximum $d(P) = d$ is attained. We shall show that *there is at least one point P in $\mathfrak{M}$ for which $d(P)$ is stationary*, i.e. for which the first derivatives of $d(u_1, u_2, \cdots, u_N)$ all vanish.

For the proof we transform the minimizing set Σ into a set Σ' by a deformation which takes the points P of Σ into points P' of Σ' in such a manner that $d(P') < d(P)$, except for the images of arbitrarily small neighborhoods of stationary points P. Hence if $\mathfrak{M}$ did not contain a stationary point we could replace Σ by another connecting set Σ' for which $d_{\Sigma'} < d$, contrary to the minimizing character of Σ.

The deformation is effected by *moving P into P' along the direction of the gradient* of the function $d(u_1, u_2, \cdots, u_N)$. Let $U: u_1, u_2, \cdots, u_N$ be a point of C and $V: v_1, v_2, \cdots, v_N$ an arbitrary vector at the point U. The first derivatives of $d(U)$ were supposed to be continuous in the domain C. Consequently for any fixed closed subdomain C^* of C and arbitrarily small δ, we can find a positive constant ϵ_δ so small that, for all points U in C^*,

$$d(U - \epsilon V) = d(U) - \epsilon V \cdot \operatorname{grad} d(U) + \epsilon\tau(\epsilon),^{14} \tag{6.8}$$

with $|\tau(\epsilon)| < \delta$, whenever $|\epsilon| < \epsilon_\delta$. We choose C^* so that it contains $\mathfrak{M}$.

Assume that everywhere in the closed set $\mathfrak{M}$

$$|\operatorname{grad} d(U)| \geq 4\alpha,$$

[14] $V \cdot \operatorname{grad} d(U)$ is the inner product of the two vectors.

α being a positive constant. Then we can choose a constant $a < 1$ so small that

$$|\operatorname{grad} d(U)| > 2\alpha$$

holds in the larger domain $\mathfrak{M}^*$ consisting of all points U whose distance to points in $\mathfrak{M}$ does not exceed a. We may assume that P_1 and P_2 do not belong to $\mathfrak{M}^*$, and that C^* contains $\mathfrak{M}^*$.

We apply (6.8) for $V = \operatorname{grad} d(U)$ to points of $\mathfrak{M}^*$ and obtain

$$(6.9) \qquad d(U - \epsilon V) \leq d(U) - 4\epsilon\alpha^2 + \epsilon\delta, \qquad 0 < \epsilon < \epsilon_\delta ;$$

if we choose $\delta = \alpha^2$ and ϵ_δ accordingly small, (6.9) becomes

$$d(U - \epsilon V) \leq d(U) - 3\epsilon\alpha^2.$$

Denote by P_0 a fixed point in $\mathfrak{M}$, by r distance from P_0. Consider all points U: $u_1, u_2, \cdots, u_N$ in $\mathfrak{M}^*$ whose distance from P_0 does not exceed a. In this sphere $r \leq a$ we replace each point U by U': $u_1', u_2', \cdots, u_N'$ according to

$$U' = U - \eta \operatorname{grad} U$$

with

$$\eta = \epsilon_\delta(a - r).$$

Then in the interior of the sphere

$$d(U') < d(U) < d,$$

while outside the sphere we set $U' = U$ and have

$$d(U) = d(U').$$

By a finite number of such spheres we can cover $\mathfrak{M}$; performing a succession of corresponding transformations $U \to U'$ we arrive at a transformation of C into itself in which all points outside $\mathfrak{M}^*$ remain unchanged. The set Σ is transformed into a compact connected set Σ' containing P_1 and P_2, and $d(U') < d$ everywhere in Σ', contrary to our assumption that Σ is a minimizing set. Hence the hypothesis $\alpha > 0$ in $\mathfrak{M}$ is absurd and Σ contains a stationary point of $d(U)$.

Let $\mathfrak{S}$ be the set of stationary points contained in $\mathfrak{M}$. It is closed because of the continuity of the derivatives of $d(U)$. Since $\mathfrak{M}$, and hence $\mathfrak{S}$, is a proper subset of Σ, there is a boundary point P_3 of $\mathfrak{S}$ in Σ. We prove

Lemma 6.2: In any neighborhood K of P_3 there exists a point P for which $d(P) < d(P_3)$.

Proof: Let Q be a point lying both in K and in the set $\Sigma - \mathfrak{S}$; then we have $d \geq d(Q)$. If $d(Q)$ were equal to d, there would be another point P in a neighborhood of Q contained in K, for which $d(P) < d(Q) = d$, since Q is not a stationary point. The lemma is proved.

The recent theory of critical points[15] has carried the analysis of stationary points much farther, by considering higher types of stationary points and establishing relations between their numbers and the topological characteristics of the domain C.

To generalize the theory of unstable stationary points to functionals in the calculus of variations one must cope with the difficulty that these functionals, in the relevant cases, are not continuous but merely semicontinuous in the respective function spaces. For the variational problem of minimal surfaces with given boundary, M. Shiffman, and independently M. Morse and C. Tompkins, have overcome this difficulty by an explicit analysis of Douglas' functional. An alternative approach is to reduce the problem of stationary values for functionals to a corresponding problem for differentiable functions of a finite number N of variables,[16] and subsequently to pass to the limit as $N \to \infty$.[17] We shall present this method for the case of simply connected minimal surfaces with piecewise smooth boundaries. The following articles deal with the first part of this program by treating the problem of unstable minimal surfaces spanned in polygons.

2. *A Modified Variational Problem.* Let γ be a simple polygonal contour with $N + 3$ vertices. We consider Dirichlet's integral $D[\mathfrak{x}]$ extended over the upper half-plane B: $v > 0$ with β: $v = 0$ as boundary. The space $\mathfrak{S}^*$ of admissible vectors consists of all vectors $\mathfrak{x}(u,v)$ continuous in $B + \beta$, with piecewise continuous first derivatives in B and finite Dirichlet integral, which map β continuously and monotonically on the polygon γ. The subspace of harmonic vectors in $\mathfrak{S}^*$ is called $\mathfrak{P}^*$. We may suppose that three fixed consecutive vertices of γ are images of three fixed points, e.g. $u = -1, 0, \infty$ on β. This normalization defines corresponding subspaces of $\mathfrak{S}^*$ and $\mathfrak{P}^*$,

[15] See Lusternik-Schnirelmann [1], [2] and M. Morse [1].

[16] See Courant [9].

[17] Shiffman [8] was the first who successfully carried out such a passage to the limit, see § 7.

which we denote by $\mathfrak{S}$ and $\mathfrak{P}$ respectively. From Chapter III, § 2, 2 we know that the subspace of $\mathfrak{P}$ defined by $D[\mathfrak{x}] \leq c$ is compact.

The other vertices of γ correspond consecutively to values $u_1, u_2, \cdots, u_N$ on β, forming the set (U). We modify the previous variational problem III to problem III-U: *To find for a fixed set (U) a vector $\mathfrak{z}$ of $\mathfrak{S}$ for which $D[\mathfrak{z}]$ is a minimum with respect to $\mathfrak{S}$.*

Problem III-U is solved exactly like Plateau's problem III in Chapter III; the solution $\mathfrak{z}$ is a vector of $\mathfrak{P}$, but, in general, does not represent a minimal surface. The value of the minimum is a function

$$d(U) = d(u_1, u_2, \cdots, u_N)$$

of the set (U) of N parameters u_i in the domain $0 < u_1 < u_2 < \cdots < u_N < \infty$. At the boundary of this domain $d(U)$ becomes infinite, as follows also from the results of Chapter III. For if $D[\mathfrak{z}]$ remained bounded as, e.g., $u_1 \rightarrow u_2$, we would have a set of admissible vectors $\mathfrak{z}$ with equally bounded Dirichlet integrals and with non-equicontinuous boundary values on β, in contradiction to lemma 3.2. The harmonic vectors $\mathfrak{z}$, depending on (U), which solve problem III-U are called *minimal vectors*; they form a subset $\mathfrak{M}$ of $\mathfrak{P}$.

By a transformation (3.12),

$$u = u' + \epsilon\Lambda,$$
$$v = v' + \epsilon\mathrm{M},$$

involving a parameter ϵ, where Λ and M are functions of u, v, and ϵ (or of u', v', and ϵ), we take the domain $B + \beta$ into itself in a one-to-one way. The functions Λ and M are supposed to be continuous in $B + \beta$ and to have piecewise continuous first derivatives absolutely bounded by a constant b; the quantities Λ and M for $\epsilon = 0$ are called λ and μ. As before in Chapter III, § 4, 1 we consider families of variations in $\mathfrak{S}^*$ of the following type: the vector $\mathfrak{x}$ of $\mathfrak{S}$ is varied into a vector $\mathfrak{z}$ of $\mathfrak{S}^*$ by the relation

$$\mathfrak{z}(u', v') = \mathfrak{x}(u, v).$$

Denoting by D^* Dirichlet's integral taken over B with respect to u', v', we have (3.16)

$$D^*[\mathfrak{z}] = D[\mathfrak{x}] + \tfrac{1}{2}\epsilon V(\mathfrak{x}, \lambda, \mu) + \epsilon^2 R,$$

where the first variation $V(\mathfrak{x},\lambda,\mu)$ of $D[\mathfrak{x}]$ is defined by (3.17)

$$V(\mathfrak{x}, \lambda, \mu) = \iint_B [p(\lambda_u - \mu_v) - q(\lambda_v + \mu_u)]\, du\, dv,$$

$$p = \mathfrak{x}_u^2 - \mathfrak{x}_v^2, \qquad q = -2\mathfrak{x}_u\mathfrak{x}_v,$$

with

$$| R | \leq c\, D[\mathfrak{x}],$$

the constant c depending only on the bound b for the first derivatives of Λ and M. If $\mathfrak{x}$ is in $\mathfrak{P}$, $p + iq = \phi(w)$ is an analytic function of $w = u + iv$ in B; in particular, the equation $\phi(w) = 0$ characterizes $\mathfrak{x}$ as a minimal surface.

The transformation (3.12) takes the system (U) into another system (U') of values $u_1', u_2', \cdots, u_N'$. In general, (3.12) will affect the points $u = -1, 0, \infty$, so that vectors of $\mathfrak{S}$ are transformed into vectors of $\mathfrak{S}^*$ no longer satisfying the three-point condition. However, by a suitable linear transformation of w, with real coefficients, superimposed on (3.12)—which leaves Dirichlet's integral and the domain B invariant—we return to a vector of $\mathfrak{S}$. Accordingly we may assume all variations normalized, so that we remain in $\mathfrak{S}$.

We define: $D[\mathfrak{x}]$ is stationary if
1) $\mathfrak{x}$ is harmonic,
2) the first variation V vanishes for all functions λ, μ or Λ, M admitted in (3.12).

Our previous remark shows that to establish the stationary character of a vector $\mathfrak{x}$ in $\mathfrak{S}$ it suffices to prove $V = 0$ for all λ, μ or Λ, M for which Λ vanishes at $u = -1, 0, \infty$ on β, so that the varied vector $\mathfrak{z}$ is also in $\mathfrak{S}$. We shall assume this normalization throughout the section.

In a sequence of lemmas, we shall analyze $d(U)$ as a function of the set (U): $u_1, u_2, \cdots u_N$. We first prove

Lemma 6.3: The minimal vector $\mathfrak{z}$ is uniquely determined by (U).

Proof: For fixed (U) the admissible vectors form a *convex* set,[18] i.e.

[18] Here is the only—but decisive—point where we use the assumption that γ is a polygon.

if two vectors $\mathfrak{z}_1$ and $\mathfrak{z}_2$ belong to (U), so does the linear combination $\mathfrak{x} = t\mathfrak{z}_1 + (1 - t)\mathfrak{z}_2, 0 \leq t \leq 1$. Setting $D[\mathfrak{z}_i] = d_i, i=1,2$, we have, by Schwarz' inequality,

$$D[\mathfrak{x}] \leq t^2 d_1 + (1 - t)^2 d_2 + 2t(1 - t) \sqrt{d_1 d_2}, \tag{6.10}$$

the equality sign holding only for $\mathfrak{z}_1 = \mathfrak{z}_2$. Suppose $\mathfrak{z}_1$ and $\mathfrak{z}_2$ were distinct minimal vectors, i.e. $d_1 = d_2 = d$; then (6.10) would imply $D[\mathfrak{x}] < d$, contrary to the minimum character of d. The lemma is proved; accordingly the space $\mathfrak{M}$ of minimal vectors is simply an N-parametric family.

The next lemma, although not indispensable for our immediate objective, will be of value later.

Lemma 6.4: An admissible vector $\mathfrak{x}$ which represents a minimal surface belongs to $\mathfrak{M}$.

Proof: On each *open* segment $S_i: u_i < u < u_{i+1}$ of β the vector $\mathfrak{x}$ is regular and harmonic, according to the principle of reflection for minimal surfaces (theorem 3.2). The first component of $\mathfrak{x}$ has on S_i a vanishing derivative with respect to v and the other two components are constant there (if the components are taken with respect to suitably chosen orthogonal axes). For a different admissible vector $\mathfrak{x} + \mathfrak{s}$ belonging to the same set (U), the last two components of $\mathfrak{s}$ vanish on S_i, so that $\mathfrak{s}\mathfrak{x}_v = 0$ everywhere on S_i.

To characterize $\mathfrak{x}$ as a member of $\mathfrak{M}$ we must show that

$$D[\mathfrak{x} + \mathfrak{s}] = D[\mathfrak{x}] + D[\mathfrak{s}] + 2D[\mathfrak{x},\mathfrak{s}] > D[\mathfrak{x}].$$

This inequality follows from the relation $D[\mathfrak{x},\mathfrak{s}] = 0$, which we obtain by integrating first over the domain B with small half-disks about the points of (U) omitted. Since $\Delta\mathfrak{x} = 0$, and since $\mathfrak{s}\mathfrak{x}_v = 0$ on S_i, Green's formula leaves only contributions from the circular arcs; these tend to zero as the radii of the small circles approach zero through properly chosen sequences.

Thirdly we show:

Lemma 6.5: The function $d(U)$ is continuous.

Proof: Consider the minimal vector $\mathfrak{z}$ belonging to (U) and denote by $\mathfrak{z}'$ the minimal vector belonging to the varied set (U'); we have

$d' = d(U') = D[\mathfrak{z}'] \leq D[\mathfrak{z}]$ and therefore, by (3.16),

$$(6.11) \qquad d(U') \leq d(U) + \tfrac{1}{2}\,\epsilon V(\mathfrak{z}, \lambda, \mu) + \epsilon^2 c\, d(U)$$

or, since $|\, V(\mathfrak{z}, \lambda, \mu) \,| \leq 4bD[\mathfrak{z}]$ by (3.18),

$$d(U') \leq d(U)(1 + 4b\epsilon + c\epsilon^2).$$

A similar inequality holds for the inverse transformation from (U') to (U); the lemma is therefore proved.

Furthermore we prove

Lemma 6.6: The minimal vector $\mathfrak{z}$ depends continuously on (U).

Proof: Let (U_n) be a sequence tending to (U), $\mathfrak{z}_n$ the corresponding minimal vectors, and $d_n = D[\mathfrak{z}_n]$ the minima of $D[\mathfrak{x}]$ in the problems III-U_n. Since $d_n \to d$ the values d_n are bounded. The vectors $\mathfrak{z}_n$ belong to a compact space, and at least a subsequence tends to a vector $\mathfrak{y}$ of $\mathfrak{P}$ corresponding to the set (U). The semicontinuity of Dirichlet's integral for harmonic vectors yields $D[\mathfrak{y}] \leq \lim d_n = d$, hence necessarily $D[\mathfrak{y}] = d$. By lemma 6.3, $\mathfrak{z}$ is uniquely determined by (U), i.e. $\mathfrak{y} = \mathfrak{z}$ is *the* minimal vector for (U) and *the* limit of the sequence $\mathfrak{z}_n$; for otherwise we could select a subsequence converging to another limit.

On the basis of the preceding lemmas a decisive conclusion can be drawn:

Lemma 6.7: The function $d(U)$ possesses continuous first derivatives with respect to u_i.

Proof: We now consider $\mathfrak{z}$ as the original vector and $\mathfrak{x}$ as obtained from the variation Λ, M; then for this variation we have

$$(6.12) \qquad d = d(U) \leq d(U') - \tfrac{1}{2}\,\epsilon V'(\mathfrak{z}', \lambda, \mu) + \epsilon^2 c\, d(U')$$

with V' defined by

$$(6.13) \qquad V'(\mathfrak{z}', \lambda, \mu) = \iint_B [p'(\lambda_u - \mu_v) - q'(\lambda_v + \mu_u)]\, du\, dv;$$

here we have written u, v as independent variables instead of u', v' and used the notation

$$p' = \mathfrak{z}_u'^2 - \mathfrak{z}_v'^2, \qquad q' = -2\mathfrak{z}_u'\mathfrak{z}_v'.$$

First we establish the relation

$$\lim_{\epsilon \to 0} V'(\mathfrak{z}', \lambda, \mu) = V(\mathfrak{z}, \lambda, \mu). \tag{6.14}$$

Given a small positive constant δ, we divide B into a closed interior domain B_1 and a boundary strip B_2 of width less than δ. Let V_1 and V_2 denote the integrals V taken over B_1 and B_2, respectively, V_1' and V_2' the corresponding integrals V'. By lemma 6.6 we have

$$\lim_{\epsilon \to 0} \mathfrak{z}' = \mathfrak{z},$$

hence the convergence of $\mathfrak{z}'$ is uniform in B_1 and by Harnack's theorem (see Chapter I, § 2), the same holds for the derivatives of $\mathfrak{z}'$. Hence we may pass to the limit under the integral sign in V_1', obtaining $V_1' \to V_1$. From (6.11) and (6.12) it follows that $D[\mathfrak{z}'] \to D[\mathfrak{z}]$. By choosing δ sufficiently small we can make $D_2[\mathfrak{z}]$, hence also $D_2[\mathfrak{z}']$, arbitrarily small uniformly in ϵ. The form of the integral V_2' shows, since the derivatives of λ and μ are bounded, that V_2' remains arbitrarily small independently of ϵ if the width δ is chosen sufficiently small. Hence we have $V' \to V$ as well.

From (6.11) it follows that $d(U')$ is bounded in ϵ for sufficiently small ϵ For $\epsilon > 0$ we infer from (6.11) and (6.12) that

$$\tfrac{1}{2} V'(\mathfrak{z}', \lambda, \mu) - \epsilon c\, d(U') \leq \frac{d(U') - d(U)}{\epsilon} \leq \tfrac{1}{2} V(\mathfrak{z}, \lambda, \mu) + \epsilon c\, d(U). \tag{6.15}$$

Hence we have, by (6.14),

$$\lim_{\epsilon \to 0} \frac{d(U') - d(U)}{\epsilon} = \tfrac{1}{2} V(\mathfrak{z}, \lambda, \mu). \tag{6.16}$$

If λ and μ are specialized in such a way that only u_i is changed (so that, in particular, $\mu = 0$), relation (6.16) establishes the existence of the partial derivative $\partial d(U)/\partial u_i$. Since V, according to the preceding arguments, depends continuously on (U) and on the values of λ in (U), it is clear that the derivatives of $d(U)$ are continuous functions of the u_i. Consequently

$$\tfrac{1}{2} V(\mathfrak{z}, \lambda, \mu) = \sum_{i=1}^{N} \lambda_i \frac{\partial d(U)}{\partial u_i}, \tag{6.17}$$

where λ_i is the value of λ at the point $u = u_i$, $v = 0$.

The continuity of the partial derivatives also follows from explicit expressions obtained if we choose the variation (3.12) as a piecewise linear transformation which leaves $v' = v$ unchanged, slightly changes one of the values u_i—e.g. takes u_2 into $u_2 + \epsilon$—and does not affect the others:

$$u' = \begin{cases} u_1 + (u - u_1)\left(\dfrac{\epsilon}{u_2 - u_1} + 1\right), & u_1 \leq u \leq u_2, \\ u_3 + (u - u_3)\left(\dfrac{\epsilon}{u_2 - u_3} + 1\right), & u_2 \leq u \leq u_3, \\ u, & u < u_1,\ u > u_3. \end{cases}$$

The function $\lambda = \Lambda$ is given by $-(u - u_1)/(u_2 - u_1)$ and $-(u - u_3)/(u_2 - u_3)$ in the respective segments, while $\mu = \mathrm{M}$ is zero. Formula (6.16) yields

$$\frac{\partial d(u_1, u_2, \cdots, u_N)}{\partial u_2} = -\int_0^\infty dv \left[\int_{u_1}^{u_2} \frac{\mathfrak{z}_u^2 - \mathfrak{z}_v^2}{u_2 - u_1}\, du + \int_{u_2}^{u_3} \frac{\mathfrak{z}_u^2 - \mathfrak{z}_v^2}{u_2 - u_3}\, du\right],$$

and this explicit expression again shows that the derivative is a continuous function of the values (U).

3. *Proof That Stationary Values of $d(U)$ Are Stationary Values for $D[\mathfrak{x}]$.* If the function $d(U)$ is stationary for a set (U), i.e. if all the derivatives $\partial d/\partial u_i$ vanish, the corresponding vector $\mathfrak{z}$ makes $D[\mathfrak{z}]$ stationary and hence represents a minimal surface. This follows immediately from the fact (proved in Chapter III, § 4, 3) that the vanishing of V for all admissible λ, μ implies $\phi(w) = 0$. From lemma 6.4 we may easily infer the *complete* equivalence of the problem of stationary values of $d(U)$ with that of minimal surfaces spanned in γ, since every minimal surface $\mathfrak{x}$ spanning γ is represented by a vector $\mathfrak{z}$ of $\mathfrak{M}$.

As a conclusion we state

Theorem 6.5: Unstable minimal surfaces spanned in γ correspond to unstable stationary points for a function $d(u_1, u_2, \cdots, u_N)$ with continuous first derivatives. The theory of the former is reduced to that of the latter.

More precisely, suppose that P_3 is a stationary point of $d(U)$ with the property established in lemma 6.2, that every neighborhood of P_3 contains points P for which $d(P) < d(P_3)$. To the coordinates (U) of such points P correspond variations of the minimal surface which decrease $D[\mathfrak{x}]$. On the other hand, there are obviously variations in $\mathfrak{S}$ which increase $D[\mathfrak{x}]$. Hence the minimal surface corresponding to P_3 represents a strict saddle point.

Since minimal surfaces furnishing strict relative minima of Dirichlet's integral obviously belong to $\mathfrak{M}$ and yield strict relative minima for $d(u_1, u_2, \cdots, u_N)$, we obtain in particular from § 6, 1

Theorem 6.6: If γ bounds two different minimal surfaces, each furnishing a strict relative minimum, there exists at least one strictly unstable minimal surface spanned in γ.

4. *Generalization.* The preceding result can be generalized by a method best explained in geometric phraseology. Let two admissible vectors $\mathfrak{y}_1(u,v)$ and $\mathfrak{y}_2(u,v)$ in $\mathfrak{S}$, with $d_1 = D[\mathfrak{y}_1] \leq D[\mathfrak{y}_2] = d_2$, be connected by a connected set Σ of admissible vectors $\mathfrak{y}(u,v)$, and let d_Σ be the least upper bound (or maximum) of $D[\mathfrak{y}]$ in Σ, or the "*elevation*" of Σ.[19] The difference $d_\Sigma - d_2$ is called the "*relative elevation*" of the connecting set Σ. The minimum or greatest lower bound of all d_Σ may be denoted by d. If $d > d_2$, i. e. $d - d_2 = h > 0$, we say that $\mathfrak{y}_1$ and $\mathfrak{y}_2$ are separated in $\mathfrak{S}$ by a "*wall*" *of relative elevation* h and *absolute elevation* d. A connecting set whose elevation equals the elevation d of the wall will be referred to as a "*minimizing*" *connecting set.* We prove first:

Lemma 6.8: Let $\mathfrak{x}_1$, $\mathfrak{x}_2$ be the harmonic vectors in $\mathfrak{P}$ with the same boundary values as $\mathfrak{y}_1$, $\mathfrak{y}_2$, respectively; then $\mathfrak{x}_1$ and $\mathfrak{x}_2$ are separated in $\mathfrak{P}$ by a wall whose elevation is at least equal to that of the wall separating $\mathfrak{y}_1$ and $\mathfrak{y}_2$ in $\mathfrak{S}$.

Proof: Let $\mathfrak{y}$ be an admissible vector in $\mathfrak{S}$, $\mathfrak{x}$ the vector in $\mathfrak{P}$ with the same boundary values. We construct the connecting set

$$\mathfrak{y}(t) = \mathfrak{x} + (1 - t)(\mathfrak{y} - \mathfrak{x}), \qquad 0 \leq t \leq 1.$$

[19] The only essential difference in the reasoning for the functional $D[\mathfrak{x}]$ in a function space from that in § 6, 1 stems from the fact that $D[\mathfrak{x}]$ is merely semicontinuous.

The vectors $\mathfrak{y}(t)$ have the same boundary values as $\mathfrak{y}$ and $\mathfrak{y}(0) = \mathfrak{y}$, $\mathfrak{y}(1) = \mathfrak{x}$. Furthermore we have

$$D[\mathfrak{y}(t)] = D[\mathfrak{x}] + (1 - t)^2 D[\mathfrak{y} - \mathfrak{x}],$$

since $D[\mathfrak{x}, \mathfrak{y} - \mathfrak{x}] = 0$ (cf. Chapter I, § 2). The vector $\mathfrak{y}(t)$ represents a deformation of $\mathfrak{y}$ into $\mathfrak{x}$, depending continuously on the parameter t, for which Dirichlet's integral decreases monotonically as t increases from 0 to 1. Such a monotonic continuous deformation will be called a "*retraction*." We denote by L_1 and L_2 the sets of surfaces $\mathfrak{y}(t)$ retracting $\mathfrak{y}_1$ into $\mathfrak{x}_1$ and $\mathfrak{y}_2$ into $\mathfrak{x}_2$, respectively. If Σ^* is a set connecting $\mathfrak{x}_1$ and $\mathfrak{x}_2$ in $\mathfrak{P}$, the elevation of Σ^* must be at least equal to the elevation d of the wall separating $\mathfrak{y}_1$ and $\mathfrak{y}_2$ in $\mathfrak{S}$; for otherwise $L_1 + \Sigma^* + L_2$ would be a set of elevation less than d connecting $\mathfrak{y}_1$ and $\mathfrak{y}_2$ in $\mathfrak{S}$. The lemma is proved; note that it remains true if γ is not assumed to be a polygon, since this assumption has not been used in the proof.

If γ is a polygon, a vector $\mathfrak{x}$ in $\mathfrak{P}$ can be still further retracted in $\mathfrak{P}$ into the vector $\mathfrak{z}$ in $\mathfrak{M}$ belonging to the same set (U) as $\mathfrak{x}$. The reasoning is the same as before: along the connecting set in $\mathfrak{P}$ given by

$$\mathfrak{x}(t) = \mathfrak{z} + (1 - t)(\mathfrak{x} - \mathfrak{z}), \qquad 0 \leq t \leq 1,$$

the values of Dirichlet's integral decrease monotonically; for

$$D[\mathfrak{x}(t)] = D[\mathfrak{z}] + (1 - t)^2 D[\mathfrak{x} - \mathfrak{z}],$$

in consequence of the relation $D[\mathfrak{z}, \mathfrak{x} - \mathfrak{z}] = 0$ which follows in the usual way from the fact that

$$D[\mathfrak{z}] < D[\mathfrak{z} + \epsilon(\mathfrak{x} - \mathfrak{z})],$$

for ϵ different from zero. We conclude, for polygonal contours γ:

Lemma 6.9: If two vectors $\mathfrak{x}_1$, $\mathfrak{x}_2$ of $\mathfrak{P}$ are separated in $\mathfrak{P}$ by a wall of elevation greater than a positive quantity A, the vectors $\mathfrak{z}_1$, $\mathfrak{z}_2$ of $\mathfrak{M}$ belonging to the same sets (U) as $\mathfrak{x}_1$, $\mathfrak{x}_2$, respectively, are separated in $\mathfrak{M}$ by a wall of elevation greater than A.

The following generalization of theorem 6.6 is now readily proved:

Theorem 6.7: If a polygon γ bounds two surfaces $\mathfrak{y}_1$, $\mathfrak{y}_2$ in $\mathfrak{S}$, separated in $\mathfrak{S}$ by a wall, γ must bound an unstable minimal surface. Note that this result does not presuppose the existence of two isolated relative minima.

Proof: We know from the preceding discussion that the two vectors $\mathfrak{z}_1$, $\mathfrak{z}_2$ in $\mathfrak{M}$ belonging to the same sets (U) as $\mathfrak{y}_1$, $\mathfrak{y}_2$, respectively, are separated by a wall in $\mathfrak{M}$. Consequently the function $d(u_1, u_2, \cdots, u_N)$ has an unstable stationary point (cf. article 1) and the polygon γ bounds an unstable minimal surface.

5. *Remarks on a Variant of the Problem and on Second Variation.* Recent investigations by I. Marx and M. Shiffman [1] have shown that the function $d(u_1, u_2, \cdots, u_N)$ can be used to classify unstable minimal surfaces. This classification depends on the second derivatives of $d(u_1, u_2, \cdots, u_N)$, whose existence and continuity could not be easily ascertained for the problem formulated in article 2. We therefore relax the conditions of that problem by dropping the requirement of monotonicity for the mapping of the segments

$$u_i \leq u \leq u_{i+1}$$

of $v = 0$ on the corresponding edge of the polygon; instead we permit the admissible vectors to map each segment onto the straight line containing the edge of the polygon, fixing only the images of the points u_i and u_{i+1} as the corresponding vertices of the polygon but allowing multiple covering and even overshooting of the ends by the image of the segment on the v-axis. This new problem will be called Problem M.

This problem (which, incidentally, does not lead readily to the generalization in the next section) allows an analysis in the following steps, which will be described in detail in the paper by Marx and Shiffman. First, it can be shown that the minimal vectors depend *differentiably* on the parameters u_1, u_2, $\cdots$, u_N. This in turn implies that the function $d(u_1, u_2, \cdots, u_N)$ has continuous *second* derivatives.

Now let (U) be a stationary set for $d(U)$. Construct the matrix of second derivatives $\partial^2 d/\partial u_i \, \partial u_j$ at this point. Then the following result is obtained: the number of negative eigenvalues of this matrix is equal to the number of negative eigenvalues of the operator

$$L[c] = -(c_{uu} + c_{vv}) + 2KWc$$

under the boundary condition $c(u, 0) = 0$. The operator $L[c]$ is the Euler expression associated with the *second variation* of the surface area:

$$\iint [c_u^2 + c_v^2 + 2KWc^2]\, du\, dv,$$

where c stands for the variation in the direction of the surface normal, and K and W are the Gauss curvature and surface element, respectively, of the minimal vector corresponding to (U).

Incidentally, for the simplified problem described in this article, it is not hard to show (see Marx, Shiffman [1]) that, for a minimal vector, $\phi(w)$ has at most poles at $u_1, u_2, \cdots u_N$ and at $-1, 0$:

$$\phi(w) = \sum_{i=1}^{N} \frac{R_i}{w - u_i} + \frac{R_{N+1}}{w + 1} + \frac{R_{N+2}}{w}.$$

Furthermore, if one substitutes this expression of $\phi(w)$ into the formula for the derivatives of d, one readily gets

$$\frac{\partial d(u_1, u_2, \cdots u_N)}{\partial u_i} = \frac{\pi}{2} R_i.$$

7. *Unstable Minimal Surfaces in Rectifiable Contours*

1. *Preparations. Main Theorem.* The preceding result will now be extended by a passage to the limit to more general classes of boundary curves γ, e.g. rectifiable curves.[20] We confine ourselves first to the more special class $\mathfrak{K}$ of *rectifiable Jordan curves* γ consisting of a *finite number of arcs with continuously varying direction*; no assumption concerning continuity or existence of the tangent is made for the end points of these arcs. A curve γ of the class $\mathfrak{K}$ has the following property:

Lemma 6.10: Let σ be an arbitrarily small constant. We can approximate the curve γ by a polygon γ_σ, and transform the whole $\mathfrak{x}$-space continuously into an $\mathfrak{x}'$-space, in such a way that an admissible vector $\mathfrak{x}(u,v)$ for the boundary γ is taken into an admissible vector $\mathfrak{x}'(u,v)$ for the boundary γ_σ satisfying the inequality

$$D[\mathfrak{x}'] < (1 + \sigma)D[\mathfrak{x}]. \tag{6.18}$$

The values of $\mathfrak{x}'$ on the u-axis β differ, for corresponding values of u, from those of $\mathfrak{x}$ by less than a quantity depending only on σ and

[20] Cf. footnote 17 on page 226.

going to zero with σ. If σ_i is a sequence of positive numbers tending to zero, the polygons γ_{σ_i} can be so constructed as to approximate γ in the Fréchet sense, the length of γ_{σ_i} tending to the length of γ. The proof of this lemma will be postponed (see § 8).

Our objective is

Theorem 6.8 (*Main Theorem*): Let $\mathfrak{x}_1$, $\mathfrak{x}_2$ be two admissible vectors spanned in γ and separated by a wall. Then the wall has finite elevation d; the vectors $\mathfrak{x}_1$ and $\mathfrak{x}_2$ are contained in a minimizing connecting set Σ which also contains a strictly unstable minimal surface $\mathfrak{x}$; and

$$D[\mathfrak{x}] = d$$

while for all $\mathfrak{y}$ in Σ

$$D[\mathfrak{y}] \leq d.$$

In particular the theorem holds if $\mathfrak{x}_1$ and $\mathfrak{x}_2$ are two separate relative minima.

Remarks: We note that from the outset we may restrict ourselves to harmonic vectors $\mathfrak{x}$, since a non-harmonic vector $\mathfrak{y}$ can be retracted into a harmonic vector with the same boundary values (cf. lemma 6.8), and since this process of retraction, if applied to all surfaces of a connected set Σ, leads to a connected set of harmonic surfaces. Likewise we may assume that all admissible vectors $\mathfrak{x}$ are subject to the three-point condition, i.e. belong to the vector space $\mathfrak{P}$. If $\mathfrak{x}_1$ or $\mathfrak{x}_2$ is not a minimal surface, we can retract it in $\mathfrak{P}$ into a minimal surface by the method of descent of Chapter III, § 7.

Proof: We recall the following results of Chapter III: the lower semicontinuous dependence of the minimum of Dirichlet's integral on the boundary curve (theorem 3.1), the compactness property of harmonic vectors as used for the proof of semicontinuity (§ 2, 3), and the continuity theorem for the area of minimal surfaces with rectifiable boundaries (§ 8, 4).[21] The proof consists of the following steps:

a) We replace γ and Σ by γ_σ and Σ_σ respectively;
b) We apply theorem 6.7 for each value of σ;
c) We pass to the limit as $\sigma \to 0$.

[21] The use of this third property is the specific idea that makes a passage to the limit possible, see Shiffman [8].

We assume a wall between the (harmonic) vectors $\mathfrak{x}_1$ and $\mathfrak{x}_2$ whose relative elevations satisfy the inequality

$$d - d_1 > d - d_2 > 2\alpha > 0.$$

The absolute elevation $d = \infty$ is not excluded *a priori*, but will be shown impossible in the course of the proof. Consider a sequence of positive numbers $\sigma = \sigma_i$ tending to zero and a corresponding approximation of γ by polygons γ_σ to which we shall apply lemma 6.10. If $\mathfrak{x}$ is a vector of $\mathfrak{P}$ spanning γ, every γ_σ bounds, by lemma 6.10, an admissible vector $\mathfrak{x}' = \mathfrak{x}_\sigma$ for which inequality (6.18) holds. By Dirichlet's Principle we can retract $\mathfrak{x}_\sigma$ into the harmonic vector with the same boundary values without invalidating (6.18). For $\sigma \to 0$ these harmonic vectors—which we again call $\mathfrak{x}_\sigma$—tend to $\mathfrak{x}$, in consequence of the last statement of the lemma. We denote by $\mathfrak{P}_\sigma$ the space of harmonic vectors in γ_σ. It may be assumed that the three fixed points on γ also lie on γ_σ and that the same three-point condition is satisfied for $\mathfrak{P}_\sigma$ as for $\mathfrak{P}$.

Since γ_σ is a polygon, we may retract every vector $\mathfrak{x}_\sigma$ of $\mathfrak{P}_\sigma$ into a vector $\mathfrak{z}_\sigma$ of the subspace $\mathfrak{M}_\sigma$ of $\mathfrak{P}_\sigma$, as in lemma 6.9. For every connected set Σ of vectors $\mathfrak{x}$ in $\mathfrak{P}$, bounded by γ, we thus have a connected set Σ_σ of vectors $\mathfrak{z}_\sigma$ in $\mathfrak{M}_\sigma$ for which

$$D[\mathfrak{z}_\sigma] \leq (1 + \sigma)D[\mathfrak{x}].$$

The boundary values of $\mathfrak{z}_\sigma$ differ from those of $\mathfrak{x}$ by less than a quantity depending only on σ and tending to zero with σ. In particular, we obtain from $\mathfrak{x}_1$, $\mathfrak{x}_2$ first $\mathfrak{x}_{1,\sigma}, \mathfrak{x}_{2,\sigma}$ and then $\mathfrak{z}_{1,\sigma}, \mathfrak{z}_{2,\sigma}$, with

$$\begin{aligned} D[\mathfrak{z}_{1,\sigma}] &= d_{1,\sigma} \leq (1 + \sigma)d_1 , \\ D[\mathfrak{z}_{2,\sigma}] &= d_{2,\sigma} \leq (1 + \sigma)d_2 . \end{aligned} \tag{6.19}$$

If $\mathfrak{x}_1$ and $\mathfrak{x}_2$ are connected by a set Σ in which $D[\mathfrak{x}] \leq a$, $\mathfrak{z}_{1,\sigma}$ and $\mathfrak{z}_{2,\sigma}$ can be connected by a set of vectors $\mathfrak{z}_\sigma$ in $\mathfrak{M}_\sigma$ for which $D[\mathfrak{z}_\sigma] \leq (1+\sigma)a$.

By the results of § 6, 4 we can connect $\mathfrak{z}_{1,\sigma}$ and $\mathfrak{z}_{2,\sigma}$ by a minimizing set Σ_σ with elevation d_σ, and consequently

$$d_\sigma \leq (1 + \sigma)a. \tag{6.20}$$

In Σ_σ we have elements $\mathfrak{t}_\sigma$ at the "top," i.e. elements for which $D[\mathfrak{t}_\sigma] = d_\sigma$, while $D[\mathfrak{z}_\sigma] \leq d_\sigma$ for all $\mathfrak{z}_\sigma$ in Σ_σ.

According to theorem 6.7, three possibilities arise: 1) $d_\sigma = d_{1,\sigma}$, 2) $d_\sigma = d_{2,\sigma}$, 3) there is a wall between $\mathfrak{z}_{1,\sigma}$ and $\mathfrak{z}_{2,\sigma}$, whose top $\mathfrak{t}_\sigma$ may be assumed to be a minimal surface. In all three cases we have a fixed bound A, independent of σ, for $D[\mathfrak{z}_\sigma]$ in Σ_σ. In the first two cases this follows from (6.19), in the third case from the isoperimetric inequality (theorem 3.5), which states that $D[\mathfrak{t}_\sigma] \leq L^2/\pi$ if we assume γ_σ so close to γ that the length of γ_σ does not exceed $2L$, L being the length of γ.

Let σ tend to zero through a sequence $\sigma_1, \sigma_2, \cdots$, and denote by δ the limit or least upper bound of d_{σ_i} as $\sigma_i \to 0$. We consider—as in § 6, 1—the totality of all sequences of vectors $\mathfrak{x}_{\sigma_i}$ in Σ_{σ_i} and denote by Σ the set of all limit vectors of such sequences. Because of the compactness of our harmonic vectors, Σ is a connected set in $\mathfrak{P}$ containing $\mathfrak{x}_1$ and $\mathfrak{x}_2$. Furthermore, by the semicontinuity of Dirichlet's integral in its dependence on the boundary, we have $D[\mathfrak{x}] \leq \delta$ for all $\mathfrak{x}$ in Σ. Since the relative elevation of Σ above d_2 was supposed to be at least a positive quantity 2α, we have

$$\delta \geq d_2 + 2\alpha.$$

Consequently (6.19) yields, for suitable large i, i.e. σ_i sufficiently small

$$d_{\sigma_i} \geq d_{1,\sigma} + \alpha, \qquad d_{\sigma_i} \geq d_{2,\sigma} + \alpha.$$

The set Σ_{σ_i} has positive relative elevation, so that the possibilities 1) and 2) above are excluded; $\mathfrak{t}_{\sigma_i}$ is a minimal surface. By the property of compactness, at least a sub-sequence of the $\mathfrak{t}_{\sigma_i}$ has as limit a minimal surface $\mathfrak{x}$ in Σ. We may assume $\mathfrak{x} = \lim_{\sigma_i \to 0} \mathfrak{t}_{\sigma_i}$ and $\delta = \lim_{\sigma_i \to 0} d_{\sigma_i}$, with $d_{\sigma_i} = D[\mathfrak{t}_{\sigma_i}]$. Since the length of γ_σ tends to that of γ, we have by theorem 3.6,

$$D[\mathfrak{x}] = \lim_{\sigma_i \to 0} d_{\sigma_i} = \delta,$$

while the Dirichlet integrals of all other vectors in Σ are of value at most δ, because of the semicontinuity. In other words, $\mathfrak{x}$ is the "top" of Σ, i.e. $\mathfrak{x}$ does not furnish a relative minimum.

It remains to ascertain that Σ with $\mathfrak{x}$ on its "top" is a minimizing set connecting $\mathfrak{x}_1$ and $\mathfrak{x}_2$ in $\mathfrak{P}$ (and consequently that $\delta = d$). Inequality (6.20) is valid for the elevation a of any connected set Σ

in $\mathfrak{P}$ containing $\mathfrak{x}_1$ and $\mathfrak{x}_2$. Letting σ_i tend to zero, we find that

$$a \geq \lim d_{\sigma_i} = \delta.$$

In other words, δ is the absolute elevation between $\mathfrak{x}_1$ and $\mathfrak{x}_2$ in $\mathfrak{P}$, i.e. $\delta = d$: our proof is complete.

2. *Remarks and Generalizations.* The result could be generalized for arbitrary rectifiable curves by essentially the same method; however, complications would arise in the approximation of γ by polygons γ_σ. These were avoided by Shiffman, who introduced a different approximation of the problem for γ by one involving only N parameters. Let the unit circle B serve as parameter domain, with polar coordinates r, θ. We subdivide γ by $N + 3$ points $A_1, A_2, \cdots, A_{N+3}$, so that for $N \to \infty$ the distance between two consecutive points will tend to zero. On the boundary β of B we mark off $N + 3$ points $Q_1, Q_2, \cdots, Q_{N+3}$ with the coordinates $\theta_1, \theta_2, \cdots, \theta_{N+3}$. The arcs A_iA_{i+1} of γ may be imbedded in convex point sets K_i (whose diameters later will shrink uniformly to zero as N tends to infinity). Vectors $\mathfrak{x}(r,\theta)$ are admissible if they are continuous and have piecewise continuous first derivatives in B. Furthermore their boundary values corresponding to the arcs Q_iQ_{i+1} are required to be in K_i, and concentric circles $r < 1$ are to be mapped by $\mathfrak{x}(r,\theta)$ on lines of lengths not exceeding the length L of γ. Finally we impose a condition equivalent to the three-point condition by selecting three fixed points on β and stipulating that the boundary sets K_i belonging to the intervals that contain these points are to contain three fixed points on γ. For a fixed set of values $Q_1, Q_2, \cdots, Q_{N+3}$ our admissible vectors form a convex manifold, and this fact again is the key to the solution of the problem $D[\mathfrak{x}] = \text{minimum} = d(\theta_1, \theta_2, \cdots, \theta_{N+3})$. The theory of the minimum problem and the result are exactly the same as in § 6. The passage to the limit, as N tends to infinity and the maximum diameter of the convex sets K_i tends to zero, can be carried out in a way similar to that of article 1. For further details, reference is made to Shiffman [8].

Another remark concerns the extension of the theory to k contours, $k > 1$. For two contours no essentially new element enters; for $k > 2$ it is necessary to make a more detailed study of the possible degenerations of the domain B (see e.g. Chapter IV, § 2, 3), since degenerated surfaces must be admitted in the proof of the theorem. Such questions are treated in papers by Shiffman [7] and Morse and Tompkins [5].

8. *Continuity of Dirichlet's Integral under Transformation of $\mathfrak{x}$-Space*

It remains to prove lemma 6.10. To this end we apply the method of transformation of the $\mathfrak{x}$-space already used in Chapter IV, § 3, 3. The curve γ is transformed into a polygon γ_σ by a finite sequence of elementary transformations of the space, called "pinching" processes, each of which affects only the immediate neighborhood of a point or a segment:

a) *δ, t-pinching about a point P.* This transformation takes vectors $\mathfrak{x}(u, v)$ into vectors $\mathfrak{y}(u, v)$ defined in the same domain B, continuously in the parameters δ, t. For convenience, we choose P as the origin. The pinching leaves fixed all straight lines through P, transforms the points of the sphere

$$r = (x_1^2 + x_2^2 + \cdots + x_m^2)^{1/2} \leq \delta$$

into the origin, and leaves the coordinates of all points outside a larger sphere $r > t\delta$ unchanged. We define

$$y_\nu = p(r)x_\nu, \qquad \nu = 1, 2, \cdots, m,$$

where

$$p(r) = \begin{cases} 0, & r < \delta, \\ \dfrac{1}{\log t} \log \dfrac{r}{\delta}, & \delta \leq r \leq t\delta, \\ 1, & r > t\delta. \end{cases}$$

For $\delta = \eta^2$ and $t = \eta^{-1}$, this transformation is identical to that used to prove lemma 4.2. As in that proof, we have

$$D[\mathfrak{y}] \leq \frac{1}{2} \iint_B (1 + \epsilon)^2(\mathfrak{x}_u^2 + \mathfrak{x}_v^2)\, du\, dv, \tag{6.21}$$

where the function $\epsilon(x_1, x_2, \cdots, x_m) = \epsilon(r)$ is given by

$$\epsilon = \begin{cases} \dfrac{1}{\log t}, & r < t\delta, \\ 0, & r \geq t\delta. \end{cases}$$

(If we suppose $\mathfrak{x}$ to be a harmonic vector, $r < t\delta$ and $r \geq t\delta$ define subregions of B with analytic boundaries.)

b) *δ, t-pinching about a straight segment S.* Without loss of generality we may choose for S the segment

$$x_2 = x_3 = \cdots = x_m = 0; \qquad |x_1| < a.$$

We define the transformation for $|x_1| \leq a$ by

$$y_1 = x_1, \qquad y_i = p(\rho)x_i, \qquad i=2,3,\cdots,m,$$

with $\rho^2 = x_2^2 + x_3^2 + \cdots + x_m^2$ and $p(\rho)$ as above; for $|x_1| > a$ the transformation is defined as δ, t-pinching about the points $x_1 = a$ and $x_1 = -a$ on the segment S. By this pinching process a cylindrical bar about S with two half-spherical ends is contracted into S.

Again we find the same inequality (6.21) as for the pinching about a point if $\epsilon = 1/\log t$ for all points of B corresponding to vectors affected by the pinching and $\epsilon = 0$ elsewhere in B.

By carrying out successively N pinching processes about different points and lines, possibly with different values $\delta = \delta_i$, $\epsilon = \epsilon_i$, we obtain from $\mathfrak{x}$ a vector $\mathfrak{y}$ with

$$D[\mathfrak{y}] \leq \frac{1}{2} \iint_B (1 + \epsilon_1)^2 \cdots (1 + \epsilon_N)^2(\mathfrak{x}_u^2 + \mathfrak{x}_v^2)\, du\, dv,$$

the ϵ_i being defined like ϵ. If at most two pinching processes affect the same point in space, i.e. if no more than two quantities ϵ_i differ from zero at the same point, we have

$$D[\mathfrak{y}] \leq (1 + \epsilon)^4 D[\mathfrak{x}],$$

where ϵ now denotes the largest value $1/\log t_i$ employed in our pinching processes.

In applying these transformations we first assume that the points of discontinuity of the tangent to γ are simple vertices in which the direction of γ has only a jump discontinuity, and at which the adjacent arcs are not tangent to each other. We inscribe in γ a polygon γ' whose vertices include the vertices of γ and whose edges are between τ and 2τ in length; the quantity τ is later to be determined as a function of σ. As a consequence of the piecewise continuity of the tangent to γ', there is a quantity $\eta(\tau)$, tending to zero with τ, such that every point of γ is at a distance less than $\eta^4\tau$ from the corresponding edge of γ'.

The deformation of the $\mathfrak{x}$-space taking γ into γ' is constructed by successive pinching processes, so balanced that they do not interfere with each other. First we cut off from the ends of each edge of γ' segments of the length $\eta^2\tau$. About the remaining middle segments S of the edges we perform δ, t-pinching processes with

$$\delta = \eta^4\tau, \qquad t = \frac{1}{\eta}.$$

These pinching processes, for sufficiently small τ, will not interfere with each other; they pull γ into γ' except for neighborhoods of the vertices. The arcs about the vertices will not extend farther than, say, $2\tau\eta^2$ from them. By additional pinching processes about the vertices with

$$\delta = 2\tau\eta^2, \qquad t = \frac{1}{\eta},$$

these arcs are transformed into the vertices while points already on the edges remain there. Hence the total effect of the pinching processes is a transformation of γ into γ'; our construction ensures that γ' is monotonically described by the resulting vector $\mathfrak{y}(u,v)$ as the point (u,v) describes β monotonically. Furthermore, for sufficiently small τ no more than two of our pinching processes will affect the same portion of the vector space. Hence, according to the preceding result,

$$D[\mathfrak{y}] \leq \left(1 - \frac{1}{\log \eta}\right)^4 D[\mathfrak{x}].$$

By making η so small that

$$\left(1 - \frac{1}{\log \eta}\right)^4 < 1 + \sigma$$

we obtain the result stated in our lemma.

Finally, to remove the preliminary restriction concerning the endpoints of the arcs of γ, we have merely to begin the construction with pinching processes about the vertices. Then γ will be taken into another curve γ^* for which our restrictive assumptions are valid, and the argument can be applied without change.[22]

[22] If in the pinching processes η is replaced by $\alpha\eta$ the result will be a transformation depending continuously on the parameter α, so that a continuous deformation takes place as α increases from 0 to 1.

Bibliography, Chapters I to VI

Alexandroff, P.

[1] (and H. Hopf). *Topologie*, Volume 1 (Chapter I), Julius Springer, 1935.

Bieberbach, L.

[1] *Über einen Riemannschen Satz aus der Lehre von der konformen Abbildung*, Sitzungsber. Berliner Math. Ges., Volume 24, 1925, p. 6 ff.

Blaschke, W.

[1] *Vorlesungen über Differentialgeometrie*, Volume 1, Julius Springer, 1929.

Bliss, G. A.

[1] *Calculus of Variations*, Open Court, 1925.

Carathéodory, C.

[1] *Über die Bergrenzung einfach zusammenhängender Gebiete*, Math. Ann., Volume 73, 1913, pp. 323–370.

Carleman, T.

[1] *Über eine isoperimetrische Aufgabe und ihre physikalischen Anwendungen*, Julius Springer, 1919.

[2] *Sur la représentation conforme des domaines multiplement connexes*, C. R. Ac. Sci. Paris, Volume 168, 1919, pp. 843–845.

[3] *Zur Theorie der Minimalflächen*, Math. Z., Volume 9, 1921, p. 160 ff.

Chen, Y. W.

[1] *Branch points, poles, and planar points of minimal surfaces in R^3*, Ann. Math., Volume 49, Oct., 1948, pp. 790–806.

Courant, R.

[1] *On the problem of Plateau*, Proc. Nat. Ac. Sci. U. S. A., Volume 22, 1936, p. 367 ff.

[2] (and D. Hilbert). *Methoden der mathematischen Physik*, Volume 2, Julius Springer, 1937. Interscience Publishers, Inc., Photolithoprint reproduction, 1943.

[3] *Plateau's problem and Dirichlet's Principle*, Ann. Math., Volume 38, 1937, pp. 679–725.

[4] *The existence of a minimal surface of least area bounded by prescribed Jordan arcs and prescribed surfaces*, Proc. Nat. Ac. Sci. U. S. A., Volume 24, 1938, pp. 97–101.

[5] *Remarks on Plateau's and Douglas' problem*, Proc. Nat. Ac. Sci. U. S. A., Volume 24, 1938, pp. 519–523.

[6] *Conformal mapping of multiply connected domains*, Duke Math. J., Volume 5, 1939, pp. 314–823.

[7] *The existence of minimal surfaces of given topological structure under prescribed boundary conditions*, Acta Math., Volume 72, 1940, pp. 51–98.

[8] (and N. Davids). *Minimal surfaces spanning closed manifolds*, Proc. Nat. Ac. Sci. U. S. A., Volume 26, 1940, pp. 194–199.

[9] (with B. Manel and M. Shiffman). *A general theorem on conformal mapping of multiply connected domains*, Proc. Nat. Ac. Sci. U. S. A., Volume 26, 1940, pp. 503–507.

[10] *Critical points and unstable minimal surfaces*, Proc. Nat. Ac. Sci. U. S. A., Volume 27, 1941, pp. 51–57.

[11] *On the first variation of the Dirichlet-Douglas integral and on the method of gradients*, Proc. Nat. Ac. Sci. U. S. A., Volume 27, 1941, pp. 242–248.

[12] *On a generalized form of Plateau's problem*, Trans. Amer. Math. Soc., Volume 50, 1941, pp. 40–47.

[13] (and H. Robbins). *What is Mathematics?* Oxford University Press, 1941.

[14] *On Plateau's problem with free boundaries*, Proc. Nat. Ac. Sci. U. S. A., Volume 31, 1945, pp. 242–246.

[15] *Über direkte Methoden in der Variationsrechnung und über verwandte Fragen*, Math. Ann., Volume 97, 1927, pp. 711–736.

[16] *Über direkte Methoden bei Variations- und Randwertproblemen*, Jber. Deutsch. Math. Verein., Volume 34, 1925, pp. 90–117.

[17] *Neue Bemerkungen zum Dirichletschen Prinzip*, J. reine angew. Math., Volume 165, 1931, pp. 248–256.

[18] *Soap film experiments with minimal surfaces*, Amer. Math. Monthly, Volume 47, 1940, pp. 168–174.

Davids, N.

[1] *Minimal surfaces spanning closed manifolds and having prescribed topological position*, Amer. J. Math., Volume 64, 1942, pp. 348–362.

Douglas, J.

[1] *Solution of the problem of Plateau*, Trans. Amer. Math. Soc., Volume 33, 1931, pp. 263–321.

[2] *The problem of Plateau*, Bull. Amer. Math. Soc., Volume 39, 1933, pp. 227–251.

[3] *Some new results in the problem of Plateau*, Journal Math. Phys., Volume 14, 1936, pp. 55–64.

[4] *Minimal surfaces of general topological structure*, Journal Math. Phys., Volume 15, 1936, pp. 105–123.

[5] *Minimal surfaces of higher topological structure*, Proc. Nat. Ac. Sci. U. S. A., Volume 24, 1938, pp. 343–353.

[6] *Green's function and the problem of Plateau*, Proc. Nat. Ac. Sci. U. S. A., Volume 24, 1938, pp. 353–360.

[7] *The most general form of the problem of Plateau*, Proc. Nat. Ac. Sci. U. S. A., Volume 24, 1938, pp. 360–364.

[8] *Minimal surfaces of higher topological structure*, Ann. Math., Volume 40, 1939, pp. 205–298.

[9] *The higher topological form of Plateau's problem*, Ann. R. Scuola Norm. Super. Pisa, Ser. II, Volume 8, 1939, pp. 1–24.

[10] *The problem of Plateau for two contours*, J. Math. Phys., Volume 10, 1931, pp. 315–359.

[11] *Remarks on Riemann's doctoral dissertation*, Proc. Nat. Ac. Sci. U. S. A., Volume 24, 1938, pp. 297–302.

[12] *Green's function and the problem of Plateau. The most general form of the problem of Plateau*, Amer. J. Math., Volume 61, 1939, pp. 545–608.

Grunsky, H.

[1] *Über die konforme Abbildung mehrfach zusammenhängender Bereiche auf mehrblättrige Kreise*, Sitzungsber. Preuss. Akad. Wiss., 1937, p. 40 ff.

Hadamard, J.

[1] *Mémoire sur le problème d'analyse relatif à l'équilibre des plaques élastiques encastrées*, Mém. savants étrangers Acad. Sci. Inst. France, Series 2, Volume 33, No. 4, 1908.

Hurwitz, A.

[1] (and R. Courant). *Funktionentheorie*, Julius Springer, 1939; photolithoprint reproduction, Interscience, New York, 1946.

von Kerékjártó, B.

[1] *Vorlesungen über Topologie*, Volume 1, Julius Springer, 1923.

Lebesgue, H.

[1] *Sur le problème de Dirichlet*, Rend. circ. mat. Palermo, Volume 24, 1907, pp. 371–402.

Lusternik, L.

[1] (and L. Schnirelmann). *Méthodes topologiques dans les problèmes variationnels*, Actualités scient. et indust., No. 188, 1934.

[2] (and L. Schnirelmann). *Functional topology and abstract variational theory*, Mémorial des sciences math., 1938.

McShane, E. J.

[1] *Parametrizations of saddle surfaces, with applications to the problem of Plateau*, Trans. Amer. Math. Soc., Volume 35, 1933, pp. 716–733.

Manel, B.

[1] *Conformal mapping of multiply connected domains on the basis of Plateau's problem*, Revista, Universidad Nacional de Tucuman, Volume 3, 1942, pp. 141–149.

Marx, I.

[1] (and M. Shiffman). A paper to be published in the near future.

Morrey, C. D.

[1] *The problem of Plateau on a Riemannian manifold*, Ann. Math., Volume 49, No. 4, 1948.

Morse, M.

[1] *The calculus of variations in the large*, Amer. Math. Soc. Colloquium Publication, Volume 18, 1934.

[2] (and C. Tompkins). *Existence of minimal surfaces of general critical type*, Ann. Math., Volume 40, 1939, p. 443 ff.

[3] (and C. Tompkins). *Existence of minimal surfaces of general critical type*, Proc. Nat. Ac. Sci. U. S. A., Volume 25, 1939, pp. 153–158.

[4] (and C. Tompkins). *The continuity of the area of harmonic surfaces as a function of the boundary representation*, Amer. Jour. Math., Volume 63, 1941, pp. 325–338.

[5] (and C. Tompkins). *Unstable minimal surfaces of higher topological structure*, Duke Math. J., Volume 8, 1941, pp. 350–375.

[6] *The critical points of functions and the calculus of variations in the large*, Bull. Amer. Math. Soc., Volume 35, 1929, pp. 38–54.

[7] *Functional topology and abstract variational theory*, Ann. Math., Volume 38, 1937, pp. 386–449.

[8] *Rank and span in functional topology*, Ann. Math., Volume 41, 1940, pp. 419–445.

Plateau, J.

[1] *Sur les figures d'équilibre d'une masse liquide sans pésanteur*, Mém. acad. roy. Belgique, New Series, Volume 23, 1849.

[2] *Statique expérimentale et théorique des liquides*, Paris, 1873.

Rado, T.

[1] *On Plateau's problem*, Ann. Math., Volume 31, 1930, pp. 457–469.

[2] *On the problem of Plateau*, Ergeb. d. Math. und ihrer Grenzgebiete, Volume 2, Julius Springer, 1933.

[3] *The isoperimetric inequality and the Lebesgue definition of surface area*, Trans. Amer. Math. Soc., Volume 61, 1947, pp. 530–555.

[4] *The problem of the least area and the problem of Plateau*, Math. Z., Volume 32, 1930, pp. 763–796.

Ritter, I. F.

[1] *Solution of Schwarz' problem concerning minimal surfaces*, Revista, Universidad Nacional de Tucuman, Volume 1, 1940, pp. 49–62.

Schwarz, H. A.

[1] *Gesammelte mathematische Abhandlungen*, Volume 1, Julius Springer, 1890.

Shiffman, M.

[1] *The Plateau problem for minimal surfaces which are relative minima*, Ann. Math., Volume 39, 1938, p. 309 ff.

[2] *The Pleateau problem for minimal surfaces which are not relative minima*, Bull. Amer. Math. Soc., Volume 44, 1938, p. 637.

[3] *The Plateau problem for minimal surfaces of arbitrary topological structure*, Amer. J. Math., Volume 61, 1939, pp. 853–882.

[4] *The Plateau problem for non-relative minima*, Ann. Math., Volume 40, 1939, pp. 834–854.

[5] *The Plateau problem for non-relative minima*, Proc. Nat. Ac. Sci. U. S. A., Volume 25, 1939, pp. 215–220.

[6] *Uniqueness theorems for conformal mapping of multiply connected domains*, Proc. Nat. Ac. Sci. U. S. A., Volume 27, 1941, pp. 137–139.

[7] *Unstable minimal surfaces with several boundaries*, Ann. Math., Volume 43, 1942, pp. 197–222.

[8] *Unstable minimal surfaces with any rectifiable boundary*, Proc. Nat. Ac. Sci. U. S. A., Volume 28, 1942, pp. 103–108.

Tonelli, L.

[1] *Sul problema della superficie limitata da un dato contorno ed avente la minima area*, Atti. R. Acad. Sci. Torino, Volume 72, 1936–37, pp. 1–11.

[2] *Sul problema di Plateau*, Rend. R. Acad. Naz. Lincei, Volume 29, 1936, pp. 333–339.

Weyl, H.

[1] *Die Idee der Riemannschen Fläche*, B. G. Teubner, Berlin, 1923.

APPENDIX

Some Recent Developments in the Theory of Conformal Mapping

By M. SCHIFFER

1. Green's Function and Boundary Value Problems

1. In Chapters II and V, Dirichlet's Principle was used to prove the existence of solutions for various boundary value problems and to demonstrate for each the possibility of several different conformal maps. Each such existence proof can, in principle, be transformed into a procedure for constructing the required solution; the difficulties involved are, however, such as to render this method impracticable. Certain fundamental solutions have, therefore, been introduced, in terms of which large classes of boundary value problems may be solved by simple computation. The whole theory of boundary value problems is thus reduced to the construction and consideration of a few fundamental solutions. Here one such fundamental solution will be considered, namely the Green's function. Various important properties of this function will be pointed out, and computational procedures for its construction will be derived from them.

Let G be again a domain in the x,y-plane bounded by k smooth curves γ_ν which form the boundary $\gamma = \sum_{\nu=1}^{k} \gamma_\nu$ of G. Let $\zeta = \xi + i\eta$ be a fixed point in G, and let $h(z, \zeta)$ be a harmonic function of the point $z = x + iy$ which has on γ the boundary values

$$h(z, \zeta) = \log |z - \zeta|, \qquad z \in \gamma \qquad \zeta \in G. \tag{A1.1}$$

According to the existence theorem of Chapter I such a function exists. We wish to study its properties and consider various applications.

We define the function

$$g(z, \zeta) = \log \frac{1}{|z - \zeta|} + h(z, \zeta) \tag{A1.2}$$

which has the three following characteristic properties:

a) $g(z, \zeta)$ is harmonic for $z \in G$, except at the point $z = \zeta$;
b) $g(z, \zeta) + \log | z - \zeta |$ is harmonic at $z = \zeta$;
c) $g(z, \zeta)$ has the boundary value 0 on γ.

The function $g(z, \zeta)$ is called the Green's function of the domain G.

Now let $u(z)$ be harmonic in G and continuously differentiable in the closed region $G + \gamma$. Then from Green's identity and the boundary behavior of the Green's function, we obtain immediately the identity

$$u(\zeta) = \frac{1}{2\pi} \int_\gamma \frac{\partial g(z, \zeta)}{\partial n_z} u(z) \, ds_z , \tag{A1.3}$$

where $\partial/\partial n$ again denotes differentiation in the direction of the interior normal. We recognize from (A1.3) that the solution of a boundary value problem with respect to the domain G can be reduced to simple integration, once the Green's function has been determined. It can be shown by a finer argument that (A1.3) holds for harmonic functions which are continuous in the closed region $G + \gamma$.

We conclude from the minimum principle that the Green's function is positive in the domain G since it becomes positively infinite at ζ and vanishes on γ; hence $\partial g(z, \zeta)/\partial n_z \geq 0$ on γ. The dependence of the Green's function on the parameter point ζ is clarified by the application of Green's identity

$$\begin{aligned} &g(\eta, \zeta) - g(\zeta, \eta) \\ &\quad = \frac{1}{2\pi} \int_\gamma \left[g(z, \zeta) \frac{\partial g(z, \eta)}{\partial n} - g(z, \eta) \frac{\partial g(z, \zeta)}{\partial n} \right] ds = 0, \end{aligned} \tag{A1.4}$$

which proves the symmetry law for the Green's function

$$g(z, \zeta) = g(\zeta, z). \tag{A1.4'}$$

Hence we may consider z as the parameter point; then $g(z, \zeta)$, as a function of ζ, will be the Green's function.

The Green's function has a very simple electrostatic interpretation. Consider the boundary γ of G as a system of grounded conducting walls and locate at the point $\zeta \in G$ a point of charge 1 exerting a field of force derived from a logarithmic potential. The charge at ζ induces on γ a countercharge with potential $h(z, \zeta)$. The Green's function will then be considered as the potential of the total field thus created.

The Green's function, being harmonic in z, may be extended to an analytic function of z by addition of its conjugate harmonic function

$$p(z, \zeta) = g(z, \zeta) + i\gamma(z, \zeta). \tag{A1.5}$$

The function p is harmonic in ζ and analytic in z; it has a logarithmic pole with residue 1 at ζ and is determined only up to an additive imaginary constant. It is not single-valued because of the logarithmic singularity which introduces periods $2\pi in$. If G is multiply connected, $p(z, \zeta)$ has, in addition, periods with respect to circuits around the boundary curves γ_ν. These periods are

$$2\pi i\omega_\nu(\zeta) = \oint_{\gamma_\nu} dp(z, \zeta) = i\oint_{\gamma_\nu} \frac{\partial\gamma(z, \zeta)}{\partial s}\, ds = i\oint_{\gamma_\nu} \frac{\partial g(z, \zeta)}{\partial n}\, ds. \tag{A1.6}$$

The period functions $\omega_\nu(\zeta)$ are harmonic in ζ and single-valued in G. From (A1.3) we conclude that $\omega_\nu(\zeta)$ has the boundary value 0 on all γ_μ, except on γ_ν where it has the value 1. The function $\omega_\nu(\zeta)$ is called the harmonic measure of the boundary component γ_ν with respect to G at the point ζ.

Again we may extend the harmonic function $\omega_\nu(\zeta)$ to an analytic function $w_\nu(\zeta)$ which is determined up to an additive imaginary constant

$$w_\nu(\zeta) = \omega_\nu(\zeta) + i\Omega_\nu(\zeta). \tag{A1.7}$$

In general, $w_\nu(\zeta)$ will possess imaginary periods if ζ describes closed curves in G. We find as the period of $w_\nu(\zeta)$ with respect to the contour γ_μ

$$-2\pi iP_{\mu\nu} = i\oint_{\gamma_\mu} \frac{\partial\omega_\nu}{\partial n}\, ds. \tag{A1.8}$$

It should be remarked that the periods $\omega_\nu(\zeta)$ and $P_{\mu\nu}$ could also be computed by extending the integrations in (A1.6) and (A1.8) over closed curves γ_ν' and γ_μ' which lie entirely in G, and are obtained from γ_ν and γ_μ by continuous deformation through the domain. This fact is an immediate consequence of Green's theorem which asserts the independence of these integrals of their path, as long as the path remains in the same homotopy class. From (A1.6) and (A1.8), we conclude

$$P_{\mu\nu} = -\frac{1}{4\pi^2}\oint_{\gamma_\mu'}\oint_{\gamma_\nu'} \frac{\partial^2 g(z, \zeta)}{\partial n_z\, \partial n_\zeta}\, ds_z\, ds_\zeta = P_{\nu\mu}, \tag{A1.9}$$

i.e. the symmetry of the matrix of the $P_{\mu\nu}$.

The constants $P_{\mu\nu}$ also have a simple electrostatic interpretation. Suppose that all conductors γ_ρ are grounded, except for γ_ν which is kept at the potential 1. Then the electrostatic field in G is described by the potential $\omega_\nu(z)$. From the elements of electrostatics, we conclude that the charge on the conductor γ_μ induced by these conditions equals

$$(\text{A1.10}) \qquad -\frac{1}{2\pi}\int_{\gamma_\mu} \frac{\partial \omega_\nu}{\partial n}\, ds = P_{\nu\mu}.$$

The constants $P_{\nu\mu}$ are characteristics of the conductor system γ; they were first considered by Maxwell and called the induction coefficients of the system [30, 42].

Consider the somewhat more general case of an electrostatic field in G where each conductor γ_ρ is kept at a constant potential, say α_ρ. The corresponding field in G is described by the potential function

$$(\text{A1.11}) \qquad \omega(z) = \sum_{\rho=1}^{k} \alpha_\rho \omega_\rho(z).$$

The energy associated with this field is defined by the integral

$$(\text{A1.12}) \qquad E = \frac{1}{4\pi}\iint_G (\text{grad } \omega)^2\, dx\, dy$$

$$= -\frac{1}{4\pi}\int_\gamma \omega \frac{\partial \omega}{\partial n}\, ds = -\frac{1}{4\pi}\sum_{\rho=1}^{k}\int_{\gamma_\rho} \omega \frac{\partial \omega}{\partial n}\, ds.$$

Here the final integral can be computed by using the fact that $\omega(z) = \alpha_\rho$ on γ_ρ and applying definition (A1.8). We conclude

$$(\text{A1.13}) \qquad E = \frac{1}{2}\sum_{\rho,\sigma=1}^{k} P_{\rho\sigma}\alpha_\rho\alpha_\sigma.$$

As immediately follows from its definition, E is always non-negative. The energy integral can only vanish if grad $\omega \equiv 0$ in G, which implies $\omega =$ constant. But in this case all boundary values α_ρ of this constant function must be equal. Conversely, according to the uniqueness theorem for the boundary value problem, if all boundary values α_ρ of $\omega(z)$ are equal, then $\omega(z)$ must be a constant. Hence we have proved:

$$(\text{A1.14}) \qquad \sum_{\rho,\sigma=1}^{k} P_{\rho\sigma}\alpha_\rho\alpha_\sigma \geq 0.$$

Equality holds only when all α_ρ are equal. Thus we find: *The matrix $(P_{\rho\sigma})$ defines a positive semi-definite quadratic form.*

Let us, in particular, assume that $\alpha_k = 0$. The quadratic form

$$\sum_{\rho,\sigma=1}^{k-1} P_{\rho\sigma}\alpha_\rho\alpha_\sigma \geq 0 \tag{A1.14'}$$

can vanish only if all α_ρ become equal to α_k, that is, also vanish. Hence the matrix $(P_{\rho\sigma})_{1\ldots k-1}$ is positive definite and has a non-vanishing determinant. Consequently, each system of linear equations

$$\sum_{\sigma=1}^{k-1} P_{\rho\sigma} u_\sigma = v_\rho, \qquad \rho = 1, 2, \cdots, k-1 \tag{A1.15}$$

has exactly one solution vector u_σ, $\sigma = 1, 2, \cdots, k - 1$. This fact will play an important role in our subsequent developments.

2. *Canonical Conformal Mappings.* In article 1 we saw that if we know the Green's function for a domain G, we are able to solve the boundary value problems with respect to this domain. We want now to show that, by means of the Green's function and the harmonic measures and induction coefficients derived from it, univalent functions in G which map this domain upon interesting types of canonical domains may be explicitly constructed.

Let us consider at first a simply connected domain with one boundary curve γ_1. We extend its Green's function to the analytic function $p(z, \zeta)$. In order to eliminate the multivaluedness arising from the logarithmic pole at ζ, we introduce the function

$$f(z, \zeta) = \exp\{-p(z, \zeta)\}, \tag{A1.16}$$

which is single-valued in G, vanishes to the first order at $z = \zeta$, and has the modulus 1 for $z \in \gamma$. This function has already been constructed in Chapter I, §7. It maps G onto the unit circle in such a manner that the point $z = \zeta$ corresponds to its center, the origin.

We can generalize this construction to the case of a multiply connected domain G. In this case, the function $p(z, \zeta)$ will have additional periods besides those which arise from the logarithmic pole at ζ. We have, however, in the functions $w_\nu(z)$ another set of multivalued functions in G and we will be able to make $p(z, \zeta)$ single-valued by addition of an appropriate linear combination of these functions. In fact, let us solve the system of linear equations

$$\sum_{\sigma=1}^{k-1} P_{\rho\sigma} u_\sigma = \omega_\rho(\zeta), \qquad \rho = 1, 2, \cdots, k-1. \tag{A1.17}$$

Introducing the matrix $(\Pi_{\rho\sigma})_{1\ldots k-1}$ which is inverse to $(P_{\rho\sigma})_{1\ldots k-1}$, we may solve this system in the form

$$u_\rho(\zeta) = \sum_{\sigma=1}^{k-1} \Pi_{\rho\sigma}\, \omega_\sigma(\zeta). \tag{A1.17'}$$

Next construct the linear combination of functions $w_\sigma(z)$

$$w(z, \zeta) = \sum_{\sigma=1}^{k-1} u_\sigma(\zeta) w_\sigma(z), \tag{A1.18}$$

which is analytic in z and has the period

$$-2\pi i \sum P_{\rho\sigma} u_\sigma(\zeta) = -2\pi i \omega_\rho(\zeta) \tag{A1.19}$$

with respect to a circuit around the boundary continuum γ_ρ, $\rho = 1, 2, \cdots, k-1$. Then, in view of (A1.6), the function

$$\log f(z, \zeta) = -\left[p(z, \zeta) + \sum_{\sigma=1}^{k-1} u_\sigma(\zeta) w_\sigma(z)\right] \tag{A1.20}$$

has no periods if z describes any circuit which is topologically equivalent to any γ_ρ, $\rho = 1, 2, \cdots, k-1$.

Now we need only inquire about the period of the function (A1.20) with respect to the contour γ_k. We remark that the path consisting of all boundary curves γ_ν, passed in the positive sense with respect to the domain G, is topologically equivalent to a small circle around the logarithmic pole ζ and leads, therefore, to the period $2\pi i$. Since a circuit around each γ_ρ, $\rho < k$, does not change the function (A1.20), it is clear that this function increases by $2\pi i$ if γ_k is described. Thus the function

$$f(z, \zeta) = \exp -\left\{p(z, \zeta) + \sum_{\sigma,\rho=1}^{k-1} \Pi_{\rho\sigma} w_\sigma(z) \omega_\rho(\zeta)\right\} \tag{A1.21}$$

is single-valued in G and has a simple zero at $z = \zeta$. In order to determine the modulus of $f(z, \zeta)$ for $z \in \gamma_\rho$, we have to consider the real part of the function (A1.20) for $z \in \gamma_\rho$. In view of the known boundary behavior of $g(z, \zeta)$ and $\omega_\sigma(z)$ on γ_ρ, we find

$$\log | f(z, \zeta) | = \begin{cases} 0 & \text{for } z \in \gamma_k, \\ -u_\rho(\zeta) & \text{for } z \in \gamma_\rho. \end{cases} \tag{A1.22}$$

This shows that $|f(z, \zeta)|$ is constant on each γ_ρ and is, in particular, 1 on γ_k.

From all the facts established about it, we conclude that the function $w = f(z, \zeta)$ maps the domain G onto a domain G_w such that the images of the boundary curves lie on circles $|w| = \exp\{-u_\rho(\zeta)\}$ and $|w| = 1$, respectively. Since the argument of $f(z, \zeta)$ does not change if z describes any of the contours γ_ρ, $\rho < k$, it is evident that the images of these γ_ρ form circular arcs on a Riemann surface over the w-plane. We cannot assert the same for the image of γ_k; in fact, if z describes γ_k its image point $w(z)$ encircles the origin in the w-plane exactly once, since its argument increases by 2π.

From the behavior of $f(z)$ on the boundary of G it follows easily that $f(z)$ has all its values inside the unit circle in the w-plane and that each point $|w| < 1$ is assumed at most once in G. In fact, let a be an arbitrary point in the complex plane which does not lie on one of the circumferences $|w| = \exp\{-u_\rho(\zeta)\}$ or $|w| = 1$. Then, according to the argument principle, the integral

$$\text{(A1.22')} \qquad \frac{1}{2\pi i}\int_\gamma \frac{f'(z)}{f(z) - a}\,dz = \frac{1}{2\pi}\int_\gamma d\arg\,[f(z) - a]$$

gives the number of times this value a is attained inside G. From our characterization of $f(z)$ on the boundary curves γ_ρ it is obvious that $1/2\pi \arg\,[f(z) - a]$ does not change if z runs over any γ_ρ with $\rho < k$; if z describes the contour γ_k the expression will change by 1 and 0, for $|a| < 1$ and $|a| > 1$, respectively. Thus $f(z)$ cannot assume any value $|a| > 1$, and attains each value $|a| < 1$ at most once. The same must clearly hold for points on the exceptional circumferences, excluded above. Thus we have proved:

The function $f(z, \zeta)$ maps the domain G univalently onto the interior of the unit circle which is slit along $k - 1$ circular arcs around the origin. The point $z = \zeta$ is mapped into the origin and the boundary continuum γ_k into the unit circumference [34] (Figure 1).

Remark: Since the numeration of the boundary continua is quite arbitrary, we have proved the existence of k different canonical maps depending on the choice of the particular boundary continuum which is to correspond to the unit circumference.

Let ζ and η be two arbitrary points in G and consider the analytic function of z

$$\text{(A1.23)} \qquad F(z; \zeta, \eta) = \frac{f(z, \zeta)}{f(z, \eta)}.$$

This function is of constant modulus on each boundary continuum of γ; it vanishes at $z = \zeta$ and has a simple pole at $z = \eta$. Its argument does not change if z describes any boundary contour of G. Hence, the same considerations as above lead to the result:

The function $F(z; \zeta,\eta)$ maps the domain G univalently onto the whole complex plane, slit along k circular arcs around the origin. The points ζ and η are transformed by this mapping into the origin and the point at infinity, respectively [34] (Figure 2).

Figure 1. Circular-slit domain in unit circle.

Figure 2. Circular-slit domain in plane.

It is easy to derive further canonical map functions from the examples already constructed. From the fact that

$$\text{(A1.24)} \qquad \mathcal{R}e\{\log f(z, \zeta)\} = \begin{cases} 0 & \text{for} \quad z \in \gamma_k, \\ -u_\rho(\zeta) & \text{for} \quad z \in \gamma_\rho, \rho < k \end{cases}$$

we obtain by differentiation with respect to ξ (putting $\zeta = \xi + i\eta$)

$$\text{(A1.25)} \qquad \mathcal{R}e\left\{\frac{\partial}{\partial \xi} \log f(z, \zeta)\right\} = \lambda_\rho(\zeta) \quad \text{for} \quad z \in \gamma_\rho, \rho \leq k,$$

where the λ_ρ are independent of z. Thus the function

$$\text{(A1.26)} \qquad \psi(z, \zeta) = -\frac{\partial}{\partial \xi} \log f(z, \zeta)$$

has a constant real part on each boundary continuum γ_ρ of γ. It has at the point $z = \zeta$ a simple pole with residue 1. Hence it maps the domain G univalently onto the whole plane slit along segments parallel to the imaginary axis. The point $z = \zeta$ corresponds in this map to the point at infinity.

Similarly, we show by differentiating (A1.24) with respect to η that

$$\text{(A1.27)} \qquad \varphi(z, \zeta) = -\frac{1}{i}\frac{\partial}{\partial \eta} \log f(z, \zeta)$$

maps the domain univalently onto the plane slit along segments parallel to the real axis. At $z = \zeta$ the function has again a simple pole with residue 1.

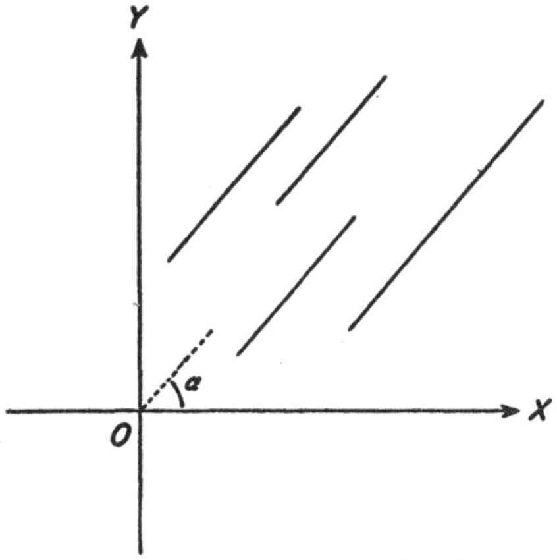

Figure 3. Parallel-slit domain in direction α.

From the functions $\varphi(z, \zeta)$ and $\psi(z, \zeta)$ infinitely many new slit mappings can be constructed. In fact, consider the combination

(A1.28) $$\varphi_\alpha(z, \zeta) = e^{i\alpha}[\cos \alpha\, \varphi(z, \zeta) - i \sin \alpha\, \psi(z, \zeta)]$$

for arbitrary α. This function admits around $z = \zeta$ a series development

(A1.29) $$\varphi_\alpha(z, \zeta) = \frac{1}{z - \zeta} + \text{regular terms,}$$

and on each boundary continuum γ_ρ we have clearly

(A1.30) $$\mathcal{I}m\, \{e^{-i\alpha}\varphi_\alpha(z, \zeta)\} = \text{constant} \qquad \text{for } z \in \gamma_\rho\,.$$

Therefore,

The function $\varphi_\alpha(z, \zeta)$ maps the domain G on the whole plane slit along straight segments which are turned at the angle α to the real axis; the point $z = \zeta$ is a simple pole with residue 1 [24, 46, 60] (Figure 3).

Finally we are led to the following construction. For every value of α, we find

(A1.31) $$\tfrac{1}{2}[\varphi_\alpha(z, \zeta) + \varphi_{\alpha+\pi/2}(z, \zeta)] = \tfrac{1}{2}[\varphi(z, \zeta) + \psi(z, \zeta)] = \Phi(z, \zeta).$$

This function is also regular in the domain G except at the point $z = \zeta$ where it has a simple pole with residue 1. We want to show that it is univalent in G. At first it is clear that each boundary contour γ_ρ of G is mapped by $w = \Phi(z, \zeta)$ into a simple convex curve in the w-plane. In fact, if z describes γ_ρ the real part of $\psi(z, \zeta)$ remains constant while the real part of $\varphi(z, \zeta)$ assumes every value in a certain interval exactly twice; this is an immediate consequence of the mapping properties of the functions φ and ψ. Hence, in view of (A1.31), we see that the image of γ_ρ is cut by each parallel to the imaginary axis at most twice. Using the fact that Φ can also be decomposed into φ_α and $\varphi_{\alpha+\pi/2}$ we prove that the image of each γ_ρ is cut by each straight line in at most two points. This establishes the asserted convexity.

This result does not in itself guarantee the univalence of the mapping; the image domain G_w of G may be spread over a Riemann surface in spite of the fact that the image of each γ_ρ is a simple curve. In this case, the domain G_w must contain winding points of the Riemann surface which correspond to the point $z \in G$ with $\Phi'(z, \zeta) = d/dz\, \Phi(z, \zeta) = 0$. We will show, however, that $\Phi'(z, \zeta)$ does not vanish in G. In fact, since $\varphi_\alpha(z, \zeta)$ is univalent in G it has everywhere a non-vanishing derivative, i.e.

(A1.32) $$\frac{\varphi'(z, \zeta)}{\psi'(z, \zeta)} \neq i \tan \alpha$$

for any choice of α and $z \in G$. At $z = \zeta$, the quotient (A1.32) has the value 1; its value cannot be imaginary for any choice of $z \in G$; hence,

(A1.32′) $$\mathcal{R}e \left\{ \frac{\varphi'(z, \zeta)}{\psi'(z, \zeta)} \right\} > 0, \qquad z \in G.$$

But $\Phi'(z, \zeta) = 0$ in G would imply

(A1.33) $$\frac{\varphi'(z, \zeta)}{\psi'(z, \zeta)} = -1,$$

which is impossible because of (A1.32′).

Thus, we have proved:

The function $\Phi(z, \zeta)$ maps G univalently onto an infinite domain of the w-plane with convex boundary curves [17, 25, 60].

We shall see later that this map is distinguished by an important extremum property. Here we want to stress the interesting fact that the formulas (A1.28) and (A1.31) permit the construction of univalent functions from simpler univalent functions by the process of addition, a quite unusual phenomenon in the theory of univalent functions. It will also be important to remember that all functions constructed so far may be expressed explicitly in terms of the Green's function and functions derived from it.

For later use, let us note the fact that, in view of (A1.26), (A1.27), and (A1.31),

$$\Phi(z,\zeta) = -\frac{1}{2}\left(\frac{\partial}{\partial\xi} - i\frac{\partial}{\partial\eta}\right)\log f(z,\zeta), \quad \zeta = \xi + i\eta. \tag{A1.34}$$

Obviously, $f(z, \zeta)$ is not analytic with respect to its parameter point ζ; it is, however, convenient to introduce the following symbolism. We define the two differential operators [71]

$$\frac{\partial}{\partial\zeta} = \frac{1}{2}\left(\frac{\partial}{\partial\xi} - i\frac{\partial}{\partial\eta}\right), \qquad \frac{\partial}{\partial\bar\zeta} = \frac{1}{2}\left(\frac{\partial}{\partial\xi} + i\frac{\partial}{\partial\eta}\right), \tag{A1.35}$$

and apply them to arbitrary differentiable functions of ξ and η. In this sense, we may put (A1.34) into the form

$$\Phi(z,\zeta) = -\frac{\partial}{\partial\zeta}\log f(z,\zeta). \tag{A1.36}$$

The function

$$m(z,\zeta) = -\log|f(z,\zeta)| = -\tfrac{1}{2}\log f(z,\zeta) - \tfrac{1}{2}\log\overline{f(z,\zeta)} \tag{A1.37}$$

is harmonic in z and has constant values on all boundary continua γ_ρ; it has a logarithmic pole with residue -1. The function $m(z, \zeta)$ is a generalization of the Green's function and will occur in later considerations. In terms of it, we may write

$$\Phi'(z,\zeta) = 2\frac{\partial^2}{\partial z\,\partial\zeta}m(z,\zeta). \tag{A1.38}$$

3. *Boundary Value Problems of Second Type and Neumann's Function.* A well-known problem in potential theory which occurs in many applications is the following: Along the boundary curves γ_ρ of the domain G the value of the normal derivative of a harmonic function is prescribed, i.e. we have

$$\frac{\partial u}{\partial n} = \mu(s), \qquad s = \text{arc length on } \gamma. \tag{A1.39}$$

The problem is to determine the harmonic function $u(z)$ in G. This is the so-called boundary value problem of the second type in potential theory.

We remark, at first, that we are not quite free in prescribing the function $\mu(s)$ on γ. For, according to Gauss' identity for harmonic functions, we have

$$\text{(A1.40)} \qquad \oint_\gamma \mu(s)\, ds = \oint_\gamma \frac{\partial u}{\partial n}\, ds = -\iint_G \Delta u\, dx\, dy = 0.$$

Thus $\mu(s)$ is subjected to the restriction that its integral over γ must be equal to zero. We will now show that if this condition is fulfilled the problem has a unique solution. In fact, assuming that the required harmonic function $u(z)$ exists in G, let us extend it to an analytic function $w(z) = u(z) + iv(z)$. According to the Cauchy-Riemann differential equations which connect the real and imaginary parts of an analytic function, we have on γ

$$\text{(A1.41)} \qquad \mu(s) = \frac{\partial u}{\partial n} = -\frac{\partial v}{\partial s}.$$

This requirement determines the boundary values of the unknown harmonic function $v(z)$ on each boundary curve γ_ρ, up to an additive constant.

We are now able to give a procedure for the construction of the function $u(z)$. First, we determine a boundary value distribution $v(s)$ which is compatible with (A1.41). Then, by means of Green's function, we construct a harmonic function $v(z)$ in G with these boundary values. We remark that

$$\text{(A1.42)} \qquad V(z) = v(z) + \sum_{\nu=1}^{k-1} \lambda_\nu \omega_\nu(z)$$

will differ on each γ_ρ, $\rho < k$, from $v(z)$ by the constant λ_ρ and is, therefore, also a permissible conjugate function according to (A1.41). We now have to determine among all functions $V(z)$ that function which has a single-valued conjugate harmonic function $u(z)$; i.e. we require, in view of (A1.8):

$$\text{(A1.43)} \quad 0 = \oint_{\gamma_\rho} \frac{\partial u}{\partial s}\, ds = \oint_{\gamma_\rho} \frac{\partial V}{\partial n}\, ds = \oint_{\gamma_\rho} \frac{\partial v}{\partial n}\, ds - 2\pi \sum_{\nu=1}^{k-1} P_{\rho\nu}\lambda_\nu.$$

But we have already proved that the system of equations

$$\text{(A1.43')} \qquad \sum_{\nu=1}^{k-1} P_{\rho\nu}\lambda_\nu = \frac{1}{2\pi}\oint_{\gamma_\rho} \frac{\partial v}{\partial n}\, ds, \qquad \rho = 1, 2, \cdots, k-1$$

has exactly one solution. Thus we can find a harmonic function $u(z)$ which is single-valued in G and satisfies on γ the boundary condition (A1.39). It is clear that this function $u(z)$ is determined only up to an additive constant.

We have just shown that *the boundary value problems of the second type can be solved by means of the Green's function* and the harmonic measures. The procedure is, however, somewhat involved; it will, therefore, be more convenient to introduce another fundamental function which, in the present problem, plays the same role as the Green's function in the boundary value problems of the first type. This function is called the Neumann's function and is defined by the following three properties:

a) $N(z, \zeta)$ is harmonic for $z \in G$, except at the point $z = \zeta$;
b) $N(z, \zeta) + \log|z - \zeta|$ is harmonic at $z = \zeta$;
c) $\partial/\partial n\, N(z, \zeta) = 2\pi/L$ for $z \in \gamma$, $L =$ total length of γ.

The meaning of condition c) is clearly understood if we remember that, since $N(z, \zeta)$ has a logarithmic pole at $z = \zeta$, necessarily

$$\text{(A1.44)} \qquad \frac{1}{2\pi}\oint_\gamma \frac{\partial}{\partial n_z} N(z, \zeta)\, ds_z = 1.$$

The simplest requirement for the normal derivative of N is that its value on γ be constant, from which, clearly, condition c) follows.

Not even with all these conditions is $N(z, \zeta)$ uniquely determined; we may still add an additive constant depending on ζ. In order to fix $N(z, \zeta)$ completely the following requirement is usually made:

$$\text{(A1.45)} \qquad \oint_\gamma N(z, \zeta)\, ds_z \equiv 0 \qquad \text{for all } \zeta.$$

Let $u(z)$ be an arbitrary harmonic function which is continuously differentiable in the closed region $G + \gamma$. We find, because of Green's identity,

$$\text{(A1.46)} \qquad \begin{aligned} u(\zeta) &= \frac{1}{2\pi}\oint_\gamma \left(u(z)\,\frac{\partial N(z, \zeta)}{\partial n} - N(z, \zeta)\,\frac{\partial u}{\partial n} \right) ds \\ &= -\frac{1}{2\pi}\oint_\gamma N(z, \zeta)\,\frac{\partial u}{\partial n}\, ds + \frac{1}{L}\oint_\gamma u\, ds. \end{aligned}$$

Since $u(\zeta)$ is only defined up to an additive constant if its normal derivative is prescribed, it is sufficient to restrict the second boundary value problem to the class of all harmonic functions which satisfy the additional requirement

$$\oint_\gamma u\, ds = 0. \tag{A1.47}$$

For all these functions, we have the representation

$$u(\zeta) = -\frac{1}{2\pi}\oint_\gamma N(z, \zeta)\frac{\partial u(z)}{\partial n}\, ds \tag{A1.46'}$$

in terms of the Neumann's function and the boundary values of the normal derivative.

There remains the problem of representing the Neumann's function explicitly in terms of the Green's function. We consider the regular harmonic function

$$N(z, \zeta) + \log | z - \zeta |$$

in G and denote by $\nu(z, \zeta)$ its conjugate function with respect to z. By the Cauchy-Riemann differential equations, we have, using the notation

$$\log (z - \zeta) = \log | z - \zeta | + i \arg (z - \zeta),$$

for $z \in \gamma$,

$$\begin{aligned} -\frac{\partial}{\partial s}\nu(z, \zeta) &= \frac{\partial}{\partial n}[N(z, \zeta) + \log | z - \zeta |] \\ &= \frac{2\pi}{L} - \frac{\partial}{\partial s}\arg (z - \zeta). \end{aligned} \tag{A1.48}$$

Choosing on each γ_ρ an arbitrary branch of the multivalued function $\arg (z - \zeta)$, we determine

$$\nu(z, \zeta) = -2\pi/L\, s + \arg (z - \zeta) + \lambda_\rho \quad \text{for } z \in \gamma_\rho,\ \lambda_k = 0, \tag{A1.48'}$$

and solve this boundary value problem by the formula [cf. (A1.3) and (A1.6)]:

$$\begin{aligned} \nu(z, \zeta) = \frac{1}{2\pi}\oint_\gamma \frac{\partial g(z, t)}{\partial n_t}\left[-\frac{2\pi}{L} s_t + \arg (t - \zeta)\right] ds_t \\ + \sum_{\nu=1}^{k-1} \lambda_\nu \omega_\nu(z). \end{aligned} \tag{A1.49}$$

Here the $\lambda_\nu(\zeta)$ are again determined by the requirement that the conjugate function of $\nu(z, \zeta)$ shall be single-valued in G. Using (A1.6) and (A1.8), we are led to the conditions

$$\oint_\gamma \frac{\partial \omega_\rho(t)}{\partial n_t}\left[-\frac{2\pi}{L}s_t + \arg(t - \zeta)\right] ds_t = 2\pi \sum_{\nu=1}^{k-1} \lambda_\nu(\zeta) P_{\nu\rho}, \qquad \rho < k \tag{A1.50}$$

which determine the $\lambda_\nu(\zeta)$ uniquely.

There exists a close relationship between the formula (A1.49), giving the conjugate of the regular part of the Neumann's function, and the following formula for the regular part of the Green's function:

$$h(z, \zeta) = \frac{1}{2\pi}\oint_\gamma \frac{\partial g(z, t)}{\partial n_t} \log|t - \zeta|\, ds_t. \tag{A1.51}$$

This equation follows directly from the fact that $h(z, \zeta)$ has on γ the boundary values $\log|z - \zeta|$. Let us assume that the γ_ρ, $\rho < k$, have been so determined that on them $\log(z - \zeta)$ is single-valued. Then, comparing the periods of the conjugates of both sides of (A1.51) with respect to a circuit around a curve γ_ρ, we obtain in view of (A1.6)

$$\omega_\rho(\zeta) = \frac{1}{2\pi}\oint_\gamma \frac{\partial \omega_\rho(t)}{\partial n_t} \log|t - \zeta|\, ds_t. \tag{A1.52}$$

Let $u(\zeta)$, $v(\zeta)$ be a pair of conjugate harmonic functions and set $w(\zeta) = u(\zeta) + iv(\zeta)$. Then by the Cauchy-Riemann equations we have, for the differential operators $\partial/\partial\zeta$ and $\partial/\partial\bar{\zeta}$ defined in (A1.35),

$$\frac{\partial u}{\partial \zeta} = i\frac{\partial v}{\partial \zeta} = \frac{1}{2}\frac{dw}{d\zeta}, \qquad \frac{\partial u}{\partial \bar{\zeta}} = -i\frac{\partial v}{\partial \bar{\zeta}}. \tag{A1.53}$$

Differentiating (A1.50) and (A1.52) with respect to ζ and bearing in mind that $\log|t - \zeta|$ and $\arg(t - \zeta)$ are conjugate, we obtain

$$i\sum_{\nu=1}^{k-1} P_{\nu\rho}\frac{\partial \lambda_\nu}{\partial \zeta} = \frac{\partial \omega_\rho}{\partial \zeta}. \tag{A1.54}$$

Introducing the analytic functions $w_\rho(\zeta)$ defined in (A1.7), we get the following simple expression for the derivatives of the coefficients $\lambda_\nu(\zeta)$:

$$i\frac{\partial \lambda_\nu}{\partial \zeta} = \frac{1}{2}\sum_{\rho=1}^{k-1} \Pi_{\nu\rho} w_\rho'(\zeta). \tag{A1.55}$$

Now, differentiating (A1.49) and (A1.51) with respect to ζ and using (A1.55), we find

$$i\frac{\partial\nu(z,\zeta)}{\partial\zeta} = \frac{\partial h(z,\zeta)}{\partial\zeta} + \frac{1}{2}\sum_{\rho,\nu=1}^{k-1}\Pi_{\nu\rho}\,\omega_\nu(z)w_\rho'(\zeta). \tag{A1.56}$$

This is an interesting identity between the derivatives of the conjugate of the Neumann's function on the one hand, and of the Green's function on the other. It is, of course, still more interesting to introduce the Neumann's function itself into the identity. This is done by again using (A1.53), this time with respect to differentiation in z. We find from (A1.56) and (A1.53) [8, 17]:

$$\frac{\partial^2 N(z,\zeta)}{\partial z\,\partial\zeta} = \frac{\partial^2 g(z,\zeta)}{\partial z\,\partial\zeta} + \frac{1}{4}\sum_{\rho,\nu=1}^{k-1}\Pi_{\nu\rho}\,w_\nu'(z)w_\rho'(\zeta). \tag{A1.57}$$

Had we differentiated the conjugate-complex identity related to (A1.56) with respect to z, we would have obtained analogously

$$\frac{\partial^2 N(z,\zeta)}{\partial z\,\partial\bar\zeta} = -\frac{\partial^2 g(z,\zeta)}{\partial z\,\partial\bar\zeta} - \frac{1}{4}\sum_{\rho,\nu=1}^{k-1}\Pi_{\nu\rho}\,w_\nu'(z)\overline{w_\rho'(\zeta)}. \tag{A1.58}$$

We have thus identified the second derivatives of the Neumann's function with simple expressions based on the Green's function. The central role of the latter becomes quite evident from all the formalism based on its properties.

We can bring (A1.57) and (A1.58) into a particularly elegant form by introducing the function $m(z,\zeta)$, defined in (A1.37). Because of (A1.21), we can write

$$m(z,\zeta) = g(z,\zeta) + \sum_{\sigma,\rho=1}^{k-1}\Pi_{\rho\sigma}\,\omega_\rho(z)\omega_\sigma(\zeta) \tag{A1.59}$$

and hence we may put (A1.57) and (A1.58) into the form

$$\frac{\partial^2 N(z,\zeta)}{\partial z\,\partial\zeta} = \frac{\partial^2 m(z,\zeta)}{\partial z\,\partial\zeta}, \qquad \frac{\partial^2 N(z,\zeta)}{\partial z\,\partial\bar\zeta} = -\frac{\partial^2 m(z,\zeta)}{\partial z\,\partial\bar\zeta}. \tag{A1.60}$$

The close relation of $\partial^2 m/\partial z\partial\zeta$ to a univalent mapping function has already been shown in (A1.38).

Let us show, finally, how the Neumann's function is connected with an important canonical mapping of the domain G. Let ζ and η be two arbitrary points in G and consider the function $N(z,\zeta) - N(z,\eta)$.

In view of the fundamental property of the Neumann's function, we have

$$\frac{\partial}{\partial n}[N(z, \zeta) - N(z, \eta)] = 0 \qquad \text{for } z \in \gamma. \tag{A1.61}$$

We may extend the above harmonic function in z to an analytic function $q(z; \zeta, \eta)$ of z. From the Cauchy-Riemann equations we can assert that the imaginary part of q remains constant on each boundary curve γ_ρ. In particular, this imaginary part will be single-valued on each γ_ρ. Further, $q(z; \zeta, \eta)$ has two logarithmic poles at ζ and η

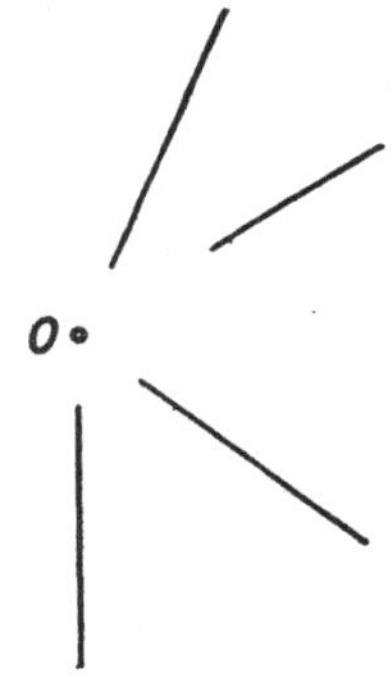

Figure 4. Radial-slit domain.

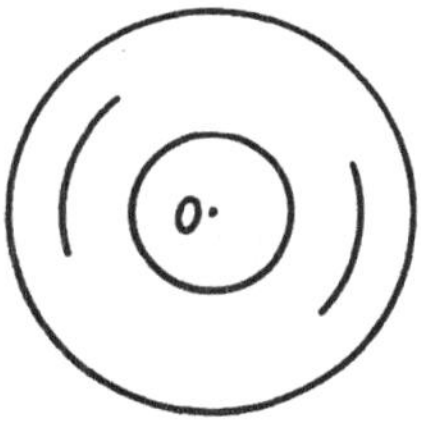

Figure 5. Ring-slit domain.

with the residues -1 and $+1$, respectively. Hence, we easily obtain the following

Theorem: The function

$$Q(z; \zeta, \eta) = \exp\{-q(z; \zeta, \eta)\} \tag{A1.62}$$

is univalent and single-valued in G and maps this domain onto the whole plane slit along linear segments, all directed towards the origin. The point $z = \zeta$ corresponds to the origin, the point $z = \eta$ to infinity (Figure 4).

These properties determine the conformal map up to a multiplicative constant. The possibility of this so-called radial-slit domain was first proved by Köbe [34]. We see here that its explicit analytical expression is given by the Neumann's function and, hence, is ultimately reducible to the Green's function of G. Numerous further canonical maps could be constructed from the Green's and Neumann's functions and the harmonic measures. We mention, for example, the mapping of G onto a circular ring slit along $k - 2$ concentric circular arcs (Figure 5). One has here the freedom to determine the two boundary curves γ_ρ which shall correspond to the full circumferences which bound the ring [34].

For further results concerning the Green's function and canonical mappings, compare [33, 38].

2. *Dirichlet Integrals for Harmonic Functions*

1. *Formal Remarks.* In the previous chapters, we considered the family of all piecewise smooth functions in G which satisfy given conditions with respect to the boundary. Within this family, we distinguished the harmonic function as that having the minimum Dirichlet integral. In this section, Dirichlet's integral will be used in a different way; we shall apply it only within the family of harmonic functions and characterize certain harmonic functions by extremum properties of Dirichlet's integral.

Let $u(z)$ and $v(z)$ be harmonic in G; applying the complex differential operator $\partial/\partial z$ defined in (A1.35), we obtain two analytic functions

$$\text{(A2.1)} \qquad f(z) = u_x - iu_y = 2\frac{\partial u}{\partial z}, \qquad g(z) = v_x - iv_y = 2\frac{\partial v}{\partial z}.$$

The Dirichlet product $D[u, v]$ may then be expressed in the form

$$\text{(A2.2)} \qquad D[u, v] = \mathfrak{Re}\left\{\iint_G f(z)\overline{g(z)}\, d\tau\right\}, \qquad d\tau = dx\, dy.$$

In order to utilize the full power of the theory of analytic functions we shall henceforth consider analytic functions $f(z)$ and $g(z)$ instead of harmonic functions $u(z)$ and $v(z)$. We regard them as vectors in an infinite-dimensional linear space $\mathfrak{F}$ over the field of complex numbers in which a metric is established by defining as the scalar product of two vectors

$$\text{(A2.3)} \qquad (f, \bar{g}) = \iint_G f\bar{g}\, d\tau.$$

In particular, we define as the norm or length of the vector the expression

$$\text{(A2.3}') \qquad \|f\| = \sqrt{(f, \bar{f})}, \quad \text{i.e.} \quad \|f\|^2 = \iint_G |f|^2\, d\tau.$$

In the following, the formal rules on integration by parts will be of importance. If there exists a function $F(z)$, single-valued and continuously differentiable in G and such that $f(z) = F'(z)$, then from Green's formula one can easily derive

$$(f, \bar{g}) = \iint_G F'(z)\overline{g(z)}\, d\tau_z \tag{A2.4}$$
$$= -\frac{1}{2i}\oint_\gamma F(z)\overline{g(z)}\, d\bar{z}, \qquad d\bar{z} = dx - i\, dy.$$

Similarly, if $g(z) = G'(z)$, we have

$$(f, \bar{g}) = +\frac{1}{2i}\oint_\gamma f(z)\overline{G(z)}\, dz. \tag{A2.4'}$$

If $f(z)$ or $g(z)$ is derived from a single-valued harmonic function $u(z)$ or $v(z)$ according to (A2.1), we may write

$$(f, \bar{g}) = -\frac{1}{i}\oint_\gamma u(z)\overline{g(z)}\, d\bar{z} = +\frac{1}{i}\oint_\gamma f(z)v(z)\, dz. \tag{A2.5}$$

For each single-valued analytic function $f(z)$ in G there exists a function $F(z)$ such that $F'(z) = f(z)$. In general, however, $F(z)$ will not be single-valued in G. For a circuit around the boundary continuum γ_ρ, the function $F(z)$ will have the period

$$[f]_\rho = \oint_{\gamma_\rho} f(z)\, dz. \tag{A2.6}$$

These periods can easily be written in the form of scalar products in our Dirichlet metric. In fact, consider the function $w_\rho'(z)$ defined in (A1.7) and apply (A2.5) to the expression

$$(f, \overline{w_\rho'}) = +\frac{1}{i}\oint_\gamma f(z)\omega_\rho(z)\, dz. \tag{A2.7}$$

Using the fact that $\omega_\rho = 1$ on γ_ρ and $\omega_\rho = 0$ elsewhere on γ, we find finally

$$[f]_\rho = +i(f, \overline{w_\rho'}) = \overline{-i(w_\rho', \bar{f})}. \tag{A2.8}$$

Consider the subspace $\mathfrak{F}_0 \subset \mathfrak{F}$ consisting of all analytic functions in G with single-valued integrals. We conclude from (A2.8) the important result:

The necessary and sufficient condition for $f(z)$ to lie in $\mathfrak{F}_0$ is that it be orthogonal to all functions $w_\rho'(z)$.

2. *The Kernels K and L.* We define next the two functions

$$K(z, \bar{\zeta}) = -\frac{2}{\pi}\frac{\partial^2 g(z, \zeta)}{\partial z\,\partial \bar{\zeta}}, \qquad L(z, \zeta) = -\frac{2}{\pi}\frac{\partial^2 g(z, \zeta)}{\partial z\,\partial \zeta}. \tag{A2.9}$$

These combinations of derivatives of the Green's function have al-already occurred in the important identities (A1.57) and (A1.58). We shall immediately recognize their significance in our Dirichlet calculus. Since we may write

$$\log | z - \zeta | = \tfrac{1}{2} \log (z - \zeta) + \tfrac{1}{2} \log (\bar{z} - \bar{\zeta}),$$

it is obvious that the singularity of the Green's function disappears under the differentiation processes leading to the function K. Hence $K(z, \bar{\zeta})$ is a regular analytic function of z and $\bar{\zeta}$ over the whole of G. The function $L(z, \zeta)$ has at $z = \zeta$ a double pole; we put

$$L(z, \zeta) = \frac{1}{\pi(z - \zeta)^2} - l(z, \zeta), \tag{A2.10}$$

where $l(z, \zeta)$ is a regular analytic function of z and ζ in G. As an illustration, consider the case where the domain G is the circle $| z | < R$; the Green's function has then the form

$$g(z, \zeta) = \log \left| \frac{R^2 - z\bar{\zeta}}{R(z - \zeta)} \right| \tag{A2.11}$$

and hence

$$K(z, \bar{\zeta}) = \frac{R^2}{\pi(R^2 - z\bar{\zeta})^2}, \quad L(z, \zeta) = \frac{1}{\pi(z - \zeta)^2}, \quad l(z, \zeta) = 0. \tag{A2.12}$$

From the definition of K and L and from the symmetry of the Green's function, we infer the symmetry laws

$$K(z, \bar{\zeta}) = \overline{K(\zeta, \bar{z})}, \qquad L(z, \zeta) = L(\zeta, z), \qquad l(z, \zeta) = l(\zeta, z). \tag{A2.13}$$

Now let $f(z)$ be analytic and continuous in $G + \gamma$. Consider the Dirichlet integral

$$\begin{aligned} (f(z), \overline{K(z, \bar{\zeta})}) &= \iint_G K(\zeta, \bar{z}) f(z)\, d\tau \\ &= -\frac{2}{\pi} \iint_G \frac{\partial^2 g(\zeta, z)}{\partial \zeta\, \partial \bar{z}} f(z)\, d\tau_z. \end{aligned} \tag{A2.14}$$

Let $G_{\zeta,\epsilon}$ be the domain obtained from G by removal of a circle of radius ϵ around the point ζ and compute first, by means of (A2.5), Dirichlet's integral over this sub-domain:

$$\iint_{G_{\zeta,\epsilon}} K(\zeta, \bar{z})f(z)\, d\tau_z = -\frac{1}{\pi i}\oint \frac{\partial g(\zeta, z)}{\partial \zeta} f(z)\, dz. \tag{A2.15}$$

The line integral on the right-hand side is to be extended over the boundary of $G_{\zeta,\epsilon}$, i.e. over γ and the circumference $|z - \zeta| = \epsilon$, the

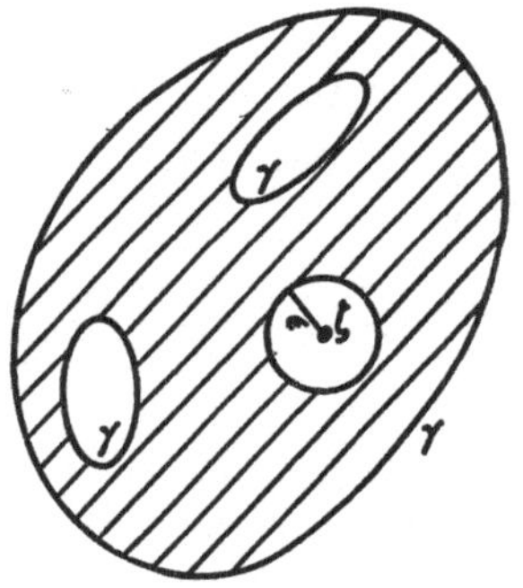

Figure 6. The domain $G_{\epsilon,\zeta}$.

latter in the negative sense. Since for $z \in \gamma$, we have $g(z, \zeta) \equiv 0$ for all $\zeta \in G$, it is clear that

$$\frac{\partial g(\zeta, z)}{\partial \zeta} = 0 \quad \text{for} \quad z \in \gamma. \tag{A2.16}$$

Thus we find in view of the behavior of the Green's function near ζ, as given by (A1.2),

$$\begin{aligned}\iint_{G_{\zeta,\epsilon}} K(\zeta, \bar{z})f(z)\, d\tau &= \frac{1}{2\pi i}\oint_{|z-\zeta|=\epsilon} \left(\frac{1}{z-\zeta} + \frac{\partial h}{\partial \zeta}\right) f(z)\, dz \\ &= f(\zeta) + O(\epsilon).\end{aligned} \tag{A2.17}$$

Let $\epsilon \to 0$; the left-hand side converges towards the Dirichlet integral (A1.14) and thus we have proved the remarkable identity [62, 72]:

$$(f(z), \overline{K(z, \bar{\zeta})}) = \iint_G K(\zeta, \bar{z})f(z)\, d\tau = f(\zeta). \tag{A2.18}$$

The function $K(\zeta, \bar{z})$ is, therefore, a reproducing kernel with respect to the integral operation (A2.18); it may also be considered as the unit matrix in the vector space $\mathfrak{F}$ with Dirichlet metric. Identity (A2.18) was first observed by Wirtinger; it has, as we shall see, numerous applications in the theory of analytic and harmonic functions.

It is now of interest to study the Dirichlet integral

$$(f(z), \overline{L(z, \zeta)}) = -\frac{2}{\pi} \iint_G \frac{\partial^2 g(z, \zeta)}{\partial \bar{z} \partial \bar{\zeta}} f(z)\, d\tau_z. \tag{A2.19}$$

We must remark at first that, because of its double pole at ζ, $L(z, \zeta)$ does not belong to the class $\mathfrak{F}$. We may, however, define the improper integral (A2.19) as the limit value for $\epsilon \to 0$ of analogous integrals extended over the subdomains $G_{\zeta,\epsilon}$. We find again by use of (A2.5)

$$\iint_{G_{\zeta,\epsilon}} f(z)\overline{L(z, \zeta)}\, d\tau_z = -\frac{1}{\pi i} \int \frac{\partial g(z, \zeta)}{\partial \bar{\zeta}} f(z)\, dz, \tag{A2.20}$$

where the integration in the line integral is to be extended over γ and $|z - \zeta| = \epsilon$. On γ the integrand vanishes again because of (A2.16) and hence, in view of (A1.2),

$$\begin{aligned} \iint_{G_{\zeta,\epsilon}} f(z)\overline{L(z, \zeta)}\, d\tau_z &= \frac{1}{2\pi i} \oint_{|z-\zeta|=\epsilon} \left[\frac{1}{\bar{z} - \bar{\zeta}} + \frac{\partial h}{\partial \bar{\zeta}}\right] f(z)\, dz \\ &= \frac{1}{2\pi i} \oint_{|z-\zeta|=\epsilon} \left[\frac{1}{\epsilon^2} (z - \zeta) + \frac{\partial h}{\partial \bar{\zeta}}\right] f(z)\, dz \\ &= O(\epsilon). \end{aligned} \tag{A2.21}$$

This result proves that the limit of the left-hand integral exists for $\epsilon \to 0$ and that it has the value 0. Thus we have proved

$$(f(z), \overline{L(z, \zeta)}) = 0, \tag{A2.22}$$

that is:

The kernel $L(z, \zeta)$ is orthogonal to all analytic functions in G.

From the representation (A2.10) follows the identity

$$\frac{1}{\pi} \iint_G \frac{f(z)}{(\bar{z} - \bar{\zeta})^2}\, d\tau_z = \iint_G f(z)\overline{l(z, \zeta)}\, d\tau_z; \tag{A2.23}$$

the improper integral on the left is to be understood in the same sense as the L-integral. This interpretation is to be given to all improper integrals which will occur. Identity (A2.23) states a very interesting fact: The function $1/\pi(z - \zeta)^2$ is a simple analytic function which becomes infinite only at the point ζ. This fact excludes it from the family $\mathfrak{F}$. The family contains, however, the function $l(z, \zeta)$ which is equivalent to the given function in the space with Dirichlet metric, having the same scalar product with all functions of $\mathfrak{F}$.

If we choose, in particular, $f(z) = K(z, \bar{\eta})$ for arbitrary $\eta \in G$, we obtain from (A2.23) and the reproducing property (A2.18) of K

$$\frac{1}{\pi} \iint_G K(z, \bar{\eta})(\bar{z} - \bar{\zeta})^{-2}\, d\tau_z = \overline{l(\zeta, \eta)}. \tag{A2.24}$$

Thus the knowledge of the K-kernel leads to an easy construction of L:

$$L(\zeta, \eta) = \frac{1}{\pi}\left[(\zeta - \eta)^{-2} - \iint_G K(\eta, \bar{z})(z - \zeta)^{-2}\, d\tau_z\right]. \tag{A2.25}$$

Since the Green's and Neumann's functions are obtainable from K and L, our construction problem for these functions is reduced to the problem of expressing the K-kernel in a convenient way.

On the other hand, the kernel K can be constructed if the L-kernel is known. To show this, we start again from the formula (A2.16)

$$\frac{\partial g(\zeta, z)}{\partial \zeta} \equiv 0 \qquad \text{for} \qquad z \in \gamma.$$

Let $z(s)$ be a parametric representation of the curve system γ and s the length parameter. Denoting dz/ds by z', we obtain from (A2.16) by differentiation

$$\frac{\partial^2 g(\zeta, z)}{\partial \zeta \partial z} z' + \frac{\partial^2 g(\zeta, z)}{\partial \zeta \partial \bar{z}} \bar{z}' = 0. \tag{A2.26}$$

Hence, using definitions (A2.9), we arrive at the result

$$L(z, \zeta) z' = -\overline{K(z, \bar{\zeta}) z'} \qquad \text{for } z \in \gamma, \zeta \in G. \tag{A2.27}$$

Consider now the Dirichlet integral

$$\begin{aligned} (l(z, \zeta), \overline{l(z, \eta)}) &= \iint_G \overline{l(z, \eta)}\, l(z, \zeta)\, d\tau_z \\ &= \frac{1}{\pi} \iint_G l(z, \zeta)(\bar{z} - \bar{\eta})^{-2}\, d\tau_z. \end{aligned} \tag{A2.28}$$

By means of (A2.4′), we establish the formula

$$\text{(A2.29)}\qquad \frac{1}{\pi}\iint_{G_{\eta,\epsilon}} l(z,\zeta)(\bar z-\bar\eta)^{-2}\,d\tau_z = -\frac{1}{2\pi i}\oint_\gamma l(z,\zeta)(\bar z-\bar\eta)^{-1}\,dz + O(\epsilon);$$

passing to the limit as $\epsilon\to 0$, we obtain, from (A2.10),

$$\text{(A2.29′)}\qquad (l(z,\zeta),\overline{l(z,\eta)}) = +\frac{1}{2\pi i}\oint_\gamma L(z,\zeta)(\bar z-\bar\eta)^{-1}\,dz - \frac{1}{2\pi^2 i}\oint_\gamma (z-\zeta)^{-2}(\bar z-\bar\eta)^{-1}\,dz.$$

Using the boundary behavior (A2.27) of the L-kernel, we can, in view of the residue theorem, write

$$\text{(A2.30)}\qquad \frac{1}{2\pi i}\oint L(z,\zeta)(\bar z-\bar\eta)^{-1}\,dz = +\overline{\frac{1}{2\pi i}\oint_\gamma K(z,\bar\zeta)(z-\eta)^{-1}\,dz} = K(\zeta,\bar\eta).$$

Let further $\tilde G$ be the complementary part of the z-plane with respect to G. Applying (A2.4′) with respect to $\tilde G$, we find

$$\text{(A2.31)}\qquad \frac{1}{2\pi^2 i}\oint_\gamma (z-\zeta)^{-2}(\bar z-\bar\eta)^{-1}\,dz = \frac{1}{\pi^2}\iint_{\tilde G}(z-\zeta)^{-2}(\bar z-\bar\eta)^{-2}\,d\tau_z = \Gamma(\zeta,\bar\eta).$$

$\Gamma(\zeta,\bar\eta)$ is a regular analytic function of ζ and $\bar\eta$ in G which can be obtained by simple integration. It will play an important role in the final construction of the K-kernel. We can use (A2.30) and (A2.31) to put (A2.29′) into the final form

$$\text{(A2.32)}\qquad \iint_G l(z,\zeta)\overline{l(z,\eta)}\,d\tau_z = K(\zeta,\bar\eta) - \Gamma(\zeta,\bar\eta).$$

This formula permits, in fact, the construction of K, once the kernel L is known.

3. *Inequalities.* We shall now apply the identities obtained above to the theory of analytic functions in G. We have for every such function the identity (A2.18). Hence, by Schwarz' inequality and the reproducing property of the kernel K, we find

$$|f(\zeta)|^2 \leq \|f\|^2 \iint_G K(\zeta, \bar{z})K(z, \bar{\zeta})\, d\tau = \|f\|^2 K(\zeta, \bar{\zeta}). \tag{A2.33}$$

This formula shows:

Each function $f \in \mathfrak{F}$ can be estimated at each point of G if its norm in the Dirichlet metric is known.

The equality sign in (A2.33) can only hold if $f(z) = \lambda K(z, \bar{\zeta})$; this shows that the K-kernel can be characterized by an extremum property within the space $\mathfrak{F}$. In fact:

Among all functions $f(z) \in \mathfrak{F}$ which have at a given point $\zeta \in G$ the value 1, *the function $K(z, \bar{\zeta})/K(\zeta, \bar{\zeta})$ has the least norm, namely $K(\zeta, \bar{\zeta})^{-1}$.*

This remark opens a new approach to the whole theory of the boundary value problem. Suppose the existence of a Green's function were unproved; the construction of the kernels K and L would then be impossible. It is always possible to define an analytic function $K(z, \bar{\zeta})$ by the above extremum problem since the class of all analytic functions in G with bounded norm is compact. This extremal function is easily shown to have the reproducing property (A2.18). If the L-kernel is defined by means of (A2.25) it becomes possible to construct Green's and Neumann's functions and verify all their properties. This approach was used by Lehto and Garabedian-Schiffer in order to give a new proof for the existence of various canonical mappings [18, 36].

Consider next the analytic function

$$f(z) = \sum_{i=1}^{N} [K(z, \overline{\zeta_i})\bar{x}_i + \lambda l(z, \zeta_i)x_i], \tag{A2.34}$$

where the ζ_i are N arbitrary points in G and the x_i arbitrary numbers. The norm of $f(z)$ is clearly non-negative. It can be easily computed by means of identity (A2.32) and the reproducing property of the

K-kernel. We find:

$$0 \leq \| f \|^2 = \sum_{i,k=1}^{N} K(\zeta_i, \bar{\zeta}_k) x_i \bar{x}_k + 2\mathcal{R}e\left\{\lambda \sum_{i,k=1}^{N} l(\zeta_i, \zeta_k) x_i x_k\right\} + |\lambda|^2 \sum_{i,k=1}^{N} [K(\zeta_i, \bar{\zeta}_k) - \Gamma(\zeta_i, \bar{\zeta}_k)] x_i \bar{x}_k. \tag{A2.35}$$

Note first that the Hermitian form [cf. (A2.31)]

$$\sum_{i,k=1}^{N} \Gamma(\zeta_i, \bar{\zeta}_k) x_i \bar{x}_k = \frac{1}{\pi^2} \iint_{\tilde{G}} \left| \sum_{i=1}^{N} \frac{x_i}{(z - \zeta_i)^2} \right|^2 d\tau_z \tag{A2.36}$$

is positive definite. Hence, we conclude from (A2.35) that

$$\sum_{i,k=1}^{N} K(\zeta_i, \bar{\zeta}_k) x_i \bar{x}_k \geq \sum_{i,k=1}^{N} \Gamma(\zeta_i, \bar{\zeta}_k) x_i \bar{x}_k > 0 \tag{A2.37}$$

is also positive definite. Since the inequality (A2.35) holds for every choice of the complex parameter λ, the discriminant condition is

$$\left| \sum_{i,k=1}^{N} l(\zeta_i, \zeta_k) x_i x_k \right|^2 \leq \sum_{i,k=1}^{N} K(\zeta_i, \bar{\zeta}_k) x_i \bar{x}_k \sum_{i,k=1}^{N} [K(\zeta_i, \bar{\zeta}_k) - \Gamma(\zeta_i, \bar{\zeta}_k)] x_i \bar{x}_k. \tag{A2.38}$$

From (A2.38) we derive the weaker result

$$\left| \sum_{i,k=1}^{N} l(\zeta_i, \zeta_k) x_i x_k \right| \leq \sum_{i,k=1}^{N} K(\zeta_i, \bar{\zeta}_k) x_i \bar{x}_k. \tag{A2.39}$$

Consider next the class of all analytic functions $f(z)$ in G which have at N arbitrarily given points $\zeta_i \in G$ the prescribed values a_i. There exists always exactly one function $\sum_{i=1}^{N} K(z, \bar{\zeta}_i)\bar{x}_i$ in this class; for the determinant of the system

$$\sum_{i=1}^{N} K(\zeta_k, \bar{\zeta}_i) \bar{x}_i = a_k \tag{A2.40}$$

does not vanish because of the positive definite character of the K-kernel. Thus every function of the above class can be written in the form

$$f(z) = \sum_{i=1}^{N} K(z, \bar{\zeta}_i) \bar{x}_i + \varphi(z), \tag{A2.41}$$

where $\varphi(z)$ vanishes at all points ζ_i. Let us now compute the norm of f; because of the reproducing property of K and because $\varphi(\zeta_i) = 0$, we obtain

$$\|f(z)\|^2 = \sum_{i,k=1}^{N} K(\zeta_i, \bar{\zeta}_k)x_i\bar{x}_k + \|\varphi\|^2 \geq \sum_{i,k=1}^{N} K(\zeta_i, \bar{\zeta}_k)x_i\bar{x}_k. \tag{A2.42}$$

The quadratic form on the right may be defined as a minimum in an appropriate interpolation problem. We recognize from this result the importance of the K-kernel in the interpolation theory for analytic functions in G.

4. *Conformal Transformations.* Let us map the domain G by a univalent function $w = \varphi(z)$ onto a domain G_w. Let $g_w(w, \omega)$ denote the Green's function of G_w and correspondingly K_w and L_w the kernels connected with this domain. It is easily seen that, if $\omega = \varphi(\zeta)$,

$$g_w(w, \omega) = g(z, \zeta), \tag{A2.43}$$

i.e. the Green's function is invariant under conformal mapping.

If we differentiate this identity with respect to z and ζ, we obtain the following transformation laws for the kernels:

$$K_w(w, \bar{\omega})\varphi'(z)\overline{\varphi'(\zeta)} = K(z, \bar{\zeta}),\ L_w(w, \omega)\varphi'(z)\varphi'(\zeta) = L(z, \zeta). \tag{A2.44}$$

Finally, we find for the transformation of $l(z, \zeta)$ the rule

$$l_w(w, \omega)\varphi'(z)\varphi'(\zeta) = \frac{1}{\pi}\left[\frac{\varphi'(z)\varphi'(\zeta)}{[\varphi(z) - \varphi(\zeta)]^2} - \frac{1}{(z - \zeta)^2}\right] + l(z, \zeta). \tag{A2.45}$$

The function

$$U(z, \zeta) = \frac{\varphi'(z)\varphi'(\zeta)}{[\varphi(z) - \varphi(\zeta)]^2} - \frac{1}{(z - \zeta)^2} \tag{A2.46}$$

is of great interest in the theory of conformal mapping. If we consider it for an arbitrary function $\varphi(z)$, we can immediately state a criterion for the univalence of $\varphi(z)$. This criterion is obviously the regularity of $U(z, \zeta)$ within the domain considered. If $z = \zeta$, U has the value

$$U(z, z) = \frac{1}{6}\left[\frac{\varphi'''}{\varphi'^2} - \frac{3}{2}\frac{\varphi''^2}{\varphi'^2}\right] = \frac{1}{6}\{\varphi, z\}, \tag{A2.47}$$

where $\{\varphi, z\}$ denotes the Schwarz-Cayley differential parameter which plays an important role in the conformal mapping of domains whose boundaries are circular arcs.

Let z_1 and z_2 be two arbitrary points in G which are mapped by $\varphi(z)$ into w_1 and w_2. Clearly, in view of (A2.44) the expression

$$\text{(A2.48)} \qquad \int_{z_1}^{z_2} \sqrt{K(z, \bar{z})} \mid dz \mid = \int_{w_1}^{w_2} \sqrt{K_w(w, \bar{w})} \mid dw \mid$$

is a *conformal invariant* and can serve to introduce in G a length concept which is unchanged by any conformal mapping. The geometry of this metric was studied by Bergman [5]. In the case of a simply connected domain G it was first considered by Poincaré.

5. *An Application to the Theory of Univalent Functions.* The theories of conformal mapping and of univalent functions (cf. [43, 68]) are closely interrelated: conformal mapping is carried out by means of univalent functions. The two theories differ, however, in their points of view and in the problems considered. While the theory of conformal mapping deals primarily with existence and behavior of canonical mappings, the central problem in the theory of univalent functions is the establishing of necessary and sufficient conditions for a given analytic function $\varphi(z)$ in G to be univalent. We shall now show that inequalities (A2.39) yield considerable information in this respect.

In fact, suppose that $w = \varphi(z)$ is univalent in G and maps this domain onto the domain G_w. Here inequalities (A2.39) must hold with respect to the functions l_w and K_w. But according to the results of the last article these functions can be expressed in terms of the given function $\varphi(z)$ and the known functions l and K connected with our basic domain G. Using formulas (A2.44) and (A2.45), we obtain the result:

In order that $\varphi(z)$ be univalent in G, it is necessary that the inequality

$$\text{(A2.49)} \qquad \left| \sum_{i,k=1}^{N} \left[\frac{1}{\pi} U(\zeta_i, \zeta_k) + l(\zeta_i, \zeta_k) \right] x_i x_k \right| \leq \sum_{i,k=1}^{N} K(\zeta_i, \bar{\zeta}_k) x_i \bar{x}_k$$

hold for arbitrary values $\zeta_i \in G$ and complex numbers x_i.

Since U is simply constructed from $\varphi(z)$ this inequality puts an infinity of necessary conditions on the function $\varphi(z)$. As an illustration let us consider the case in which G is the unit circle and let $N = 1$. We obtain, because of (A2.12) and (A2.47), the necessary condition

$$\text{(A2.50)} \qquad \mid \{\varphi, z\} \mid \leq \frac{6}{(1 - \mid z \mid^2)^2},$$

which is closely related to the distortion theorem for this class of univalent functions. Nehari has shown conversely that if a function $\varphi(z)$ satisfies in $|z| < 1$ the inequality

$$|\{\varphi, z\}| \leq \frac{2}{(1 - |z|^2)^2}, \tag{A2.51}$$

we can assert that $\varphi(z)$ is univalent in $|z| < 1$. This condition is, however, not necessary [29, 44].

It can also be shown that a countable subset can be selected from the inequalities (A2.49) which as a whole represents a sufficient condition for the univalence of the function $\varphi(z)$. Such a set of inequalities was given first by Grunsky and is intimately related to inequality (A2.49) [26].

6. *Discontinuities of the Kernels.* By definition, the kernel $L(z, \zeta)$ has a double pole for $z = \zeta$. The kernel $K(z, \bar{\zeta})$ is continuous for $z \in G + \gamma$ as long as $\zeta \in G$; however, if ζ lies on γ, the kernel K also becomes infinite for $z = \zeta$. In the case of the circle this is quite obvious, in view of (A2.12). That this is generally so may be deduced from the fact that the homogeneous integral equation

$$f(z) + \lambda \iint_G K(z, \bar{\zeta}) f(\zeta)\, d\tau_\zeta = 0 \tag{A2.52}$$

has the value $\lambda = -1$ as an eigenvalue of infinite order, since all analytic functions $f(z)$ satisfy this equation because of the reproducing property of K. Hence, by virtue of general theorems on integral equations, $K(z, \bar{\zeta})$ cannot be a bounded function in the closed region $G + \gamma$ (cf. [14]).

We prove now the remarkable

Theorem: The function $l(z, \zeta)$ is continuous in both variables in the closed region $G + \gamma$ if the boundary curves γ_ρ everywhere possess three continuous derivatives with respect to the length parameter.

This statement is immediately verified for the circle by means of (A2.12). The result can be extended by conformal mapping to all simply connected domains by applying the transformation law (A2.45) and remarking that, if $\varphi(z)$ denotes the function mapping the unit circle onto the simply connected domain G, our assumption on the curves γ_ρ guarantees the continuity of $U(z, \zeta)$ in the closed region $|z| \leq 1$. There remains to prove, therefore, only the regularity of the l-kernel for multiply connected domains. Let γ_1 be one fixed boundary curve. It divides the plane into two domains; we call the one which contains G the domain G_1. Let $g_1(z, \zeta)$ be its Green's function. The function $g_1(z, \zeta) - g(z, \zeta)$ is harmonic everywhere in G and vanishes, by construction, on γ_1. Hence, we may represent it by Green's formula (A1.3) and obtain

$$g_1(z, \zeta) - g(z, \zeta) = \frac{1}{2\pi} \int_{\gamma-\gamma_1} \frac{\partial g(z, t)}{\partial n_t} [g_1(t, \zeta) - g(t, \zeta)] \, ds_t. \tag{A2.53}$$

Differentiating with respect to z and ζ, we obtain in obvious notation

$$l_1(z, \zeta) - l(z, \zeta) = \frac{1}{\pi^2} \int_{\gamma-\gamma_1} \frac{\partial^2 g(z, t)}{\partial z \partial n_t} \left[\frac{\partial g_1(t, \zeta)}{\partial \zeta} - \frac{\partial g(t, \zeta)}{\partial \zeta}\right] ds_t. \tag{A2.54}$$

This formula already contains the stated result. For if ζ and z lie near or on γ_1, the integrand in (A2.54) remains continuous, since the integration point t does not run over γ_1. The function $l_1(z, \zeta)$ is continuous in the closed simply connected region G_1; hence $l(z, \zeta)$ is continuous near and on γ_1. But, since γ_1 might have been any boundary curve of γ, our result is proved.

The continuity of $l(z, \zeta)$ in $G + \gamma$ is of great value for the theory of the Green's function. As a first application let us consider formula (A2.32). We know now that the left-hand side is continuous in the closed region $G + \gamma$. Thus, we can infer the continuity of the expression $K(z, \bar{\zeta}) - \Gamma(z, \bar{\zeta})$ in this closed region. The significance of this result lies in the following fact: the kernel $K(z, \bar{\zeta})$ is a certain derivative of the Green's function of G and depends on the domain G in a highly transcendental way; the Γ-function, however, is obtained by integrating an elementary function over the complement of G and can in effect be computed by integration. No solution of boundary value problems is necessary for its construction; we express this fact by terming Γ as a geometric quantity. Thus our result shows that the K-kernel has on γ the same order of infinity as the geometric quantity Γ.

7. *An Eigenvalue Problem.* We consider the positive definite Hermitian kernel $K(z, \bar{\zeta}) - \Gamma(z, \bar{\zeta})$ (cf. (A2.37)), which is uniformly bounded in G, and study the homogeneous integral equation

$$\varphi_\nu(z) = \lambda_\nu^2 \iint_G [K(z, \bar{\zeta}) - \Gamma(z, \bar{\zeta})] \varphi_\nu(\zeta) \, d\tau_\zeta. \tag{A2.55}$$

Because of the enumerated properties of the kernel this equation has positive eigenvalues λ_ν^2 and eigenfunctions $\varphi_\nu(z)$. We want to show that we have always $\lambda_\nu^2 \geq 1$. In fact, we derive from (A2.55), (A2.31), and the reproducing property of the kernel:

$$\text{(A2.56)} \quad \iint_G |\varphi_\nu(z)|^2 \, d\tau_z = \lambda_\nu^2 \iint_G |\varphi_\nu(z)|^2 \, d\tau_z - \lambda_\nu^2 \iint_{\tilde{G}} |\chi_\nu(z)|^2 \, d\tau,$$

where

$$\text{(A2.57)} \qquad \chi_\nu(z) = \frac{1}{\pi} \iint_G \frac{\overline{\varphi_\nu(\zeta)}}{(z-\zeta)^2} \, d\tau_\zeta, \qquad z \in \tilde{G}.$$

The equation (A2.56) proves our statement. Furthermore it shows that $\lambda_\rho > 1$ except for functions $\varphi_\nu(z)$ whose transform $\chi_\nu(z)$ vanishes identically in the complement $\tilde{G}$ of G. Such functions can, however, be easily characterized. We remark first that, because of (A2.55) and the continuity properties of the kernel $K - \Gamma$, each $\varphi_\nu(z)$ is continuous in $G + \gamma$. Hence, integrating by parts by means of (A2.4), we may transform (A2.57) into

$$\text{(A2.58)} \qquad \begin{aligned} \chi_\nu(z) &= \frac{1}{2\pi i} \oint_\gamma \overline{\varphi_\nu(\zeta)}\,(\zeta - z)^{-1} \overline{d\zeta} \\ &= \frac{1}{2\pi i} \oint_\gamma \overline{\varphi_\nu(\zeta)\zeta'^2}(\zeta - z)^{-1} \, d\zeta \end{aligned}$$

since $d\bar{\zeta} = \bar{\zeta}' \, ds$ and $|\zeta'|^2 = 1$. The same formula represents also an analytic function $\chi_\nu(z)$ for $z \in G$. The function $\chi_\nu(z)$ has a discontinuity if the argument point z crosses the boundary γ. According to a classical result of Plemelj [45], we have for any $z \in \gamma$ the jump condition

$$\text{(A2.59)} \quad \lim_{z_i \to z} \chi_\nu(z_i) = \lim_{z_e \to z} \chi_\nu(z_e) + \overline{\varphi_\nu(z) z'^2}, \qquad z_i \in G,\ z_e \in \tilde{G}.$$

If we assume now $\chi_\nu(z) \equiv 0$ in $\tilde{G}$, we find that the analytic function $\chi_\nu(z)$ in G has on γ the boundary values $\overline{\varphi_\nu(z) z'^2}$. Let

$$\text{(A2.60)} \qquad \Phi(\zeta) = \int^\zeta [\varphi_\nu(z) + \chi_\nu(z)] \, dz.$$

For two points ζ_1 and ζ_2 on the same boundary continuum γ_ρ, we have

$$\Phi(\zeta_1) - \Phi(\zeta_2) = \int_{\zeta_1}^{\zeta_2} [\varphi_\nu z' + \chi_\nu z'] \, ds \tag{A2.60'}$$
$$= \int_{\zeta_1}^{\zeta_2} [\varphi_\nu z' + \overline{\varphi_\nu z'}] \, ds,$$

if we chose the arc of γ_ρ between ζ_1 and ζ_2 as our path of integration. The right-hand term is real; this shows that $\Phi(\zeta)$ has constant imaginary parts on each γ_ρ. Hence, subtracting from $\Phi(\zeta)$ an appropriate combination of functions $iw_\rho(\zeta)$, we can obtain an analytic function which has on γ the imaginary part zero. By the maximum principle, this function has a vanishing imaginary part everywhere in G and is therefore a real constant. Thus we have proved

$$\varphi_\nu(\zeta) + \chi_\nu(\zeta) = \Phi'(\zeta) = i \sum_{\rho=1}^{k} \alpha_\rho w_\rho'(\zeta), \qquad \alpha_\rho \text{ real.} \tag{A2.61}$$

Similarly we can show

$$\varphi_\nu(\zeta) - \chi_\nu(\zeta) = \sum_{\rho=1}^{k} \beta_\rho w_\rho'(\zeta), \qquad \beta_\rho \text{ real.} \tag{A2.61'}$$

Thus we conclude finally

$$\varphi_\nu(z) = \sum \tfrac{1}{2}(\beta_\rho + i\alpha_\rho) w_\rho'(z). \tag{A2.62}$$

We have proved:

All eigenfunctions of (A2.55) with eigenvalue 1 are linear combinations of $w_\rho'(z)$.

It is easily seen that all such combinations are really eigenfunctions of (A2.55) with eigenvalue 1. This result illustrates again the importance of the harmonic measures and their many-sided role in the theory.

Let us multiply (A2.55) by $\overline{w_\rho'(z)}$ and integrate over G. Using the symmetry property of the kernel and the fact that $w_\rho'(z)$ is an eigenfunction of (A2.55) with $\lambda_\nu^2 = 1$, we obtain

$$\iint_G \varphi_\nu(z)\overline{w_\rho'(z)} \, d\tau_z = \lambda_\nu^2 \iint_G \overline{w_\rho'(\zeta)}\varphi_\nu(\zeta) \, d\tau_\zeta . \tag{A2.63}$$

Hence, if $\lambda_\nu^2 > 1$

$$(\varphi_\nu, \overline{w_\rho'}) = \iint_G \varphi_\nu \overline{w_\rho'} \, d\tau = 0 \tag{A2.63'}$$

which proves, by virtue of (A2.8), the

Theorem: Each eigenfunction $\varphi_\nu(z)$ of (A2.55) with eigenvalue > 1 lies in $\mathfrak{F}_0$, i.e. has a single-valued integral in G.

8. *Kernel Functions for the Class* $\mathfrak{F}_0$. We may subtract from the kernel $K(z, \bar{\zeta})$ a combination $\sum t_{\rho\sigma} w'_\rho(z)\overline{w'_\sigma(\zeta)}$ such that it becomes orthogonal to all functions $w'_\rho(z)$. This corrected kernel $K_0(z, \bar{\zeta})$ will then lie in the class $\mathfrak{F}_0$, i.e. will have a single-valued integral in G. It will still have the reproducing property within the class $\mathfrak{F}_0$ since all elements of this class are orthogonal to the corrective term. Using results of section 1, we can give a very elegant interpretation for $K_0(z, \bar{\zeta})$. In fact, we showed in (A1.37) that the kernel $-\partial^2 m/\partial z\partial\bar{\zeta}$ has a single-valued (and also univalent) integral $- \frac{1}{2}\partial/\partial\zeta \log f(z, \zeta)$ and lies, therefore, in $\mathfrak{F}_0$. By (A1.58) and (A1.60) we connected this kernel with $-\partial^2 g(z, \zeta)/\partial z\partial\bar{\zeta}$. But the kernels differ only by a combination of w'_ρ; hence, we have necessarily

$$K_0(z, \bar{\zeta}) = -\frac{2}{\pi}\frac{\partial^2 m}{\partial z\partial\bar{\zeta}} = \frac{2}{\pi}\frac{\partial^2 N(z, \zeta)}{\partial z\partial\bar{\zeta}}. \tag{A2.64}$$

We may also introduce the corresponding L_0-kernel for $\mathfrak{F}_0$ and define

$$L_0(z, \zeta) = -\frac{2}{\pi}\frac{\partial^2 N(z, \zeta)}{\partial z\partial\zeta} = -\frac{2}{\pi}\frac{\partial^2 m}{\partial z\partial\zeta}. \tag{A2.65}$$

A completely analogous theory for the kernel functions in this new function space is possible, the central role being played now by the Neumann's instead of the Green's function.

We may study the solutions of the integral equation

$$\varphi_\nu(z) = \lambda_\nu^2 \iint_G [K_0(z, \bar{\zeta}) - \Gamma(z, \bar{\zeta})]\varphi_\nu(\zeta)\, d\tau_\zeta. \tag{A2.66}$$

Since K_0 and Γ both lie in $\mathfrak{F}_0$, we conclude first that also $\varphi_\nu(z) \in \mathfrak{F}_0$. But then its scalar product with $K_0 - \Gamma$ is the same as with $K - \Gamma$. Hence the integral equations (A2.55) and (A2.66) have all eigenvalues $\lambda_\nu^2 > 1$ and the corresponding $\varphi_\nu(z)$ in common. Equation (A2.66) has no eigenvalue 1, since the functions $w'_\nu(z)$ do not lie in $\mathfrak{F}_0$.

Consider the identity

$$K_0(z, \bar{w}) - \iint_G [K_0(z, \bar{\zeta}) - \Gamma(z, \bar{\zeta})]K_0(\zeta, \bar{w})\, d\tau_\zeta = \Gamma(z, \bar{w}), \tag{A2.67}$$

which follows immediately from the reproducing property of K_0 in $\mathfrak{F}_0$. We may conceive of this identity as an inhomogeneous integral equation for $K_0(z, \bar{w})$ with the positive definite Hermitian kernel $K_0 - \Gamma$ which is continuous in the closed region $G + \gamma$. Since its lowest eigenvalues are greater than 1, we may solve this integral equation by the well known Neumann series based on the reciprocal kernel [14]. In fact, let

$$\text{(A2.68)} \qquad \Delta^{(\nu)}(z, \bar{\zeta}) = \iint_D \Delta^{(\nu-1)}(z, \bar{w})[K_0(w, \bar{\zeta}) - \Gamma(w, \bar{\zeta})]\, d\tau_w$$

$$\Delta^{(0)}(z, \bar{\zeta}) = K_0(z, \bar{\zeta}).$$

Then we have

$$\text{(A2.69)} \qquad K_0(z, \bar{w}) = \sum_{\nu=0}^{\infty} \iint_D \Delta^{(\nu)}(z, \bar{\zeta})\Gamma(\zeta, \bar{w})\, d\tau_\zeta .$$

If we define analogously

$$\text{(A2.70)} \qquad \Gamma^{(\nu)}(z, \bar{\zeta}) = \iint_D \Gamma^{(\nu-1)}(z, \bar{w})\Gamma(w, \bar{\zeta})\, d\tau_\zeta$$

$$\Gamma^{(1)}(z, \bar{\zeta}) = \Gamma(z, \bar{\zeta}),$$

we may easily compute the series (A2.69) in the form of a double sum

$$\text{(A2.71)} \qquad K_0(z, \bar{w}) = \sum_{\nu=0}^{\infty} \left\{ \sum_{\mu=0}^{\nu} (-1)^\mu \binom{\nu}{\mu} \Gamma^{(\mu+1)}(z, \bar{w}) \right\}.$$

We have expressed the kernel K_0 in terms of a series each of whose elements can be obtained by elementary integration, and thus have a practically and numerically important procedure for constructing a kernel function [9].

After having constructed the kernel K_0 for the class $\mathfrak{F}_0$, we can easily find the kernel K for the wider class $\mathfrak{F}$. In fact, let a_ρ be a point in that component of $\tilde{G}$ which is bounded by the curve γ_ρ. Then the function

$$\text{(A2.72)} \qquad W_\rho'(z) = (z - a_\rho)^{-1} - \iint_G K_0(z, \bar{\zeta})(\zeta - a_\rho)^{-1}\, d\tau_\zeta$$

is regular in G and, by construction, orthogonal to all functions of the class $\mathfrak{F}_0$. Its integral has the period $2\pi i$, if z describes the contour γ_ρ and is single-valued otherwise. Thus the function $W_\rho(z)$ is a combination of functions $w'_\rho(z)$, whose coefficients can easily be computed from the mentioned periodicity conditions. But knowing K_0 and the w'_ρ we can easily construct the kernel K.

9. *Comparison Theory*. An interesting generalization of the development of the kernel K_0 can be given which also is of value for more efficient calculations in practical problems. Let $\mathfrak{G}$ be a domain in the complex plane which contains our original domain G, and is also bounded by k smooth curves $\mathfrak{C}_\rho$ each of which can be deformed continuously through $\mathfrak{G}$ into the corresponding boundary curve γ_ρ of G. Let $\mathfrak{g}(z, \zeta)$, $\mathfrak{K}(z, \zeta)$ and $\mathfrak{L}(z, \zeta)$ denote the corresponding Green's function and kernels with respect to the larger domain.

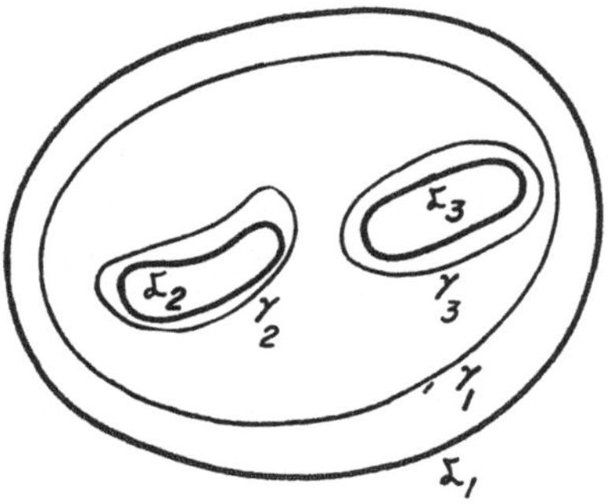

Figure 7. Comparison domains.

We remark that the function

$$\Lambda(z, \zeta) = \mathfrak{L}(z, \zeta) - L(z, \zeta) \tag{A2.73}$$

is again regular analytic in G and continuous in the closed region $G + \gamma$. This follows immediately from the result obtained for the l-kernel in article 6. In fact, $\Lambda(z, \zeta)$ is a clear generalization of this l-kernel and becomes identical with it in view of (A2.12) if we choose $\mathfrak{G}$ as a circle. Let us now compute the integral (cf. (A2.22))

$$\iint_G \Lambda(z, \zeta)\overline{\Lambda(z, \eta)}\, d\tau_z = \iint_G \mathfrak{L}(z, \zeta)\, \overline{\Lambda(z, \eta)}\, d\tau_z\,, \tag{A2.74}$$

which is analogous to (A2.32). Integrating by parts according to (A2.5) and using (A2.73) and formula (A2.27), we transform the above integral into

$$\text{(A2.75)}\qquad \begin{aligned}&\frac{1}{\pi i}\oint_\gamma \frac{\partial \mathfrak{g}(z,\zeta)}{\partial \zeta}\overline{[\mathfrak{L}(z,\eta) - L(z,\eta)]\,dz}\\ &= \frac{1}{\pi i}\oint_\gamma \frac{\partial \mathfrak{g}(z,\zeta)}{\partial \zeta}\overline{\mathfrak{L}(z,\eta)}\,d\bar z + \frac{1}{\pi i}\oint_\gamma \frac{\partial \mathfrak{g}(z,\zeta)}{\partial \zeta} K(z,\bar\eta)\,dz.\end{aligned}$$

Since $\mathfrak{L}$ and $\partial\mathfrak{g}/\partial\zeta$ are both bounded in the domain $G^* = \mathfrak{G} - G$ the first of the two right-hand integrals can be retransformed by (A2.5) into an area integral over G^*(ζ and η lying in G). Thus, we find

$$\text{(A2.76)}\qquad \begin{aligned}&\frac{1}{\pi i}\oint_\gamma \frac{\partial \mathfrak{g}(z,\zeta)}{\partial \zeta}\overline{\mathfrak{L}(z,\eta)\,dz}\\ &= -\iint_{G^*} \mathfrak{L}(z,\zeta)\overline{\mathfrak{L}(z,\eta)}\,d\tau_z = -B(\zeta,\bar\eta).\end{aligned}$$

Clearly, $B(\zeta,\bar\eta)$ is a positive definite Hermitian kernel in G of the same nature as Γ. We consider $\mathfrak{L}(z,\zeta)$ and $\mathfrak{K}(z,\bar\zeta)$ as known quantities and want to express K and L in terms of them; hence B is again to be considered as an elementary geometric quantity, obtainable by simple integration.

The second integral of (A2.75) can be transformed by integration by parts if we remark that $\partial g(z,\zeta)/\partial\zeta = 0$ on γ, and that we may write this integral in the form

$$\text{(A2.77)}\qquad \begin{aligned}&\frac{1}{\pi i}\oint_\gamma \frac{\partial[\mathfrak{g}(z,\zeta) - g(z,\zeta)]}{\partial \zeta} K(z,\bar\eta)\,dz\\ &= -\iint_G [\mathfrak{K}(\zeta,\bar z) - K(\zeta,\bar z)]K(z,\bar\eta)\,d\tau_z.\end{aligned}$$

Using the reproducing property of K, we easily evaluate (A2.77), and, combining (A2.76) with (A2.77), finally obtain the identity

$$\text{(A2.78)}\qquad \iint_G \Lambda(z,\zeta)\overline{\Lambda(z,\eta)}\,d\tau_z = K(\zeta,\bar\eta) - \mathfrak{K}(\zeta,\bar\eta) - B(\zeta,\bar\eta).$$

This result contains valuable information on the change of the kernel K if the fundamental domain G is changed. We recognize from (A2.78) that the kernel

$$\text{(A2.79)}\qquad K(z,\bar\zeta) - \mathfrak{K}(z,\bar\zeta) - B(z,\bar\zeta) = M(z,\bar\zeta)$$

is positive definite, i.e. that $\sum M(z_i, \bar z_k)x_i\bar x_k \geqq 0$ for any choice of points $z_i \in G$ and complex values x_i. This proves in particular:

The positive Hermitian form $\sum K(z_i, \bar{z}_k)x_i\bar{x}_k$ decreases monotonically if the basic domain G is increased. This result permits easy estimates for the K-kernel using domains of comparison that either lie entirely in G or contain G.

Let us define for each function $f(z) \in \mathfrak{F}$ two transforms

$$(A2.80) \qquad \begin{aligned} T_1 f(z) &= \iint_G \mathfrak{K}(z, \bar{\zeta}) f(\zeta)\, d\tau_\zeta, \\ T_2 f(z) &= \iint_G \mathfrak{L}(z, \zeta) \overline{f(\zeta)}\, d\tau_\zeta. \end{aligned}$$

Since $\mathfrak{L}(z, \zeta)$ has a double pole at $z = \zeta$ the last integral is improper; it is, however, easily verified that it represents an analytic function of z in G and G^*. This function is discontinuous if z crosses the boundary γ of G. In fact, integration by parts leads (for functions $f(z)$ which are continuous in $G + \gamma$) to

$$(A2.81) \qquad \begin{aligned} T_2 f(z) &= \frac{1}{\pi i} \oint_\gamma \frac{\partial \mathfrak{g}(z, \zeta)}{\partial z} \overline{f(\zeta)\, d\zeta} \\ &= \frac{1}{\pi i} \oint_\gamma \left(\frac{1}{2} \frac{1}{\zeta - z} + \cdots \right) \overline{f(\zeta)\, d\zeta}. \end{aligned}$$

Thus we conclude from Plemelj's theorem that $T_2 f(z)$ jumps by the amount $\overline{f(z) z'^2}$ if z crosses the boundary curve γ. For $z \in G$, we may also write $T_2 f(z)$ as the proper integral

$$(A2.82) \qquad T_2 f(z) = \iint_G \Lambda(z, \zeta)\, \overline{f(\zeta)}\, d\tau_\zeta$$

because of the identity (A2.22). $T_1 f(z)$ is obviously analytic over all of $\mathfrak{G}$.

Multiplying identity (A2.78) with $\overline{f(\zeta)} f(\eta)$ and integrating over G with respect to each variable, we obtain, in view of (A2.76) and the reproducing property of the $\mathfrak{K}$-kernel with respect to $\mathfrak{G} = G + G^*$,

$$(A2.83) \qquad \begin{aligned} \iint_G |\, T_2 f(z)\, |^2\, d\tau_z &= \iint_G |\, f(z)\, |^2\, d\tau_z \\ &\quad - \iint_{G+G^*} |\, T_1 f(z)\, |^2\, d\tau_z - \iint_{G^*} |\, T_2 f(z)\, |^2\, d\tau_z. \end{aligned}$$

Thus we have the remarkable fact that for each function $f(z)$ analytic in G two transforms $T_1 f$ and $T_2 f$ can be defined in $\mathfrak{G}$ which have the same norm sum with respect to their domain of definition: in obvious notation, we may write

$$\text{(A2.84)} \qquad \| T_1 f \|_{\mathfrak{G}}^2 + \| T_2 f \|_{\mathfrak{G}}^2 = \| f \|_G^2 .$$

Again we set up an integral equation,

$$\text{(A2.85)} \qquad \varphi_\nu(z) = \lambda_\nu^2 \iint_G M(z, \bar{\zeta}) \varphi_\nu(\zeta) \, d\tau_\zeta$$

($M(z, \bar{\zeta})$ being defined by (A2.79))

and investigate its eigenvalues. Multiplying (A2.85) by $\overline{\varphi_\nu(z)}$ and integrating over G, we obtain, in view of (A2.79) and (A2.80),

$$\text{(A2.86)} \quad \| \varphi_\nu \|_G^2 = \lambda_\nu^2 \left[\| \varphi_\nu \|_G^2 - \| T_1 \varphi_\nu \|_{\mathfrak{G}}^2 - \iint_{G^*} | T_2 \varphi_\nu |^2 \, d\tau \right].$$

This clearly proves that each $\lambda_\nu^2 \geqslant 1$; we can even state that $\lambda_\nu^2 = 1$ is only possible if

$$\text{(A2.87)} \qquad T_1 \varphi_\nu \equiv 0 \quad \text{in} \quad \mathfrak{G}, \qquad T_2 \varphi_\nu \equiv 0 \quad \text{in} \quad G^*.$$

It is easy to show that, under our assumptions with respect to $\mathfrak{G}$, no such function φ_ν can exist except for $\varphi \equiv 0$. In fact, from $T_2\varphi_\nu \equiv 0$ in G^* and the known discontinuity behavior of $T_2\varphi_\nu$ on γ, we conclude that $T_2\varphi_\nu(z)$ represents in G an analytic function with boundary values $\overline{\varphi_\nu(z) z'^2}$. From this very fact we concluded in section 7 that $\varphi_\nu(z)$ must be a linear combination of functions $w_\rho'(z)$, say

$$\text{(A2.88)} \qquad \varphi_\nu(z) = \sum_{\rho=1}^{k-1} c_\rho w_\rho'(z).$$

On the other hand, the identity $T_1\varphi_\nu = 0$ implies

$$\text{(A2.89)} \qquad \iint_G \mathfrak{K}(z, \bar{\zeta}) \, \varphi_\nu(\zeta) \, d\tau_\zeta = 0, \quad z \in \mathfrak{G},$$

and by scalar multiplication with any function analytic in $\mathfrak{G}$ we can show that $\varphi_\nu(z)$ is orthogonal to all analytic functions in $\mathfrak{G}$. Let $\mathfrak{w}_\rho'(z)$ be the derivatives of the harmonic measures in $\mathfrak{G}$, analogous to the $w_\rho'(z)$. We find

$$\text{(A2.90)} \qquad \sum_{\rho=1}^{k-1} c_\rho \iint_G w_\rho' \cdot \mathfrak{w}_\sigma' \, d\tau = 0, \qquad \sigma = 1, 2, \cdots, k-1.$$

According to (A2.8), $(1/i)(w_\rho', \mathfrak{w}_\sigma') = \mathfrak{P}_{\rho\sigma}$ represents the period scheme of the functions $\mathfrak{w}_\sigma(z)$ with respect to circuits around the boundary curves γ_ρ which are equivalent to circuits around the $\mathfrak{C}_\rho$. Thus (A2.90) may be written in the form

$$\text{(A2.90}'\text{)} \qquad \sum_{\rho=1}^{k-1} c_\rho \mathfrak{P}_{\rho\sigma} = 0, \qquad \sigma = 1, 2, \cdots, k-1.$$

In section 1 we proved that the matrix $(\mathfrak{P}_{\rho\sigma})_{1\cdots k-1}$ has a non-vanishing determinant. Hence, all c_ρ must equal zero and $\varphi_\nu(z) \equiv 0$. We have therefore proved:

The integral equation (A2.85) *has all eigenvalues* >1.

We now consider the inhomogeneous integral equation

$$\text{(A2.91)} \qquad \begin{aligned} I(z, \bar{\zeta}) &= \mathfrak{K}(z, \bar{\zeta}) + B(z, \bar{\zeta}) \\ &= K(z, \bar{\zeta}) - \iint_G M(z, \bar{w})K(w, \bar{\zeta})\, d\tau_w \end{aligned}$$

for the kernel $K(z, \bar{\zeta})$. The function $I(z,\bar{\zeta})$ is a known quantity by our assumption that all quantities connected with $\mathfrak{G}$ are to be considered as given. We may solve this equation also by an infinite series since again the eigenvalues of the kernel are greater than 1. Introducing the iterated kernels

$$\text{(A2.92)} \quad I^{(\nu)}(z, \bar{\zeta}) = \iint_G I^{(\nu-1)}(z, \bar{w})I(w, \bar{\zeta})\, d\tau_w, \qquad I^{(1)}(z, \bar{\zeta}) = I(z, \zeta)$$

we obtain using the methods of article 8 the series development [66]

$$\text{(A2.93)} \qquad K(z, \bar{\zeta}) = \sum_{\mu=0}^{\infty} \left\{ \sum_{\mu=0}^{\infty} (-1)^\mu \binom{\nu}{\mu} I^{(\mu+1)}(z, \bar{\zeta}) \right\}.$$

We remark that the ν-th term in this series can be written in the form

$$\text{(A2.94)} \quad \sum_{\mu=0}^{\infty} (-1)^\mu \binom{\nu}{\mu} I^{(\mu+1)}(z, \bar{\zeta}) = \iint_G M^{(\nu)}(z, \bar{w})I(w, \bar{\zeta})\, d\tau_w$$

where $M^{(\nu)}(z, \bar{\zeta})$ is the ν-th iterated kernel of the original kernel $M(z, \bar{\zeta})$ of the integral equation. Our data on $\mathfrak{G}$ do not give $M(z, \bar{\zeta})$ since it still contains the kernel $K(z, \bar{\zeta})$. But, by means of it, we can easily interpret the significance of the series development (A2.93).

In fact, let us assume that the boundary curves $\mathfrak{C}_\rho$ and γ_ρ of the two compared domains $\mathfrak{G}$ and G are at least three times continuously differentiable with respect to their length parameter, and that a correspondence between the points of each $\mathfrak{C}_\rho$ and each γ_ρ can be established such that the maximum distance between any two corresponding points is less than ϵ. We will express this fact by saying that the system γ of curves lies in an ϵ-neighborhood of the system $\mathfrak{C}$. It can then easily be shown that the difference $\Lambda(z, \zeta) = \mathfrak{L}(z, \zeta) - L(z, \zeta)$ between the corresponding L-kernels is of the order of magnitude $O(\epsilon)$ uniformly in G. Hence, assuming $\mathfrak{G}$ to have a finite area, i.e. not to contain the point at infinity, we conclude from (A2.78) that the kernel $M(z, \bar{\zeta})$ has the order of magnitude $O(\epsilon^2)$ and the ν-th iterated $M^{(\nu)}(z, \bar{\zeta})$ is $O(\epsilon^{2\nu})$. Thus we recognize that the series development (A2.93) for the K-kernel proceeds in powers of the

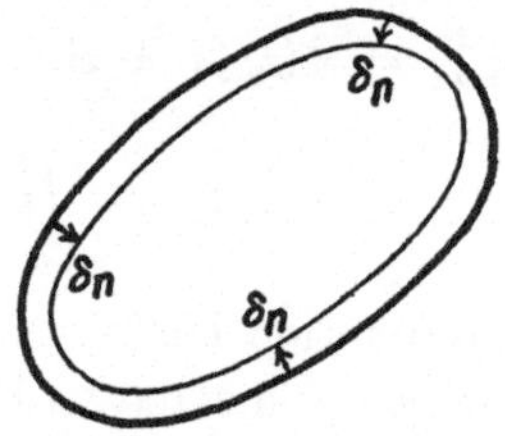

Figure 8. Variation of γ.

ϵ-measure of distance between the boundary curves. In particular, the following result is an immediate consequence of (A2.93):

$$\begin{aligned} K(z, \bar{\zeta}) &= I(z, \bar{\zeta}) + O(\epsilon^2) \\ &= \mathfrak{K}(z, \bar{\zeta}) + \iint_{G^*} \mathfrak{L}(z, w)\overline{\mathfrak{L}(\zeta, w)}\, d\tau_w + O(\epsilon^2). \end{aligned} \tag{A2.95}$$

We may transform (A2.95) in order to bring it into the usual form of functional analysis. We assume that the correspondence $\mathfrak{C} \leftrightarrow \gamma$ has been established in the following form: if one proceeds from any point on $\mathfrak{C}$ along the interior normal with respect to $\mathfrak{C}$ one arrives just at the corresponding point on γ. We denote the distance between the two corresponding points by δn. The variation δn is a continuous function on $\mathfrak{C}$ of order $O(\epsilon)$. By means of it we may express (A2.95) in the equivalent form

$$\begin{aligned} \delta\mathfrak{K}(z, \bar{\zeta}) &= K(z, \bar{\zeta}) - \mathfrak{K}(z, \bar{\zeta}) \\ &= \oint_{\mathfrak{C}} \mathfrak{L}(z, w)\overline{\mathfrak{L}(\zeta, w)}\, \delta n\, ds + O(\epsilon^2). \end{aligned} \tag{A2.96}$$

Formula (A2.96) describes the sensitivity of the functional $\mathfrak{K}(z, \bar{\zeta})$ with respect to infinitesimal deformations of the boundary $\mathfrak{C}$. Various conclusions with respect to $\mathfrak{K}(z, \bar{\zeta})$ can be derived from this representation. The variation formula (A2.96) determines the change of $\mathfrak{K}$ with the shift δn of the infinitely many points on $\mathfrak{C}$, just as the differential

$$dF = \sum_{i=1}^{N} \frac{\partial F}{\partial x_i}\, dx_i$$

describes the corresponding change of a function $F(x_1, x_2, \cdots, x_N)$ with the shift dx_i of its N variables x_i. Analogous variation formulas for $\mathfrak{L}$, $\mathfrak{g}$, $\mathfrak{w}$ etc. may be derived from (A2.96). We shall return to this problem in section 3.

10. *An Extremum Problem in Conformal Mapping.* Consider the class $\mathfrak{S}(z_0)$ of all functions $f(z)$ which are univalent in G and regular except for a pole at $z = z_0$ with residue 1. These functions map G onto a domain which contains the point at infinity and is bounded by k continua corresponding to the different γ_ρ. Let $f(z)$ be continuously differentiable in $G + \gamma$ except at $z = z_0$. Then the expression

$$A(f) = \frac{1}{2i} \oint_\gamma f\, d\bar{f} \tag{A2.97}$$

measures the area of the complements of the image domain of G. By its nature this expression is non-negative. There arises the question of an upper bound for $A(f)$ in the class $\mathfrak{S}(z_0)$.

In order to answer this question we remark that the function $\Phi(z, z_0)$ defined in (A1.31) and (A1.36), and discussed in section 1, article 2 belongs to the class $\mathfrak{S}(z_0)$. It has the following simple relation to the L-kernel (cf. (A1.38), (A1.59))

$$\Phi'(z, z_0) = -\pi L(z, z_0) + \frac{1}{2} \sum_{\rho,\sigma=1}^{k-1} \Pi_{\rho\sigma} w_\rho'(z) w_\sigma'(z_0). \tag{A2.98}$$

From this fact, the formula (A2.4) for integration by parts, and the identities (A2.8) and (A2.10) we conclude, for every function $\varphi(z)$ which is continuously differentiable in $G + \gamma$,

$$\frac{1}{2i} \oint_\gamma \Phi(z, z_0)\, d\overline{\varphi(z)} = -(\Phi'(z, z_0), \overline{\varphi'(z)}) = 0. \tag{A2.99}$$

Now every function in the class $\mathfrak{S}(z_0)$ may be written in the form

$$f(z) = \Phi(z, z_0) + \varphi(z), \tag{A2.100}$$

where $\varphi(z)$ is regular in G. Hence we obtain

$$A(f) = A(\Phi) + \frac{1}{2i} \oint_{\gamma} \varphi \, d\bar{\varphi} = A(\Phi) - \|\varphi'\|^2. \tag{A2.101}$$

We have proved [17, 25, 60]:

Among all univalent functions in G with a pole at z_0 with residue 1, *the function $\Phi(z, z_0)$ maps G onto the domain with maximum complementary area.*

11. *Mapping onto a Circular Domain.* Let G be bounded by a curve system γ which is three times continuously differentiable with respect to its length parameter. We conclude, from (A2.27) and the symmetry properties (A2.13) of the kernels, that

$$\begin{aligned} L(z, \zeta)z'\zeta' &= \overline{-K(z, \bar{\zeta})z'\bar{\zeta}'} = -K(\zeta, \bar{z})\zeta'\bar{z}' \\ &= \overline{L(z, \zeta)z'\zeta'} = \text{real} \end{aligned} \tag{A2.102}$$

if z and ζ both lie on the boundary γ. Hence, we find from (A2.10)

$$\mathcal{I}m\left\{\frac{1}{\pi}\frac{z'\zeta'}{(z-\zeta)^2}\right\} = \mathcal{I}m\{l(z, \zeta)z'\zeta'\}, \qquad z, \zeta \in \gamma. \tag{A2.103}$$

We may also put $\zeta = z(s + \Delta s)$, $z = z(s)$ and let $\Delta s \to 0$. The right-hand side of (A2.103) remains continuous because of the continuity of the l-function. The limit of the left-hand side is easily computed by repeated application of the mean value theorem of calculus. We obtain

$$\mathcal{I}m\{l(z, z)z'^2\} = \frac{1}{6\pi}\frac{d}{ds}\left(\frac{1}{\rho(s)}\right), \tag{A2.104}$$

where $\rho(s)$ is the radius of curvature of γ at the point with the length parameter s.

If, in particular, G is a domain bounded by circles of the type discussed in Chapter V, §2, we have $l(z, z)z'^2 = \text{real}$ on γ. Functions $g(z)$ which are analytic in G and have the property that $z'^2 g(z)$ is real on the boundary play an important role in the theory of conformal mapping; they are called quadratic differentials with respect to the domain G considered. It is easily shown that each quadratic differential which is regular in G can be represented in the form

$$g(z) = \sum_{\rho,\sigma=1}^{k-1} \lambda_{\rho\sigma} w_\rho'(z) w_\sigma'(z), \qquad \lambda_{\rho\sigma} = \text{real}. \tag{A2.105}$$

Thus we find that, for a circular domain,

$$l(z, z) = \sum_{\rho,\sigma=1}^{k-1} \lambda_{\rho\sigma} w_\rho'(z) w_\sigma'(z). \tag{A2.106}$$

Now every domain G can be mapped by a univalent function $z^* = f(z)$ onto a circular domain G^*. According to (A2.45) and (A2.47) we have the relation

$$l^*(z^*, z^*) f'(z)^2 = \frac{1}{6\pi} \{f, z\} + l(z, z). \tag{A2.107}$$

Using the formula (A2.106) which is valid for G^*, and the fact that $\omega_\nu(z) = \omega_\nu^*(f(z))$ follows from $w_\nu'(z) = w_\nu^{*\prime}(z^*) f'(z)$, we obtain

$$\frac{1}{6\pi} \{f, z\} + l(z, z) = \sum_{\rho,\sigma=1}^{k-1} \lambda_{\rho\sigma} w_\rho'(z) w_\sigma'(z). \tag{A2.108}$$

This formula represents a differential equation in G for the function mapping $f(z)$ onto the circular domain. The constants $\lambda_{\rho\sigma}$ in this equation are not known and must be fitted in such a way that $f(z)$ is single-valued in G. The functions l and w_ρ', however, are known if the Green's function for G is given. We have here established a simple connection between the Green's function and the mapping on circular domains.

12. *Orthonormal Systems.* Suppose a system of analytic functions $\{u_\nu(z)\}$ is known in G such that

$$(u_\nu, \bar{u}_\mu) = \iint_G u_\nu(z) \overline{u_\mu(z)}\, d\tau_z = \begin{cases} 0, & \text{for } \nu \neq \mu, \\ 1, & \text{for } \nu = \mu, \end{cases} \tag{A2.109}$$

and that every function $f(z)$ with finite norm $\| f \|$ can be developed into a series

$$f(z) = \sum_{\nu=1}^{\infty} \alpha_\nu u_\nu(z) \tag{A2.110}$$

in terms of the orthonormal system, which converges uniformly in each closed subdomain of G. The coefficients α_ν are given by the Fourier formulas

$$\alpha_\nu = (f, \bar{u}_\nu) = \iint_G f(z) \overline{u_\nu(z)}\, d\tau_z. \tag{A2.111}$$

In particular, the kernel $K(z, \zeta)$ can be developed into such a series and because of the reproducing property we can easily determine all its Fourier coefficients. We find

$$K(z, \bar{\zeta}) = \sum_{\nu=1}^{\infty} u_\nu(z)\overline{u_\nu(\zeta)}. \tag{A2.112}$$

This remarkable connection between the K-kernel and a complete orthonormal system in G permits a computation of the kernel, once a complete orthonormal system of the required property has been determined. The kernel $K(z, \bar{\zeta})$ was first studied by Bergman and Bochner in connection with the formula (A2.112) which was considered as the definition of this kernel [4, 5, 7, 13]. This definition can easily be extended to the case of functions of several complex variables, analytic in a certain domain G, and useful applications of the kernel K can be made in this theory [6].

Similar developments for fundamental functions in orthogonal series were developed by Zaremba in the theory of partial differential equations [73, 74]. Related to this approach is also the theory of orthogonal polynomials and their connection with conformal mapping, as developed by Szegö [69], compare also [15]. An abstract theory of reproducing kernels was developed by Aronszajn [3].

3. *Variation of the Green's Function*

1. *Hadamard's Variation Formula.* In the foregoing sections we considered the dependence of the Green's function $g(z, \zeta)$ for a fixed domain G on its variables z and ζ. It is, however, far more important to study the dependence of the Green's function upon the domain G or, in other words, upon the boundary curves γ. In fact, the actual construction of the Green's function for an arbitrarily given domain G involves infinite series of iterated integrals and is very complicated. It is, therefore, far more convenient to study the Green's function in those simple cases in which it can be expressed in closed form, and to extend the information thus obtained to more general domains by a continuous deformation of the boundary γ. If the behavior of the Green's function for slight deformations of the domain is known, numerous conclusions about its fundamental properties can be obtained without any actual construction.

The first investigation of this kind is due to Hadamard [27, 28] who proceeded in the following manner: Let G be a domain bounded by k twice continuously differentiable curves γ_ρ. Let us define on each γ_ρ a positive continuously differentiable function $\varphi(s)$ of the arc length; we erect at each point of γ the interior normal and plot on it the segment $\delta n = \epsilon\varphi(s)$. If ϵ is sufficiently small, the endpoints of the normal segments will form a system of continuously differentiable curves γ_ρ^* which will determine a domain G^* lying entirely within G. Let $g^*(z, \zeta)$ and $g(z, \zeta)$ denote the Green's functions for G^* and G, respectively.

Consider the expression

$$\text{(A3.1)} \qquad \mathfrak{S}(z, \zeta) = g(z, \zeta) - \frac{1}{2\pi} \oint_\gamma \frac{\partial g(t, z)}{\partial n_t} \frac{\partial g(t, \zeta)}{\partial n_t} \delta n_t \, ds_t$$

$$= g(z, \zeta) - \epsilon r(z, \zeta),$$

which is harmonic in G except for the pole at $z = \zeta$. For $z \in \gamma$ it has the boundary value $-(\partial g(z, \zeta)/\partial n_z)\, \delta n_z$ (cf. (A1.3)) and is thus negative on the whole boundary of G. Since $\delta n_z = O(\epsilon)$, $\mathfrak{S}(z, \zeta)$ will be very small on γ and the locus $\mathfrak{S}(z, \zeta) = 0$ will, therefore, be very near γ. Let z^* be the point corresponding to z on γ^*; it lies on the normal to γ at z, at a distance δn_z. Hence, by Taylor's theorem

$$g(z^*, \zeta) - g(z, \zeta) = g(z^*, \zeta) = \frac{\partial g(z, \zeta)}{\partial n_z} \delta n_z + O(\epsilon^2).$$

Thus, we find for every $\zeta \in G$ (note that $\epsilon r(z, \zeta) = (\partial g(z, \zeta)/\partial n_z) \delta n_z$ on γ)

$$\text{(A3.2)} \qquad \mathfrak{S}(z^*, \zeta) = \frac{\partial g(z, \zeta)}{\partial n_z} \delta n_z - \epsilon r(z, \zeta) + O(\epsilon^2) = O(\epsilon^2),$$

where the term $O(\epsilon^2)$ can be estimated uniformly for ζ in each closed subdomain of G.

The function $g^*(z, \zeta) - \mathfrak{S}(z, \zeta)$ is regular harmonic in G^* because the logarithmic infinities cancel by subtraction, and is also $O(\epsilon^2)$ on γ^* since $g^*(z, \zeta)$ vanishes there. Hence, using the maximum principle for this difference considered as a harmonic function of z, we obtain

$$\text{(A3.3)} \quad g^*(z, \zeta) - g(z, \zeta) = -\frac{1}{2\pi} \oint_\gamma \frac{\partial g(t, z)}{\partial n_t} \frac{\partial g(t, \zeta)}{\partial n_t} \delta n_t \, ds_t + O(\epsilon^2)$$

and the remainder term $O(\epsilon^2)$ can be estimated uniformly for $z \in G^*$ and ζ in a fixed closed subdomain of G.

Using the notation of functional analysis, we may write (A3.3) in the form [28, 32, 37]:

$$\delta g(z, \zeta) = -\frac{1}{2\pi} \oint_\gamma \frac{\partial g(t, z)}{\partial n_t} \frac{\partial g(t, \zeta)}{\partial n_t} \delta n_t \, ds_t, \tag{A3.4}$$

where $\delta g(z, \zeta)$ denotes the difference of the two Green's functions in G^* up to a harmonic function of both variables, which can be estimated as $O(\epsilon^2)$ uniformly in each closed subdomain of G^*.

Let us draw some conclusions from (A3.4). Since $g(z, \zeta) = 0$ for $z \in \gamma$ and > 0 in G, it is clear that $\partial g(z, \zeta)/\partial n_z \geq 0$ for $z \in \gamma$. Hence we see that the Green's function decreases with decreasing domain, i.e. for $\delta n > 0$ on γ. This is also an immediate consequence of the maximum principle, since $g^*(z, \zeta) - g(z, \zeta) < 0$ on γ^*; however, (A3.4) permits us to estimate the speed of decrease in its dependence on δn.

The right-hand side of (A3.4) remains well defined even if z and ζ coincide. This becomes clear when we observe that the singular part of the Green's functions is the same for both domains and that, evidently, $\delta g(z, \zeta) = \delta h(z, \zeta)$, where $h(z, \zeta)$ is the regular part of the Green's function as defined in (A1.2). We conclude from (A3.4) the formula

$$\delta[h(z, z) + h(\zeta, \zeta) - 2h(z, \zeta)] = -\frac{1}{2\pi} \oint_\gamma \left[\frac{\partial g(t, z)}{\partial n_t} - \frac{\partial g(t, \zeta)}{\partial n_t}\right]^2 \delta n_t \, ds_t. \tag{A3.5}$$

This shows:

The combination $h(z, z) + h(\zeta, \zeta) - 2h(z, \zeta)$ *decreases with decreasing domain.*

This result is not quite obvious and is our first significant result from the variation formula. Inversely, if we increase the domain G this combination of the Green's functions will increase. For a circle of radius R we have

$$h(z, \zeta) = \log \frac{|R^2 - z\bar{\zeta}|}{R}. \tag{A3.6}$$

Let us enclose the whole domain G in a very large circle of radius R; because of the monotonicity just proved, we find the inequality

$$h(z, z) + h(\zeta, \zeta) - 2h(z, \zeta) \leq \log \frac{(R^2 - |z|^2)(R^2 - |\zeta|^2)}{|R^2 - z\bar{\zeta}|^2}. \tag{A3.7}$$

Letting $R \to \infty$, we finally obtain the inequality

$$2h(z, \zeta) \geq h(z, z) + h(\zeta, \zeta). \tag{A3.8}$$

This example illustrates how finite inequalities can be obtained from the infinitesimal variation formula (A3.4).

We now wish to extend Hadamard's formula (A3.4) to include the case in which neither of the two compared domains G and G^* lies within the other, but an intercrossing of the boundary curves is allowed. This means that negative and positive values of δn are considered. For this purpose we consider a third domain G_0 which contains both G and G^* in such a way that the normal distance of each boundary γ and γ^* from the boundary γ_0 of G_0 is $O(\epsilon)$. Let δn_0 and δn_0^* be the normal distances of γ and γ^* from γ_0, respectively. We can always coordinate G_0 with G in such a way that the angle between the normals at corresponding points of γ and γ_0 is also $O(\epsilon)$. We may, then, assert that $\delta n = \delta n_0^* - \delta n_0 + O(\epsilon^2)$. Now we apply Hadamard's formula (A3.3), first with respect to the domains G_0 and G, and then with respect to the domains G_0 and G^*. Taking the difference of the two results, we obtain

$$\text{(A3.9)} \qquad \begin{aligned} &g^*(z, \zeta) - g(z, \zeta) \\ &\qquad = -\frac{1}{2\pi} \oint_{\gamma_0} \frac{\partial g_0(t_0, \zeta)}{\partial n_t} \frac{\partial g_0(t_0, z)}{\partial n_t} \delta n_t \, ds_t + O(\epsilon^2), \end{aligned}$$

and this estimate holds uniformly in each closed subdomain within G. Since, as $\epsilon \to 0$, the curve system γ_0 tends to γ in such a way that the normals tend towards each other, we can assert that

$$\frac{\partial g_0(t_0, z)}{\partial n_t} \to \frac{\partial g(t, z)}{\partial n_t}$$

uniformly for z in any closed subdomain of G. This shows that Hadamard's variation formula (A3.4) still holds in the general case considered.

It is easy to derive from (A3.3) the variation formula for the harmonic measures $\omega_\rho(\zeta)$. In fact, let γ_ρ' be a curve which lies inside G and G^* and is homologous to γ_ρ and γ_ρ^*. We have, in view of (A1.6) and (A3.3),

$$\text{(A3.10)} \qquad \begin{aligned} \omega_\rho^*(\zeta) - \omega_\rho(\zeta) &= \frac{1}{2\pi} \oint_{\gamma_\rho'} \left(\frac{\partial g^*(z, \zeta)}{\partial n_z} - \frac{\partial g(z, \zeta)}{\partial n_z} \right) ds \\ &= -\frac{1}{2\pi} \oint_{\gamma} \frac{\partial \omega_\rho(t)}{\partial n_t} \frac{\partial g(t, \zeta)}{\partial n_t} \delta n_t \, ds_t + O(\epsilon^2), \end{aligned}$$

where $O(\epsilon^2)$ can be estimated uniformly for all points ζ in a fixed closed subdomain of G. Using next the definition (A1.8) for the induction coefficients $P_{\rho\sigma}$, we obtain from (A3.10) by integration over a curve γ'_σ lying in G and homologous to γ_σ and γ^*_σ

$$\begin{aligned} P^*_{\rho\sigma} - P_{\rho\sigma} &= -\frac{1}{2\pi}\oint_{\gamma'_\sigma}\left(\frac{\partial\omega^*_\rho}{\partial n} - \frac{\partial\omega_\rho}{\partial n}\right) ds \\ &= \frac{1}{2\pi}\oint_{\gamma}\frac{\partial\omega_\rho}{\partial n}\frac{\partial\omega_\sigma}{\partial n}\,\delta n\, ds + O(\epsilon^2). \end{aligned} \tag{A3.11}$$

Formulas (A3.10) and (A3.11) now permit us to compute with ease the variations of all combinations of functions g, ω_σ etc., which were studied in section 1.

We make the following application of (A3.11). Let x_ρ, $\rho = 1, 2, \cdots, k - 1$, be an arbitrary set of real numbers. We consider the variation of the non-negative quadratic form $\sum P_{\rho\sigma}x_\rho x_\sigma$. In view of (A3.11), we find

$$\delta\left(\sum_{\rho,\sigma=1}^{k-1} P_{\rho\sigma}x_\rho x_\sigma\right) = \frac{1}{2\pi}\oint_\gamma\left(\sum_{\rho=1}^{k-1} x_\rho\frac{\partial\omega_\rho}{\partial n}\right)^2\delta n\, ds. \tag{A3.12}$$

This proves the

Theorem: The quadratic form $\sum P_{\rho\sigma}x_\rho x_\sigma$ increases monotonically with decreasing domain, i.e. $\delta n > 0$.

Our result permits an estimate of this expression, by means of the corresponding forms for convenient domains of comparison with known Green's functions.

Hadamard's formula can be written in the following form: Consider the function $p(t, z)$, analytic in t, which possesses $g(t, z)$ as real part. If t varies along γ, obviously $p(t, z) = i\rho(t, z)$, where ρ is a real-valued function. Thus, differentiating with respect to the length parameter s on γ, we obtain

$$p'(t, z)t' = i\frac{\partial\rho}{\partial s}, \qquad p'(t, z) = \frac{\partial p(t, z)}{\partial t}. \tag{A3.13}$$

By the Cauchy-Riemann differential equations we further have the relation $(\partial g/\partial n) = -(\partial\rho/\partial s)$ between the partial derivatives of the real and imaginary part of $p(t, z)$. Hence we can, in (A3.4), replace $\partial/\partial n\; g(t, z)$ by $ip'(t, z)t'$ or by $\overline{ip'(t, z)t'}$.

Thus we find, in view of $|t'|^2 = 1$,

$$\delta g(z, \zeta) = -\frac{1}{2\pi} \oint_\gamma \overline{p'(t, z)} p'(t, \zeta) \delta n_t \, ds_t \tag{A3.14}$$
$$= -\frac{1}{2\pi} \oint_\gamma p'(t, z) \overline{p'(t, \zeta)} \delta n_t \, ds_t.$$

The variational equation (A3.14) may be differentiated with respect to each variable without losing its validity. For, as is easily seen, if a harmonic function is $O(\epsilon^2)$ uniformly in a given domain, the same is true of all its derivatives in each closed subdomain. Using the identity

$$p'(t, z) = 2(\partial/\partial t) g(t, z),$$

we derive from (A3.14)

$$\delta K(z, \bar{\zeta}) = \oint_\gamma K(z, \bar{t}) K(t, \bar{\zeta}) \delta n_t \, ds_t = \oint_\gamma L(z, t) \overline{L(t, \zeta)} \delta n_t \, ds_t, \tag{A3.15}$$

$$\delta L(z, \zeta) = \oint_\gamma K(z, \bar{t}) L(t, \zeta) \delta n_t \, ds_t = \oint_\gamma L(z, t) K(\zeta, \bar{t}) \delta n_t \, ds_t. \tag{A3.15'}$$

We see from the elegant form of the formulas (A3.15) and (A3.15′) again that the particular derivatives K and L of the Green's function play a distinguished role in the theory. The second formula (A3.15) is identical with the variation formula (A2.96) derived in a very different way.

If one has a variation formula for some functional it is quite natural to consider extremum problems. The variation formula permits one to compare the value of this expression for neighboring domains and to characterize the extremal domain by the requirement that under each variation of the domain the expression must change in the same direction. However, in order to apply the variation formula, we must be sure, beforehand, that it is applicable in the case of the extremal domain. To derive Hadamard's variation formula, it was necessary to assume that the boundary γ of a domain G was at least continuously differentiable. Otherwise, we would not even have been able to write down the formula (A3.4) which contains normal derivatives of the Green's function on the boundary γ. But since we cannot assert that the required extremal domain has such a boundary, Hadamard's formula will at best be of heuristic value for guessing the right extremal domain, but will never lead to its unambiguous determination. For the latter purpose, we must derive a variation formula which is valid in the most general domain possible. Such a formula can indeed be established as we will show in the next article.

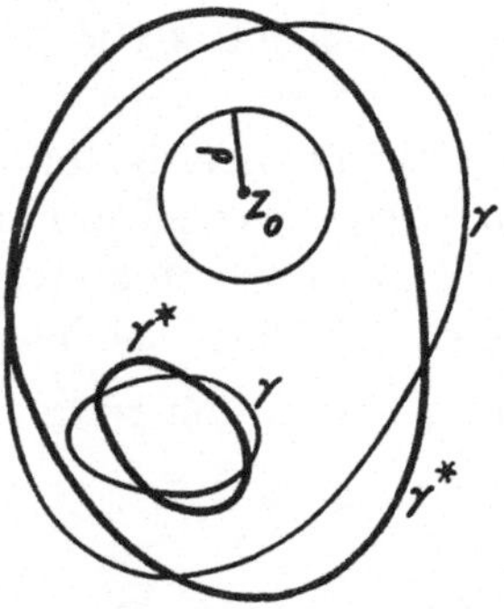

Figure 9. Interior variation.

2. *Interior Variations*. Let G again be bounded by k continuously differentiable curves γ_ρ ; let z_0 be a fixed point in G and consider the following function in G:

$$\text{(A3.16)} \quad z^* = z + e^{2i\varphi}\rho^2(z - z_0)^{-1}, \qquad 0 < \rho, \quad 0 \leq \varphi < \pi.$$

This function maps the exterior of the circle $|z - z_0| = \rho$ onto the whole plane slit along the rectilinear segment

$$\langle z_0 - 2\rho e^{i\varphi}, z_0 + 2\rho e^{i\varphi} \rangle$$

of length 4ρ with center z_0. If ρ is small enough, $|z - z_0| = \rho$ lies entirely in G, and $z^*(z)$ is, therefore, univalent on the boundary γ of the domain G and maps it into a boundary γ^* of a new domain G^*; G^* will differ by very little from G if ρ is small enough. We want to compute the variation of the Green's function for this particular variation of G into G^*.

Again let $g^*(z, \zeta)$ denote the Green's function of G^*. The function

$$d(z, \zeta) = g^*(z^*, \zeta^*) - g(z, \zeta)$$

is harmonic in G outside of the circle $|z - z_0| \leq \rho$ and vanishes on γ. Let us assume that ζ and η lie in G outside the circle $|z - z_0| \leq \rho$, and apply Green's identity to the functions $d(z, \zeta)$ and $g(z, \eta)$ with respect to the domain G_ρ which is bounded by the curve system γ and the circumference $|z - z_0| = \rho$. In view of the vanishing of $d(z, \zeta)$ and $g(z, \eta)$ on γ, we obtain

$$\text{(A3.17)}\quad d(\eta, \zeta) = \frac{1}{2\pi} \oint_{|z-z_0|=\rho} \left[d(z, \zeta) \frac{\partial g(z, \eta)}{\partial n_z} - g(z, \eta) \frac{\partial d(z, \zeta)}{\partial n_z} \right] ds_z .$$

Since the function $g(z, \zeta)$ is regular for $|z - z_0| < \rho$, we can also put (A3.17) into the form

$$\text{(A3.18)}\quad \begin{aligned} &g^*(\eta^*, \zeta^*) - g(\eta, \zeta) \\ &= \frac{1}{2\pi} \oint_{|z-z_0|=\rho} \left[g^*(z^*, \zeta^*) \frac{\partial g(z, \eta)}{\partial n_z} - g(z, \eta) \frac{\partial g^*(z^*, \zeta^*)}{\partial n_z} \right] ds_z . \end{aligned}$$

We have thus expressed $d(\eta, \zeta)$ by an integral over a small interior circle in G. In order to evaluate this integral, we observe that

$$|z - z_0| = \rho \quad \text{and} \quad |z^* - z_0| = |z - z_0 + e^{2i\varphi}\rho^2(z - z_0)^{-1}| = O(\rho)$$

on the above path of integration. We can, therefore, develop all terms of the right-hand integral in power series about the point z_0 and evaluate the integral up to first order terms. The error term will depend only on the derivatives of $g(z, \eta)$ and $g^*(z, \zeta^*)$ near the point z_0 and can, therefore, be estimated uniformly for all domains G and G^* which contain a fixed subdomain including z_0. Introducing the analytic function $p^*(z, \zeta)$ which has the real part $g^*(z, \zeta)$, we may write

$$\text{(A3.19)}\quad \begin{aligned} g^*(z^*, \zeta^*) &= g^*(z_0, \zeta^*) \\ &+ \Re\{[(z - z_0) + e^{2i\varphi}\rho^2(z - z_0)^{-1}]p^{*\prime}(z_0, \zeta^*)\} + O(\rho^2) \end{aligned}$$

and similarly

$$\text{(A3.19')}\quad g(z, \eta) = g(z_0, \eta) + \Re\{(z - z_0)p'(z_0, \eta)\} + O(\rho^2).$$

Furthermore, putting $z - z_0 = re^{i\tau}$ and remarking that on the circle $|z - z_0| = \rho$ we have $\partial/\partial n = \partial/\partial r$, we find that

$$\text{(A3.20)}\quad \frac{\partial}{\partial n_z} g^*(z^*, \zeta^*) = \Re\{(e^{i\tau} - e^{2i\varphi} e^{-i\tau})p^{*\prime}(z_0, \zeta^*)\} + O(\rho),$$

$$\text{(A3.20')}\quad \frac{\partial}{\partial n_z} g(z, \eta) = \Re\{e^{i\tau} p'(z_0, \eta)\} + O(\rho).$$

Introducing all these expressions into (A3.18), we obtain, after an elementary calculation,

$$\text{(A3.21)}\quad g^*(\eta^*, \zeta^*) - g(\eta, \zeta) = \Re\{e^{2i\varphi}\rho^2 p^{*\prime}(z_0, \zeta^*)p'(z_0, \eta)\} + O(\rho^3).$$

Inside each closed subdomain of G, we can make use of the fact that $p^{*\prime}(z_0, \zeta^*) = p'(z_0, \zeta) + O(\rho^2)$; furthermore, we can introduce in (A3.19) the Taylor series development

$$g^*(\eta^*, \zeta^*) = g^*(\eta, \zeta) + \mathcal{R}e\{e^{2i\varphi}\rho^2 p'(\eta, \zeta)(\eta - z_0)^{-1} + e^{2i\varphi}\rho^2 p'(\zeta, \eta)(\zeta - z_0)^{-1}\} + O(\rho^3). \tag{A3.22}$$

Thus we finally obtain the

Theorem: If the domain G is transformed by the variation (A3.16) into the domain G^*, the corresponding Green's functions vary according to the formula [58, 59, 61]:

$$g^*(\eta, \zeta) - g(\eta, \zeta) = \mathcal{R}e\left\{e^{2i\varphi}\rho^2\left[p'(z_0, \zeta)p'(z_0, \eta) - \frac{p'(\zeta, \eta)}{\zeta - z_0} - \frac{p'(\eta, \zeta)}{\eta - z_0}\right]\right\} + O(\rho^3). \tag{A3.23}$$

This formula has been derived under the assumption that the domain G considered possesses a smooth boundary. We remark, however, that the formula (A3.16) defines in a unique way a variation for any domain G which contains the point z_0 in its interior. No use is made here of the concept of normal shift which would have implied a smooth boundary. The final result (A3.23) contains on its right-hand side a term which is defined for every domain G containing the points z_0, ζ, and η in its interior. The error term $O(\rho^3)$ can be estimated uniformly for all domains which contain a fixed subdomain which must in turn include z_0, ζ, and η. From these observations it follows that the formula (A3.23) holds for the most general domain G in the complex plane. For such a domain we define the Green's function as the limit of the corresponding Green's functions of a sequence of domains G_n with smooth boundaries converging towards G. If now we make a variation (A3.16) of G into G^*, the domains G_n of the sequence are varied into a sequence G_n^* with smooth boundaries which converges to G^*. Since, for the difference $g_n^* - g_n$, formula (A3.23) is valid with a uniform error term $O(\rho^3)$, we recognize that (A3.23) must still hold in the limit for the variation of G into G^*.

3. *Application to the Coefficient Problem for Univalent Functions.* Let us now illustrate by an important example the use of the variation formula (A3.23). We consider the family $\mathfrak{U}$ of functions $f(z)$ which are

regular analytic and univalent in the unit circle $|z| < 1$, and have there the series development

$$f(z) = z + \sum_{n=2}^{\infty} a_n z^n . \tag{A3.24}$$

It is known that the n-th coefficient a_n of each function $f \in \mathfrak{U}$ satisfies the inequality $|a_n| < en$ [39]. Furthermore, since the family $\mathfrak{U}$ is compact it is evident that for each value n there exists at least one function $f_\iota(z)$ for which the n-th coefficient has the largest possible modulus. It is well known that the function

$$f(z) = \frac{z}{(1 - z)^2} = z + \sum_{n=2}^{\infty} nz^n \tag{A3.25}$$

belongs to the class $\mathfrak{U}$ and is an extremal function for $n = 2$ and 3 [12, 40]. It is conjectured that the same function $f(z)$ is an extremal function for each value of the integer n. We will show how the formula (A3.23) gives a partial answer to this question.

Consider a fixed value n and suppose that the extremum function $w = f(z)$ maps the unit circle upon a domain G in the w-plane. By normalization (A3.24), the domain G contains the point $w = 0$. Let $z = \Phi(w)$ be the inverse function of $f(z)$. We observe that the Green's function of the domain G has the form

$$g(w, \omega) = \log \left| \frac{1 - \overline{\Phi(\omega)}\Phi(w)}{\Phi(w) - \Phi(\omega)} \right| \tag{A3.26}$$

and, in particular, since $\Phi(0) = 0$,

$$g(w, 0) = \log \left| \frac{1}{\Phi(w)} \right| = \log \frac{1}{|z|} . \tag{A3.26'}$$

Thus the Green's function of G stands in a simple relationship to the inverse of the extremal univalent function and we may apply our variational formula for $g(w, \omega)$ in order to characterize $f(z)$. We perform on the domain G the variation

$$w^* = w + e^{2i\varphi}\rho^2(w - w_0)^{-1} \tag{A3.16'}$$

which transforms G into a domain G^* of the w-plane. By means of (A3.23), we can compute the Green's function $g^*(w, 0)$ of the new domain. Using (A3.26), we obtain

$$g^*(w, 0) = g(w, 0) + \mathcal{R}e\left\{e^{2i\varphi}\rho^2\left[\frac{\Phi'(w_0)^2}{\Phi(w_0)[\Phi(w_0) - \Phi(w)]}\right.\right.$$

$$\text{(A3.27)} \qquad + \frac{\Phi'(w_0)^2\overline{\Phi(w)}}{\Phi(w_0)[1 - \overline{\Phi(w)}\Phi(w_0)]} + \frac{\Phi'(w)}{\Phi(w)(w - w_0)}$$

$$\left.\left. + \frac{1}{w_0\Phi(w)} - \frac{\overline{\Phi(w)}}{w_0}\right]\right\} + O(\rho^3).$$

Now let $\Phi^*(w) = \lambda w + b_2w^2 + \cdots, \lambda > 0$, map the domain G^* into the unit circle. Then we clearly have $g^*(w, 0) = -\log|\Phi^*(w)|$ and hence, extending both sides of (A3.27) to analytic functions of w, we obtain

$$\log \Phi^*(w) = \log \Phi(w) - e^{2i\varphi}\rho^2\left[\frac{\Phi'(w_0)^2}{\Phi(w_0)[\Phi(w_0) - \Phi(w)]}\right.$$

$$\text{(A3.28)} \qquad \left. + \frac{\Phi'(w)}{\Phi(w)(w - w_0)} + \frac{1}{w_0\Phi(w)}\right]$$

$$- e^{-2i\varphi}\rho^2\left[\frac{\overline{\Phi'(w_0)}^2\Phi(w)}{\overline{\Phi(w_0)}[1 - \overline{\Phi(w_0)}\Phi(w)]} - \frac{\Phi(w)}{\bar{w}_0}\right] + O(\rho^3).$$

Let us take the exponential function of both sides and replace $\Phi(w)$ by z, $\Phi(w_0)$ by z_0, w by $f(z)$, w_0 by $f(z_0)$ and $\Phi'(w)$ by $f'(z)^{-1}$. We find

$$\Phi^*(w) = z - e^{2i\varphi}\rho^2\left[\frac{z}{z_0 f'(z_0)^2(z_0 - z)}\right.$$

$$\text{(A3.29)} \qquad \left. + \frac{1}{[f(z) - f(z_0)]f'(z)} + \frac{1}{f(z_0)}\right]$$

$$- e^{-2i\varphi}\rho^2\left[\frac{z^2}{\bar{z}_0\overline{f'(z_0)}^2(1 - \bar{z}_0 z)} - \frac{z^2}{\overline{f(z_0)}}\right] + O(\rho^3).$$

Further let $f^*(z) = (1/\lambda)z + \cdots$ map the unit circle onto the domain G^*. From $f^*(\Phi^*(w)) = w = f(z)$, we conclude

$$f(z) = f^*(z) - e^{2i\varphi}\rho^2\left[\frac{zf'(z)}{z_0 f'(z_0)^2(z_0 - z)}\right.$$

$$\text{(A3.30)} \qquad \left. + \frac{1}{f(z) - f(z_0)} + \frac{f'(z)}{f(z_0)}\right]$$

$$- e^{-2i\varphi}\rho^2\left[\frac{z^2 f'(z)}{\bar{z}_0\overline{f'(z_0)}^2(1 - \bar{z}_0 z)} - \frac{z^2 f'(z)}{\overline{f(z_0)}}\right] + O(\rho^3).$$

Thus we have obtained a new univalent function $f^*(z)$ which is very near the extremal $f(z)$. In order to compute the power series development of $f^*(z)$ around the origin we introduce a sequence of polynomials $S_r(t)$ by the following definition:

$$\frac{f(z)^2}{1 - tf(z)} = \sum_{n=2}^{\infty} S_n(t) z^n. \tag{A3.31}$$

The coefficients of the polynomials $S_n(t)$ are constructed in elementary rational fashion from the first coefficients of the function $f(z)$. We next compute from (A3.30) the n-th coefficient a_n^* of $f^*(z)$ and obtain, after elementary transformations,

$$\frac{a_n^*}{a_1^*} = a_n + e^{2i\varphi}\rho^2 A_n(z_0) + e^{-2i\varphi}\rho^2 \overline{B_n(z_0)} + O(\rho^3), \tag{A3.32}$$

with

$$\begin{aligned} A_n(z_0) = {} & \frac{(n+1)a_{n+1} - 2a_2 a_n}{f(z_0)} - \frac{1}{f(z_0)^2} S_n\left(\frac{1}{f(z_0)}\right) \\ & + \frac{1}{z_0^2 f'(z_0)^2}\left[(n-1)a_n + \frac{(n-1)a_{n-1}}{z_0}\right. \\ & \left. + \frac{(n-2)a_{n-2}}{z_0^2} + \cdots + \frac{2a_2}{z_0^{n-2}} + \frac{1}{z_0^{n-1}}\right] \end{aligned} \tag{A3.33}$$

and

$$\begin{aligned} B_n(z_0) = {} & \frac{1}{z_0 f'(z_0)^2}[(n-1)\bar{a}_{n-1} + (n-2)\bar{a}_{n-2} z_0 \\ & + (n-3)\bar{a}_{n-3} z_0^2 + \cdots + z_0^{n-2}] - \frac{(n-1)\bar{a}_{n-1}}{f(z_0)}. \end{aligned} \tag{A3.33'}$$

It is now clear that the function $f^*(z)\cdot 1/a_1^*$ is again of the class $\mathfrak{U}$. Hence we have

$$\left|\frac{a_n^*}{a_1^*}\right| \leq |a_n|$$

for any choice of w_0 and φ and ρ. Since $e^{i\alpha}f(e^{-i\alpha}z)$ is of the same class $\mathfrak{U}$ as $f(z)$, we may assume without loss of generality that $a_n > 0$. Thus we have the extremum condition for $f(z)$:

$$\mathcal{R}e\{e^{2i\varphi}\rho^2 A_n(z_0) + e^{-2i\varphi}\rho^2\overline{B_n(z_0)} + O(\rho^3)\} \leq 0. \tag{A3.34}$$

Dividing by ρ^2 and letting $\rho \to 0$, we obtain (note that $\mathcal{R}e\{\overline{U}\} = \mathcal{R}e\{U\}$)

$$\mathcal{R}e\{e^{2i\varphi}[A_n(z_0) + B_n(z_0)]\} \leq 0 \tag{A3.35}$$

and, since $e^{2i\varphi}$ is quite arbitrary, we arrive at the necessary extremum condition

$$A_n(z_0) + B_n(z_0) \equiv 0 \quad \text{for} \quad |z_0| < 1. \tag{A3.36}$$

We find therefore, in view of (A3.33) and (A3.33′), the following differential equation for $f(z)$:

$$\begin{aligned} \frac{z^2 f'(z)^2}{f(z)^2} S_n\left(\frac{1}{f(z)}\right) - \frac{(n+1)a_{n+1} - 2a_2 a_n - (n-1)\bar{a}_{n-1}}{f(z)} z^2 f'(z)^2 = \\ \left(\frac{1}{z^{n-1}} + \frac{2a_2}{z^{n-2}} + \cdots + \frac{(n-1)a_{n-1}}{z}\right. \\ \left. + (n-1)a_n + (n-1)a_{n-1} z + \cdots + z^{n-1}\right). \end{aligned} \tag{A3.37}$$

This result can still be considerably simplified by the following remark. Had we subjected the extremal domain G to the simple variation $w^* = w + e^{2i\varphi}\rho^2$ which is in fact a conformal mapping, we would have obtained instead of (A3.27) the identity

$$\begin{aligned} g^*(w, 0) &= g(w, 0) \\ &+ \mathcal{R}e\left\{e^{2i\varphi}\rho^2\left[\frac{\varphi'(w)}{\varphi(w)} - \frac{1}{\varphi(w)} + \overline{\varphi(w)}\right]\right\} + O(\rho^3). \end{aligned} \tag{A3.38}$$

which follows from the conformal invariance of the Green's function $g^*(w^*, \omega^*) = g(w, \omega)$. Carrying through the same considerations for this new type of variation, we would have obtained the extremum condition [41]

$$(n+1)a_{n+1} = 2a_2 a_n + (n-1)\bar{a}_{n-1}\,. \tag{A3.39}$$

This leads to the final differential equation for the extremal function [50, 55, 59]:

$$\begin{aligned} \frac{z^2 f'(z)^2}{f(z)^2} S_n\left[\frac{1}{f(z)}\right] &= \frac{1}{z^{n-1}} + \frac{2a_2}{z^{n-2}} + \cdots + \frac{(n-1)a_{n-1}}{z} \\ &+ (n-1)a_n + (n-1)\bar{a}_{n-1} z + \cdots + z^{n-1}. \end{aligned} \tag{A3.40}$$

The right-hand side of (A3.40) is a rational function of z which is real for $|z| = 1$. We recognize that $f(z)$ is still analytic for $|z| = 1$, except for those points where $f(z)$ becomes a root of $S_n(1/f) = 0$. Let $w(t) = f(e^{it})$ be the parametric representation of the boundary curve belonging to the extremum domain. Then $w'(t) = ie^{it}f'(e^{it})$ will be a tangential vector at the point $w(t)$. In view of (A3.40), we find

$$\text{(A3.41)} \qquad \frac{w'^2}{w^2} S_n\left(\frac{1}{w}\right) = \text{real}$$

as the differential equation for the boundary of the extremal domain. By appropriate choice of the parameter we can bring (A3.41) into the more usual form

$$\text{(A3.41}'\text{)} \qquad \frac{w'^2}{w^2} S_n\left(\frac{1}{w}\right) \pm 1 = 0.$$

Thus the boundary of the extremal domain G consists of a finite number of analytic arcs. We may continue the study of the extremal domain by Hadamard's method and prove, for example, that the plus sign holds always in (A3.41′). For further details see [55, 56, 57].

It is easily verified that, for every $n \geq 2$, the function (A3.25) satisfies the differential equation (A3.40). It can be shown that, for $n = 2$, it is the only function in $\mathfrak{U}$ which does so; hence $|a_2| \leq 2$ is proved. In the case $n = 3$ we can show [50] that every solution of (A3.40) satisfies $|a_3| \leq 3$; hence $|a_3| \leq 3$ is proved for all functions (A3.24) and again (A3.25) appears as the extremal function. A corresponding proof for $n > 3$ has not yet been established, which leaves the question of max $|a_n|$ for $n > 3$ still open. We see that the method of interior variation leads to differential equations as necessary conditions for extremal functions. The proof that the necessary conditions determine the extremal function uniquely and the establishment of sufficient extremum conditions can, of course, never be obtained by variational methods which are just a comparison with the neighboring functions of the class considered. The knowledge of the necessary extremum conditions reduces, however, the manifold of possible extremal functions in such a strong way that the sufficiency proof becomes in many cases quite simple. For further applications of the method of interior variations, see [17, 20, 49, 51–54, 65].

The coefficient problem for univalent functions in the unit circle was attacked first by Löwner by a variational method [40], which is closely related to the methods of the last two articles (cf. [49, 65]).

4. *Boundary Variations.* There is another method of solving extremum problems in conformal mapping by variations, a method which is particularly useful in the theory of multiply connected domains. We illustrate it by treating the following coefficient problem for univalent functions. Consider a fixed domain G in the z-plane which contains the point at infinity. Let $\mathfrak{B}$ be the class of all uni-

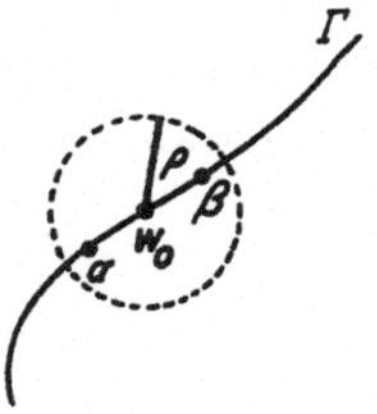

Figure 10. Boundary variation.

valent functions in G which have, near infinity, the series development

$$f(z) = z + a_0 + \frac{a_1}{z} + \frac{a_2}{z^2} + \cdots. \tag{A3.42}$$

It can again be shown that the coefficients a_ν of all $f \in \mathfrak{B}$ are bounded for fixed ν, and the question of the maximum $| a_n |$ arises again.

Let $w = f(z)$ map G onto a domain Δ in the w-plane with boundary Γ. Let $w_0 \in \Gamma$ and α, β be two points on the same component of Γ at a distance $< \rho$ from w_0. We can then easily construct the function

$$\Phi(w) = (\beta - \alpha)\left[\log\frac{w - \alpha}{w - \beta}\right]^{-1}, \tag{A3.43}$$

which is univalent in Δ. This function has in the domain $| w - w_0 | > \rho$ the series development

$$\Phi(w) = w - \frac{\alpha + \beta}{2} - \frac{(\alpha - \beta)^2}{12(w - w_0)} + O(\rho^3) \tag{A3.44}$$

where the remainder term can be estimated uniformly in each closed subdomain of $| w - w_0 | > \rho$.

In view of the obvious principle that a univalent function of a univalent function is again univalent, we have in

$$f^*(z) = \Phi[f(z)] = f(z) - \frac{\alpha + \beta}{2} - \frac{(\alpha - \beta)^2}{12(f(z) - w_0)} + O(\rho^3) \tag{A3.45}$$

a new function of the class $\mathfrak{B}$. Its n-th coefficient a_n^* will be of the form

$$a_n^* = a_n - \frac{(\alpha - \beta)^2}{12} T_n(w_0) + O(\rho^3) \tag{A3.46}$$

where the polynomials $T_n(w)$ are defined by the generating function

$$\frac{1}{f(z) - w} = \sum_{n=1}^{\infty} T_n(w) z^{-n} \tag{A3.47}$$

and depend in an elementary rational way upon the coefficients of $f(z)$.

Suppose now that $f(z)$ is a univalent function in $\mathfrak{B}$ with maximum $\mathcal{R}e\{e^{i\mu}a_n\}$. Since our basic domain G need no longer be symmetric each choice of μ leads to a separate extremum problem. Then, for every point $w_0 \in \Gamma$, we must necessarily have

$$\mathcal{R}e\{e^{i\mu}(\alpha - \beta)^2 T_n(w_0)\} + O(\rho^3) \geq 0, \tag{A3.48}$$

whence after division by ρ^2

$$\mathcal{R}e\left\{e^{i\mu}\left(\frac{\alpha - \beta}{\rho}\right)^2 T_n(w_0)\right\} + O(\rho) \geq 0. \tag{A3.48'}$$

We recognize from (A3.48′) that all points of Γ near w_0 must approximate a straight line through w_0 with angle $e^{i\tau}$ such that

$$\mathcal{R}e\{e^{i(2\tau+\mu)}T_n(w_0)\} \geq 0. \tag{A3.49}$$

This shows that Γ has no interior points; if it has a tangent at w_0 it must have the direction $e^{i\tau}$. Refining our argument, we can, in fact, prove that the extremum property of $f(z)$ implies the existence of a tangent to Γ at each point w_0 where $T_n(w_0) \neq 0$. We omit this rather lengthy proof but condense its result [56] in the following

Lemma: Let Γ_1 be a continuum in the w-plane. Suppose that there exists an analytic function $s(w) \not\equiv 0$ such that for every univalent function $\Phi(w)$ which has a representation

$$\Phi(w) = w + A_0 + \frac{A_1\rho^2}{w - w_0} + O(\rho^3) \tag{A3.50}$$

uniformly in each closed subdomain of $| w - w_0 | > \rho$, the inequality

$$\mathcal{R}e\{A_1\rho^2 s(w_0)\} + O(\rho^3) \geq 0 \tag{A3.51}$$

is fulfilled. Then Γ_1 is an analytic curve which can be expressed in such a parametric form $w = w(t)$ that

$$w'(t)^2 s[w(t)] + 1 = 0, \qquad w'(t) = dw/dt. \tag{A3.52}$$

Since for every function (A3.50) our previous reasoning would have led to the condition (A3.51), with $s(w_0) = -e^{i\mu}T_n(w_0)$, the lemma proves, for our particular problem, that the extremal domain Δ is bounded by analytic curves satisfying the differential equation

$$w'(t)^2 T_n(w) - e^{-i\mu} = 0. \tag{A3.53}$$

Let us apply this result to the particular case $n = 1$. In this case, we easily find $T_1(w) \equiv 1$ and, hence, the differential equation for all boundary curves of the extremal domain:

$$w'(t)^2 = e^{-i\mu}, \quad \text{i.e.} \quad w(t) = e^{-\frac{1}{2}i\mu}t + \text{constant}. \tag{A3.53'}$$

This shows that the extremal domain is bounded by parallel rectilinear segments with the direction $e^{-\frac{1}{2}i\mu}$.

Since the existence of an extremal domain is guaranteed by the compactness of the family $\mathfrak{B}$, our necessary condition for this domain yields an existence proof for the above types of slit mappings. It is obvious that every extremum problem in conformal mapp ng which can be treated by means of the preceding lemma, and f^lor which the existence of an extremum follows from the compactness of $\mathfrak{B}$, leads to an existence theorem for a mapping onto a domain bounded by analytic slits which satisfy a differential equation (A3.52).

The possibility of mapping a multiply connected domain onto a parallel-slit domain was proved by means of the above extremum problem by de Possel [46, 47] and Grötzsch [24]. It should also be remarked that the most straightforward proof of Riemann's mapping theorem for simply connected domains is based on a similar extremum method (cf. [12]). This method of establishing canonical maps by considering extremum problems within the family of univalent functions in G is carefully to be distinguished from the method of Dirichlet's Principle in which the same existence proofs are obtained by extremizing certain expressions within a much wider family of functions in G.

Let us now treat the following problem, due to Grötzsch [22]. Consider a domain G which contains the point at infinity and is bounded by k continua γ_ρ. Each function $f(z) \in \mathfrak{B}$ maps G onto a new domain G^+ in the w-plane bounded by continua γ_ρ^+. Let $d(\gamma_\rho)$ be the maximum distance of any two points on C_ρ, i.e. the span of C_ρ. We ask for the maximum of $d(\gamma_1^+)$ for all mappings (A3.42).

It is easily seen that there is an extremal mapping for which $d(\gamma_1^+)$ is really a maximum; let a, b be that pair of points on γ_1^+ for which $|a - b| = d(\gamma_1^+)$ is attained. We can always find a function $\Phi(w)$ of the type (A3.50) for $w_0 \in \gamma^+$ which is univalent in G^+, so that $\Phi[f(z)]$ will also be of the class $\mathfrak{B}$. The points a, b go by the mapping into

$$a_1 = a + A_0 + \frac{A_1 \rho^2}{a - w_0} + O(\rho^3), \tag{A3.54}$$

$$b_1 = b + A_0 + \frac{A_1 \rho^2}{b - w_0} + O(\rho^3).$$

But, from the assumed extremum property of G^+, we have clearly

$$|a_1 - b_1| = |a - b| \left| 1 - \frac{A_1 \rho^2}{(a - w_0)(b - w_0)} + O(\rho^3) \right| \le |a - b|. \tag{A3.55}$$

Thus, from the requirement

$$\mathcal{R}e \left\{ \frac{A_1 \rho^2}{(a - w_0)(b - w_0)} \right\} + O(\rho^3) \geq 0, \tag{A3.55'}$$

we conclude by virtue of our lemma that all boundary curves γ_ρ^+ are analytic and satisfy the differential equation

$$w'^2 (a - w)^{-1} (b - w)^{-1} + 1 = 0. \tag{A3.56}$$

This differential equation can be immediately integrated to

$$w = \tfrac{1}{2}(a + b) + \tfrac{1}{4}(a - b)(c_\rho e^{it} + 1/c_\rho e^{-it}), \qquad t = \text{real}, \tag{A3.57}$$

where the constant of integration c_ρ may have different values on each continuum γ_ρ^+. In particular, we have $c_1 = 1$ on γ_1^+, since γ_1^+ contains the points a and b. Thus γ_1^+ is just the straight segment connecting these two points. All other continua γ_ρ^+, $\rho > 1$, are easily seen to be arcs of ellipses with a and b as their foci. Hence we obtain the

Theorem: Each multiply connected domain can be mapped onto the whole plane slit along confocal elliptic arcs. One of the boundaries can be prescribed to correspond to the segment connecting the foci.

We might also formulate the following very similar problem: We consider all domains G^+ with boundary γ^+ obtained from G by maps of the class $\mathfrak{B}$ and ask for the maximal span $d(\gamma_i^+, \gamma_k^+)$: i.e. we consider pairs of points $a \in \gamma_i^+$ and $b \in \gamma_k^+$ and seek, for fixed i and k, the maximum distance $|a - b|$ under all maps $f \in \mathfrak{B}$. We may repeat just the same considerations as above leading from formula (A3.54) to formula (A3.57) only with the understanding that $a \in \gamma_i^+$ and $b \in \gamma_k^+$. Again we find from (A3.57) that all curves γ_ρ^+ are arcs of ellipses with foci a and b. In our new problem we conclude, therefore, that γ_i^+ and γ_k^+ are straight segments lying in the segment $\langle a, b \rangle$. We have proved [22]:

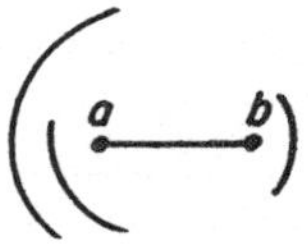

Figure 11. Elliptic slit system with foci a and b containing segment $\langle a, b \rangle$.

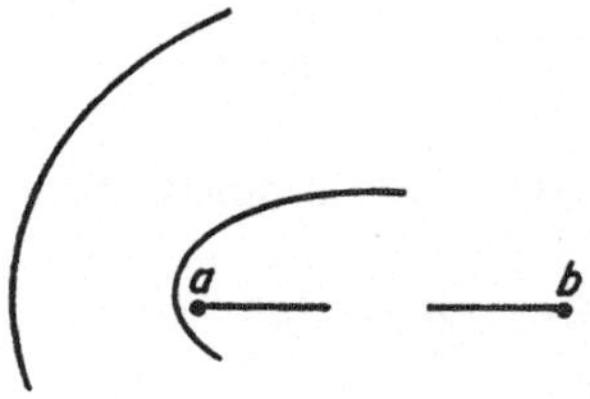

Figure 12. Elliptic slit system with foci a and b and two slits on line ab.

Each multiply connected domain can be mapped onto the whole plane slit along confocal elliptic arcs. Two boundary continua can be prescribed to be straight segments containing the foci and lying in the segment between them.

In particular:

Each doubly connected domain can be mapped onto the whole plane slit along two collinear straight segments.

The reader will have no difficulties in devising various other extremum problems and solving them analogously. Each such exercise will provide him with an existence theorem for a certain type of conformal mapping.

5. *Lavrentieff's Method.* In the preceding articles, extremum problems with respect to univalent functions in a given domain were studied. The main difficulty was in each case the proof that the

boundary of the extremal domain has certain continuity properties. For this reason one is obliged either to use interior variation methods or to apply the deep lemma of the previous article. In various cases, however, one can use a reasoning first developed by Lavrentieff [19, 35] in order to show that the extremal domain has analytic boundaries; once this fact is established one has little difficulty in characterizing the extremal domain in more detail, for example by use of Hadamard's formula which then becomes applicable.

We want to develop Lavrentieff's reasoning in an extremum problem which is in itself of interest. Consider a simply connected domain G which contains the point at infinity; there exists a univalent function $w = f(z)$ in G which has at infinity the development (A3.42) and which maps G upon the circular domain $|w| > R$. The number R is a characteristic measure of G, called the *mapping radius* of G.

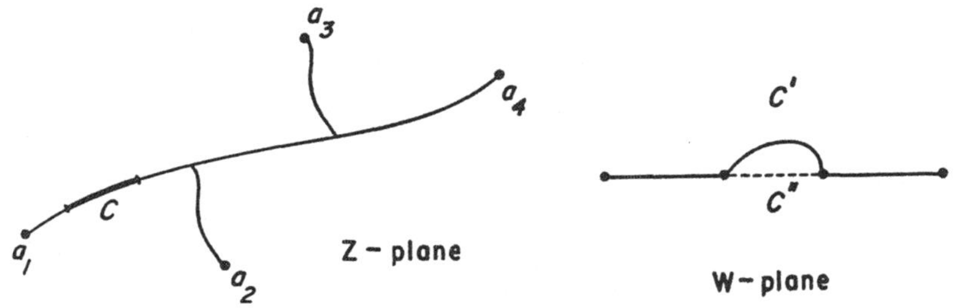

Figure 13. Lavrentieff's method.

It is immediately verified that each domain which is obtained from G by a map with a function of the normalization (A3.42) has the same mapping radius as G. By the transformation

$$t = w + R^2/w \tag{A3.58}$$

we can map the circle $|w| > R$ onto the t-plane slit along the segment $\langle -2R, 2R \rangle$. This domain has also the mapping radius R; on the other hand, one of our results in article 4 showed that the boundary curve γ^+ of this new domain has the largest span $d(\gamma^+)$ among all curves γ^+ resulting from a mapping (A3.42) of γ. Hence we obtain the inequality (due to Bieberbach [11])

$$d(\gamma) \leqslant 4R, \tag{A3.59}$$

connecting span and mapping radius. Equality in (A3.59) can only hold if γ is a segment of the length $4R$. We may also invert this result in the following way: Let two arbitrary points a, b be given and consider all continua which contain them. Each such continuum γ determines a simply connected infinite domain with a certain mapping radius; we will call this radius also the mapping radius belonging to the continuum γ. The span $d(\gamma)$ is clearly $\geq |b - a|$ and for the straight segment connecting b with a we have exactly $d(\gamma) = |b - a|$. For the mapping radii $R(\gamma)$ we have, therefore, the inequality

$$|b - a| \leq d(\gamma) \leq 4R(\gamma) \tag{A3.60}$$

and equality can hold only if γ is the segment $\langle a, b \rangle$. Thus we have proved:

The continuum through a and b with least mapping radius is the straight segment $\langle a, b \rangle$.

Grötzsch proposed the following more general question [23]: Let a_i, $i = 1, 2, \cdots, n$, be n given points in the complex plane; determine a continuum γ containing them with the least possible mapping radius. We will treat this problem by Lavrentieff's method.

Consider an extremal continuum γ; assume that we can decompose it into two separate continua γ_1 and γ_2 by deleting from γ a subcontinuum C which does not contain any of the points a_i. The continua γ_1 and γ_2 determine an infinite doubly connected domain D. According to the last result of the preceding article there exists in D a univalent function $w = f(z) = z + a_0 + a_1/z + \cdots$ which maps D onto the whole plane cut along the two collinear straight segments γ_1' and γ_2'. The continuum C is mapped into a continuum C' which connects the segments γ_1' and γ_2' and forms with them the image γ' of the extremal continuum γ. The continuum γ' has still the mapping radius of γ. If C' is not the straight segment C'' between γ_1' and γ_2', we may replace C' by C''; this replacement will, in view of our last theorem, decrease the mapping radius of γ'. The change, in view of the univalent correspondence $w = f(z)$, will create a new continuum γ in the z-plane still containing all the points a_i but with a smaller mapping radius. But this would contradict the assumed extremum nature of γ; the above decrease in the mapping radius of γ' was impossible, i.e. C' was already the straight segment connecting γ_1' and γ_2'. We know that C is obtained from C' by the inverse mapping $z = f^{(-1)}(w)$ which is analytic in the w-plane outside of γ_1' and γ_2'. Hence C is the analytic image of a straight segment, i.e. analytic.

Thus each subcontinuum C of γ which has the property that its deletion splits γ into two separate continua is an analytic arc. If we had known that γ consists of finitely many Jordan arcs, it would have followed immediately that all these arcs are analytic. This would have opened the way to an easy further investigation which we shall not continue here. Thus Lavrentieff's method permitted us to conclude, from the Jordan character of γ, its analyticity. However, the proof that the continuum γ consists of finitely many Jordan arcs requires a finer topological consideration which is nearly as deep as the proof of the general lemma on boundary variation of article 4.

In the classical calculus of variations one always makes great use of the principle that each subarc of an extremal curve is itself an extremal curve. This principle does not, in general, hold for the type of extremum problem with which we are concerned here. Lavrentieff's method, however, shows that sometimes this fruitful principle of the calculus of variations may be applied.

6. *Method of Extremal Length.* In the above articles we have developed the treatment of extremum problems by methods of variations. By considering appropriate problems, we were able to prove the existence of mappings on certain canonical domains. We showed, in fact, that the mappings belonging to the extremal functions led to domains with analytic boundary curves which satisfy certain differential equations. There arises, however, the converse question: does every such domain really solve a certain extremum problem of the type considered? This question is, in general, very hard to answer by variational considerations and here the importance of an alternate method becomes evident, namely the method of extremal length.

This method goes back to Faber [16], Grötzsch [21], and Rengel [48] and was brought into its present elegant form by Ahlfors and Beurling [1], [2], [10]. It is particularly useful as a method of verification; i.e. if it is conjectured that a certain domain has some extremum property the method frequently enables us to prove the correctness of this guess. We want again to illustrate the method by an example.

Consider a circular ring bounded by the circumferences $|z| = 1$ and $|z| = R$. We introduce in this ring a positive weight function $\rho(x, y)$ and determine, by means of it, a metric $ds^2 = \rho^2 |dz|^2$. Our only requirement on the weight factor ρ is that the total area of the ring be 1, i.e.

$$\int_0^{2\pi} \int_1^R \rho^2 r\, dr\, d\varphi = 1, \qquad z = re^{i\varphi} = x + iy. \tag{A3.61}$$

A curve C in the annulus has the length

$$l_C = \oint_C \rho\, d|z| = \int_C \rho\sqrt{r^2\, d\varphi^2 + dr^2}, \tag{A3.62}$$

and we may ask for the minimum of this length among all closed curves C which are topologically equivalent to the circumference $z = e^{i\varphi}$, $0 \leq \varphi \leq 2\pi$. The minimum number $\mu[\rho]$ depends on ρ. We shall try to choose ρ in such a way as to maximize this quantity. We consider at first the weight function $\rho_0(x, y) = (2\pi \log R)^{-1/2}(1/r)$. The constant in this definition has been chosen so that (A3.61) is fulfilled. We obtain for l_C the following estimate:

$$l_C \geq \int_C r\rho_0\, d\varphi = \left(\frac{2\pi}{\log R}\right)^{1/2}. \tag{A3.63}$$

If we choose for C any circle $|z| = c$, $1 \leq c \leq R$, the inequality becomes an equality. Thus, in this particular metric, we have $\mu[\rho_0] = (2\pi/\log R)^{-1/2}$.

Let now $\rho(x, y)$ be an arbitrary weight function satisfying (A3.61). We start with the obvious inequality

$$\int_0^{2\pi}\int_1^R (\rho - \rho_0)^2 r\, dr\, d\varphi \geq 0. \tag{A3.64}$$

Using (A3.61) twice, we can transform (A3.64) into

$$2 \geq 2(2\pi \log R)^{-1/2}\int_0^{2\pi}\int_1^R \rho r\, d\varphi\, \frac{dr}{r}. \tag{A3.65}$$

In view of the definition of $\mu[\rho]$, we have,

$$\int_0^{2\pi} \rho r\, d\varphi \geq \mu[\rho], \tag{A3.66}$$

since the left-hand integral measures the length of the circle $|z| = r$ in the ρ-metric. Hence, the inequality (A3.65) yields

$$(2\pi \log R)^{1/2} \geq \mu[\rho] \log R. \tag{A3.67}$$

Finally we obtain

$$\mu[\rho] \leq \left(\frac{2\pi}{\log R}\right)^{1/2} = \mu[\rho_0]. \tag{A3.68}$$

We repeat our result in the following form:

Among all weight functions ρ satisfying (A3.61), *the function* $\rho_0 = (2\pi \log R)^{-1/2}(1/r)$ *leads to the greatest value for the minimal length. This maximum is* $(2\pi/\log R)^{1/2}$.

Let now G be a doubly connected domain which has been obtained from the circular ring by a map $w = f(z)$. Let $w = u + iv$; each integral $\iint \rho_w \, du \, dv$ over a part of G may be expressed in the form $\iint \rho_w \mid f'(z) \mid^2 dx \, dy$ over the corresponding part of the annulus. Thus each weight function ρ_w in G gives rise to a weight function

$$\rho_z = \rho_w \mid f'(z) \mid^2 \tag{A3.69}$$

in the annulus and vice versa.

Each doubly connected domain G can be mapped onto a circular ring of the above form, i.e. bounded by two circumferences $\mid z \mid = 1$ and $\mid z \mid = R$. The number R is characteristic for G and is called its modulus. We can define it now by the following minimum property:

Consider all metrics based on a weight function ρ_w which give the total area 1 for the domain G. The minimum length for all closed irreducible curves in G and for any weight function ρ_w is less than or equal to $(2\pi/\log R)^{1/2}$ and this maximum is attained for that function ρ_w which is connected with ρ_0 by (A3.69).

In fact, we may transfer the whole question from the domain G into the annulus by means of the transformation formula (A3.69). Each metric in G determines a corresponding metric in the annulus such that corresponding curves have equal length; hence, the minima obtained in both domains must be the same. Thus we obtain the possibility of defining the modulus R of G by means of an extremum problem formulated with respect to G. We remark that a very elegant characterization of the map of G onto the annulus has been given, although the possibility of such a map has been presupposed.

Let us draw some conclusions from our result:

a) It is impossible to map two domains with different moduli onto each other. For, under conformal mapping, the metrics could be transplanted from one domain into the other; hence the extremum of the minimum lengths must be the same. But this shows that the moduli must coincide.

b) Let G' and G be two doubly connected domains such that $G' \supset G$ and that each closed irreducible curve in G is also irreducible in G'. In this case, we can assert that the modulus of G' is greater than the modulus of G. In fact, each weight function ρ' in G' can be made a permissible weight function for G after multiplication with a constant factor $\lambda > 1$ which guaranteed that, for the metric $\lambda\rho' = \rho$,

G has the total area 1. The length of each irreducible curve in G in the ρ-metric is λ times its length in the ρ'-metrix and hence definitely larger than the minimal length in G'; thus, a fortiori, this will hold for their maxima. But this implies $R' > R$ for the moduli.

We can invert our statement on the minimal length for doubly connected domains in the following way: Let ρ be a weight function in G such that, for each irreducible closed curve in G,

$$\oint_C \rho \, ds \geq 1. \tag{A3.70}$$

Then G will have in the corresponding ρ-metric the area

$$\iint_G \rho^2 \, dx \, dy \geq \frac{1}{2\pi} \log R. \tag{A3.71}$$

To apply this result we let G be a doubly connected domain bounded by two curves C_1 and C_2. We draw another irreducible closed curve Γ, and thus divide G into two doubly connected domains G_1 and G_2 bounded by the pairs of curves C_1, Γ and C_2, Γ, respectively. We introduce in G the extremal weight function ρ which satisfies (A3.70) and gives G its minimal area. Applying now the preceding result to G_1 and G_2, we conclude:

$$\iint_{G_1} \rho^2 \, dx \, dy \geq \frac{1}{2\pi} \log R_1, \qquad \iint_{G_2} \rho^2 \, dx \, dy \geq \frac{1}{2\pi} \log R_2, \tag{A3.72}$$

where R_1 and R_2 denote the moduli of G_1 and G_2, respectively. For ρ is clearly a permissible weight function in G_1 and G_2. Adding both inequalities and using the fact that ρ is the extremal weight function for G, we obtain

$$\log R \geq \log R_1 + \log R_2, \tag{A3.73}$$

an interesting superadditivity property for the logarithms of the moduli.

The method of minimal length can also easily be extended to the case of domains with higher connectivity. One introduces here again a positive weight-function ρ such that $\iint \rho^2 \, dx \, dy$ and $\int \rho \, ds$ measure area and length in G. We now submit ρ to the following restrictions. Let γ_ν' be a curve in G which is topologically equivalent to the boundary curve γ_ν of γ; we require

$$\iint_{\gamma'_\nu} \rho\, ds \geq a_\nu, \qquad \nu = 1, 2, \cdots, k, \tag{A3.74}$$

where the a_ν are a fixed given set of positive numbers. We ask for the minimum

$$\text{Min} \iint_G \rho^2\, dx\, dy = M(a_1, a_2, \cdots, a_k). \tag{A3.75}$$

This function $M(a_1, a_2, \cdots, a_k)$ is, of course a conformal invariant; hence the equality of $M(a_1, a_2, \cdots, a_k)$ for two given k-fold connected domains G is a necessary condition for their conformal equivalence. Recently, Jenkins has proved that in the case of a triply connected domain this condition is also sufficient [31]. The method of extremal length has been applied successfully in various problems of conformal mapping [21–23, 31, 48]. If we choose suitable *side conditions* the extremal length coincides with important functionals of the domain, just as in our illustrative problem it represented the modulus of a doubly connected domain. The extremal property thus exhibited leads to useful inequalities for the functionals considered. We mention here as example the beautiful result of Teichmueller obtained in the same way [70]:

Every function (A3.24) which maps the unit circle onto a slit domain in the w-plane such that the boundary slit satisfies the differential equation (A3.41′), solves some extremum problem relative to the first $n - 1$ coefficients $a_2, a_3, \cdots, a_n$.

Thus the method of extremal length permits us to show that the necessary conditions for the coefficient problem, obtained by variational methods, are sufficient conditions for some similar coefficient problem. It is only the vagueness of Teichmueller's theorem with respect to the exact nature of the coefficient problem which separates us from the complete solution of the general coefficient problem.

7. *Concluding Remarks.* For the sake of brevity, we developed the theory of the fundamental solutions and the kernels only in the case of domains imbedded in the complex plane. We might have treated the more general case of domains on arbitrary Riemann surfaces, as studied in the previous chapters. All arguments are still applicable except for one important point which may serve to clarify the relative merits of the various methods of existence proof considered.

We pointed out that in the case of a plane domain the existence of reproducing kernels can be established by certain extremum methods within the family of analytic functions in this domain; the existence of the Green's function can, therefore, be inferred without the use of Dirichlet's Principle. This is, however, impossible in the general case of a domain on a Riemann surface. In this case we are not sure *a priori* that an analytic function exists in the domain. We have to start with the family of piecewise smooth functions and only Dirichlet's Principle permits the selection of analytic functions from this family. Once the existence of such functions is ensured, we may continue the theory by the method of the fundamental functions and the kernels.

Similarly, the method of existence proofs by extremal methods within the family of univalent functions in the domain is applicable only after the existence of univalent functions is established. In each plane domain we always have the function z as an example of a univalent function, but this does not hold in the general case.

Thus three types of existence proofs for boundary value problems by extremum methods have been considered:

a) The Dirichlet Principle working with the widest class of all piecewise smooth functions with finite Dirichlet integral over the domain.

b) The kernel method using the subclass of all analytic functions with finite Dirichlet integral.

c) The extremum method within the narrowest class of univalent functions in the domain.

It is clear that the method working in the wider class is of greater generality, while the method in the narrower class, if at all applicable, is of greater convenience in operation.

Finally, we want to point out that the method of fundamental solutions and kernels can be easily extended to a wide class of partial differential equations of elliptic type and stands there in an analogous relation to Dirichlet's Principle as in the case of analytic functions and the Laplace equation, studied here [8, 67].

Bibliography to Appendix

Ahlfors, L.

[1] *Untersuchungen zur Theorie der konformen Abbildung und ganzen Funktionen*, Acta Soc. Sci. Fennicae, Nova Ser. A., Volume 1, 1930.

[2] (and A. Beurling). *Invariants conformes et problèmes extrémaux*, i xi ème Congrès des Mathématiciens Scandinaves, 1946.

Aronszajn, N.

[3] *La théorie des noyaux reproduisants et ses applications*, *I*, Proc. Cambridge Phil. Soc., Volume 39, 1943, pp. 133–153.

Bergman, S.

[4] *Über die Entwicklung der harmonischen Funktionen der Ebene und des Raumes nach Orthogonalfunktionen*, Math. Ann., Volume 86, 1922. pp. 238–271.

[5] *Partial differential equations, Advanced Topics*, Brown University, Providence, 1941.

[6] *Sur les fonctions orthogonales de plusieurs variables complexes*, Interscience Publishers, Inc., New York, 1941.

[7] *Complex orthogonal functions and conformal mapping*, Mathematical Surveys, 1950.

[8] (and M. Schiffer). *A representation of Green's and Neumann's functions in the theory of partial differential equations of second order*, Duke Math. J., Volume 14, 1947, pp. 609–638.

[9] (and M. Schiffer). *Kernel functions and conformal mapping*, Duke Math. J., Volume 17, 1950.

Beurling, A.

[10] *Études sur un problème de majoration*, Thesis, Upsala, 1933.

Bieberbach, L.

[11] *Über einige Extremalprobleme im Gebiete der konformen Abbildung*, Math. Ann., Volume 77, 1916, pp. 153–172.

[12] *Lehrbuch der Funktionentheorie*, II, B. G. Teubner, Berlin, 1931.

Bochner, S.

[13] *Über orthogonale Systeme analytischer Funktionen*, Math. Z., Volume 14, 1922, pp. 180–207.

Courant, R.

[14] (and D. Hilbert). *Methoden der mathematischen Physik*, 2 volumes, Berlin, 1931–1937; Interscience Publishers, Inc., Photolithoprint reproduction, 1943.

Daniell, P. J.

[15] *Orthogonal potentials*, Phil. Mag. (7), Volume II, 1926, pp. 247–258.

Faber, G.

[16] *Über den Hauptsatz aus der Theorie der konformen Abbildung*, Münchener Ber., 1922, pp. 91–100.

Garabedian, P. R.

[17] (and M. Schiffer]). *Identities in the theory of conformal mapping*, Trans. Amer. Math. Soc., Volume 65, 1949, pp. 187–238.

[18] (and M. Schiffer). *On existence theorems of potential theory and conformal mapping*, Ann. Math., Volume 51, 1950.

Golusin, G.

[19] *Interior problems of the theory of schlicht functions*, Uspehi Matem. Nauk, Volume 6, 1939, pp. 26–89. Translated by T. C. Doyle, A. C. Schaeffer, and D. C. Spencer for Office of Naval Research, 1947.

[20] *Method of variation in the theory of conform representation*, Rec. Math., Moscou, Nova Ser.; I, Volume 19, 1946, pp. 203–236; II, Volume 21, 1947, pp. 83–117; III, Volume 21, 1947, pp. 119–132.

Grötzsch, H.

[21] *Über einige Extremalprobleme der konformen Abbildung*, Leipziger Ber. (Abdruck aus den Berichten der Math. Phys. Kl. der sächsischen Academie der Wissenschaften zu Leipzig), Volume 80, 1928; I, pp. 367–376; II, pp. 497–502.

[22] *Über die Verzerrung bei schlichter konformer Abbildung mehrfach-zusammenhängender schlichter Bereiche*, Leipziger Ber.: I, Volume 81, 1929, pp. 38–47; II, Volume 81, 1929 pp. 217–221; III, Volume 83, 1931, pp. 283–297.

[23] *Über ein Variationsproblem der konformen Abbildung*, Leipziger Ber., Volume 82, 1930, pp. 251–263.

[24] *Über das Parallelschlitztheorem der konformen Abbildung schlichter Bereiche*, Leipziger Ber., Volume 84, 1932, pp. 15–36.

Grunsky, H.

[25] *Neue Abschätzungen zur konformen Abbildung ein- und mehrfach zusammenhängender Bereiche*, Schr. Math. Inst. u. Inst. angew, Math. Univ. Berlin, Volume 1, 1932, pp. 94–140.

[26] *Koeffizientenbedingungen für schlicht abbildende meromorphe Funktionen*, Math. Z., Volume 45, 1939, pp. 29–61.

Hadamard, J.

[27] *Mémoire sur le problème d'analyse relatif à l'équilibre des plaques élastiques encastrées*, Mém. Savants étrangers Acad. Sci. Inst. France, Series 2, Volume 33, No. 4, 1908, pp. 1–128.

[28] *Leçons sur le calcul des variations*, Hermann, Paris, 1910.

Hille, E.

[29] *Remarks on a paper by Z. Nehari*, Bull. Amer. Math. Soc., Volume 55 1949, pp. 552–553.

Jeans, J.

[30] *The mathematical theory of electricity and magnetism*, University Press, Cambridge, 1925.

Jenkins, J. A.

[31] *Some problems in conformal mapping*, Trans. Amer. Math. Soc., Volume 67, 1949, pp. 327–350.

Julia, G.

[32] *Sur une équation aux dérivées fonctionelles liée à la représentation conforme,* Ann. Ecole Norm. Sup. (3), Volume 39, 1922, pp. 1–28.

[33] *Leçons sur la représentation conforme des aires multiplements connexes,* Gauthier-Villars, Paris, 1934.

Köbe, P.

[34] *Abhandlungen zur Theorie der konformen Abbildung,* IV, Acta Math., Volume 41, 1918, pp. 305–344.

Lavrentieff, M.

[35] *On the theory of conformal mapping,* Trav. Inst. phys.-math. Stekloff, Volume 5, 1934, pp. 159–245.

Lehto, O.

[36] *Anwendung orthogonaler Systeme auf gewisse funktionentheoretische Extremal- und Abbildungsprobleme,* Ann. Acad. Sci. Fennicae, Ser. A, I, 59, 1949.

Lévy, P.

[37] *Leçons d' analyse fonctionelle,* Gauthier-Villars, Paris, 1922.

Lichtenstein, L.

[38] *Neuere Entwicklung der Potentialtheorie. Konforme Abbildung.* Encykl. d. math. Wissenschaften, Leipzig, IIC3, 1918.

Littlewood, J. E.

[39] *On inequalities in the theory of functions,* Proc. London Math. Soc. (2), Volume 23, 1924, pp. 481–519.

Löwner, K.

[40] *Untersuchungen über schlichte konforme Abbildungen des Einheitskreises,* I, Math. Ann., Volume 89, 1923, pp. 103–121.

Marty, F.

[41] *Sur le module des coefficients de MacLaurin d'une fonction univalente,* C. R. Acad. Sci. Paris, Volume 198, 1934, pp. 1569–1571.

Maxwell, J. C.

[42] *A treatise on electricity and magnetism,* I, Oxford, 1892.

Montel, P.

[43] *Leçons sur les fonctions univalentes ou multivalentes,* Gauthier-Villars, Paris, 1933.

Nehari, Z.

[44] *The Schwarzian derivative and schlicht functions,* Bull. Amer. Math. Soc., Volume 55, 1949, pp. 545–551.

Plemelj, J.

[45] *Ein Ergänzungssatz zur Cauchyschen Integraldarstellung analytischer Funktionen, Randwerte betreffend,* Monatsh. f. Math. u. Phys., Volume 19, 1908, pp. 205–210.

de Possel, R.

[46] *Zum Parallelschlitztheorem unendlichvielfach zusammenhängender Gebiete,* Göttinger Nachrichten, 1931, pp. 199–202.

[47] *Sur la représentation conforme d'un domaine a connexion infinie sur un domaine à fente parallèles,* J. de Math. (9), Volume 18, 1939, pp. 285–290.

Rengel, E.

[48] *Über einige Schlitztheoreme der konformen Abbildung*, Schr. Math. Inst. u. Inst. angew. Math. Univ. Berlin, Volume 1, 1933, pp. 141–162.

Schaeffer, A. C.

[49] (with M. Schiffer and D. C. Spencer). *The coefficient regions of schlicht functions*, Duke Math. J., Volume 16, 1949, pp. 493–527.

[50] (and D. C. Spencer). *The coefficients of schlicht functions*, Duke Math. J., Volume 10, 1943, pp. 611–635.

[51] (and D. C. Spencer). *The coefficients of schlicht functions*, II, Duke Math. J., Volume 12, 1945, pp. 107–125.

[52] (and D. C. Spencer). *The coefficients of schlicht functions*, III, Proc. Nat. Ac. Sci. U.S.A., Volume 32, 1946, pp. 111–116.

[53] (and D. C. Spencer). *The coefficients of schlicht functions*, IV, Proc. Nat. Ac. Sci. U.S.A., Volume 35, 1948, pp. 143–150.

[54] (and D. C. Spencer). *A variational method in conformal mapping*, Duke Math. J., Volume 14, 1947, pp. 949–966.

[55] (and D. C. Spencer). *Coefficient regions for schlicht functions*, Amer. Math. Soc. Colloquium Series, 1950.

Schiffer, M.

[56] *A method of variation within the family of simple functions*, Proc. London Math. Soc. (2), Volume 44, 1938, pp. 432–449.

[57] *On the coefficients of simple functions*, Proc. London Math. Soc. (2), Volume 44, 1938, pp. 450–452.

[58] *Sur la variation de la fonction de Green de domaines plans quelconques*, C. R. Acad. Sci. Paris, Volume 209, 1939, pp. 980–982.

[59] *Variation of the Green function and theory of p-valued functions*, Amer. J. Math., Volume 65, 1943, pp. 341–360.

[60] *The span of multiply connected domains*, Duke Math. J., Volume 10, 1943, pp. 209–216.

[61] *Hadamard's formula and variation of domain functions*, Amer. J. Math., Volume 68, 1946, pp. 417–448.

[62] *The kernel function of an orthonormal system*, Duke Math. J., Volume 13, 1946, pp. 529–540.

[63] *An application of orthonormal functions in the theory of conformal mapping*, Amer. J. Math., Volume 70, 1948, pp. 147–156.

[64] *On various types of orthogonalization*, Duke Math. J., Volume 17, 1950.

[65] (and D. C. Spencer). *The coefficient problem for multiply-connected domains*, Ann. Math., 1950.

[66] (and D. C. Spencer). *Lectures on conformal mapping and extremal methods*, Princeton, 1949–1950.

[67] (and G. Szegö). *Virtual mass and polarization*, Trans. Amer. Math. Soc., Volume 67, 1949, pp. 130–205.

Spencer, D. C.

[68] *Some problems in conformal mapping*, Bull. Amer. Math. Soc., Volume 53, 1947, pp. 417–439.

Szegö, G.

[69] *Orthogonal polynomials*, Amer. Math. Soc. Colloquium Series, Volume 23, 1939.

Teichmueller, O.

[70] *Ungleichungen zwischen den Koeffizienten schlichter Funktionen*, Sitzungsber. Preuss. Akad. Wiss. Math. Phys. Kl., 1938, pp. 363–375, Berlin.

Wirtinger, W.

[71] *Zur formalen Theorie der Funktionen von mehr komplexen Veränderlichen*, Math. Ann., Volume 97, 1927, pp. 357–375.

[72] *Über eine Minimalaufgabe im Gebiete der analytischen Funktionen*, Monatsh. Math. u. Phys., Volume 39, 1932, pp. 377–384.

Zaremba, S.

[73] *L'équation biharmonique et une classe remarquable de fonctions fondamentales harmoniques*, Bull. Int. Acad. Sci. Cracovie, 1907, pp. 147–196.

[74] *Sur le calcul numérique des fonctions demandées dans le problème de Dirichlet et le problème hydrodynamique*, Bull. Int. Acad. Sci. Cracovie, 1909, I, pp. 125–195.

INDEX

Supplementary Notes (1977)

Since the last printing of this book, a number of the unsolved problems mentioned in section 6 of Chapter III have been completely solved and substantial progress has been achieved on the others.[1]

We mention first the question of boundary regularity. That is, if the boundary of a minimal surface is analytic or of some degree of smoothness, how regular is the minimal surface? Lewy [8] proved that if the boundary is analytic then the surface is analytic up to the boundary. Later on a number of authors beginning with S. Hildebrandt [7] (see [9] p. 284 for additional references) proved that if the boundary is of class $C^{m,\alpha}$, $m \geq 1$, the surface is also of class $C^{m,\alpha}$ up to the boundary.

Next consider the question of the existence of branch points for solutions of Plateau's problem. It has been shown that when the number of space dimensions is three, the minimal surfaces having least area in the problems of Plateau and of Douglas (for orientable surfaces) are free of branch points (Osserman [11], Gulliver [4], [5], Alt [1], [2]). Consider a branched surface such as is constructed in section 6 of chapter III (p. 123). Choose one branch line and modify the topology of the surface so that points adjacent to the branch line are only considered to be neighbors if they both lie above or both below the plane. The new surface is topologically two disks attached at a point; a single disk may be used as a parameter domain by pinching along a diameter and twisting (see section 2 of [11] for details). The area is unchanged. But the new surface has an edge in its interior, so that it, and hence the original surface, can not have smallest area. Since this procedure may be carried out in any neighborhood of the branch point, the example in Courant [4] is incorrect. Concerning the structure of a minimal surface at a simple branch point it has been shown (Gulliver [5]) there are at least three branch lines all approaching the branch point in a continuously differentiable fashion and making equal angles.

Finally we note some recent results on the number of solutions of Plateau's problem. We mention the uniqueness theorem of Nitsche [10] for analytic boundaries of total curvature at most 4π; the theorem of Tomi [12] on the finiteness of the number of least area solutions for analytic boundaries, and the theorem of Böhme and Tromba [3] on the generic finiteness of the number of solutions (not necessarily least area) of the classical problem of Plateau.

---

[1] An entirely new approach to Plateau's problem has been developed by Reifenberg, Federer, Fleming and Almgren in which one seeks to minimize two dimensional Hausdorff measure among very general classes of geometric objects. We refer the interested reader to the books of Almgren (Plateau's Problem. New York: W.A. Benjamin Inc. 1966) and Federer (Geometric Measure Theory. New York, Heidelberg, Berlin: Springer 1969).

Bibliography

1. Alt, H.W.: Verzweigungspunkte von H. Flächen, I. Math. Z. **127** (1972), 333 - 362.

2. ——: Verzweigungspunkte von H. Flächen, II. Math. Ann. **201** (1973), 33 - 55.

3. Böhme, R.; Tromba, A.J.: The number of solutions to the classical Plateau problem is generically finite, to appear.

4. Courant, R.: On a generalized form of Plateau's problem. Trans. Amer. Math Soc. **30** (1941), 40 - 47.

5. Gulliver, R.D.: Regularity of minimizing surfaces of prescribed mean curvature. Ann of Math (2) **97** (1973), 275 - 305.

6. ——: Branched immersions of surfaces and reduction of topological type I. Math. Z. **145** (1975), 267 - 288.

7. Hildebrandt, S.: Boundary behavior of minimal surfaces. Arch. Rat. Mech. Anal. **35** (1969), 47 - 82.

8. Lewy, H.: On the boundary behavior of minimal surfaces. Proc. Nat. Acad. Sci. USA **37** (1951), 103 - 110.

9. Nitsche, J.CC.: Vorlesungen über Minimalflächen. Berlin, Heidelberg, New York: Springer 1975.

10. ——: A new uniqueness theorem for minimal surfaces, Arch. Rat. Mech. Anal. **52** (1973), 319 - 329.

11. Osserman, R.: A proof of the regularity everwhere of the classical solution to Plateau's problem. Ann of Math (2) **91** (1970), 550 - 569.

12. Tomi, F.: On the local uniqueness of the problem of least area, Arch. Rat. Mech. Anal. **52** (1973), 312 - 318.